Musical Acoustics, Neurocognition and Psychology of Music
Musikalische Akustik, Neurokognition und Musikpsychologie

AF787476

Hamburger Jahrbuch für Musikwissenschaft

Herausgegeben vom
Musikwissenschaftlichen Institut
der Universität Hamburg

Redaktion:
Arne von Ruschkowski

Band 25

PETER LANG
Frankfurt am Main · Berlin · Bern · Bruxelles · New York · Oxford · Wien

Rolf Bader (ed./Hrsg.)

Musical Acoustics, Neurocognition and Psychology of Music

Musikalische Akustik, Neurokognition und Musikpsychologie

Current Research in Systematic Musicology at the Institute of Musicology, University of Hamburg

Aktuelle Forschung der Systematischen Musikwissenschaft am Institut für Musikwissenschaft, Universität Hamburg

PETER LANG
Internationaler Verlag der Wissenschaften

Bibliographic Information published by the Deutsche Nationalbibliothek
The Deutsche Nationalbibliothek lists this publication in the Deutsche Nationalbibliografie; detailed bibliographic data is available in the internet at http://dnb.d-nb.de.

Gratefully acknowledging financial support by the University of Hamburg.
Gedruckt mit Unterstützung der Universität Hamburg.

ISSN 0342-8303
ISBN 978-3-631-59991-4

© Peter Lang GmbH
Internationaler Verlag der Wissenschaften
Frankfurt am Main 2009
All rights reserved.

All parts of this publication are protected by copyright. Any utilisation outside the strict limits of the copyright law, without the permission of the publisher, is forbidden and liable to prosecution. This applies in particular to reproductions, translations, microfilming, and storage and processing in electronic retrieval systems.

www.peterlang.de

Table of Content
Inhaltsverzeichnis

Institute of Musicology, University of Hamburg

Institut für Musikwissenschaft, Universität Hamburg

Preface

Systematic research in music knows many supplementary approaches to the topic, with their specific disciplinary methods, views and questions. A comprehensive understanding of music is only possible by the interdependencies, completions and influences of these different disciplines. So music can be investigated via instrument and room acoustics, the physiology of the ear and the brain, through psychology of perception, musical logic and music theory, but also by investigating musical meaning, research in emotions and movements, as well as by working on social, ethnical and political aspects. Only when dealing with all these disciplines, a extensive knowledge und understanding of music can be achieved, which again acts back into music, by inventing algorithms for sound production and manipulation in music production, by constructing a music psychological and logical music theory used in music performance and composition, in research concerning noise, for understanding and improving social structures and systems or in cultural exchange through the understanding of music and culture of foreign ethnic groups.

The present volume presents some of the current research, inventions, trends, and discussions at the Institute of Musicology at the University of Hamburg in the field of Systematic Musicology. It is one of a two volume release, where the second volume is in the field of Historic Musicology. The generation change, which happened in recent years at the Institute made a presentation of recent work reasonable in analogy to the first volume of this publication 'Hamburger Jahrbuch für Musikwissenschaft'.

So all authors of this volume are professors, scientific stuff or students of the Hamburg Institute. The international cooperations of the research presented here with universities, societies and regular conference presentations all around the world appear indirectly here, through the treated topics, which are mostly located around 'hot spots' of current international research. Briefly mentioned here shall be the *International Cooperative in Systematic and Cognitive Musicology* with its *International Summer School for Systemtatic Musicology* established since many years, regular preparation of sessions in the fields of musical acoustics and music psychology at the *American Acoustical Society*, the *European Acoustical Association* and the *Deutschen Gesellschaft für Akustik*, as well as the *European Society for the Cognitive Science of Music*, the *International Council for Traditional Music*, the *Society for Ethnomusicology* and many others.

The contributions to this volume consider the topics of research in Systematic Musicology mentioned above. The 'Acoustic Camera' developed in the Institute over the last two years for visualization of sound fields, radiation of musical instruments as well as measuring the area of vibration of these instruments is presented in the paper of Bader, Münster, Richter and Timm. Devices of this kind are commercially very successful nowadays in many fields of acoustics, where most of them are qualitatively not appropriate for musical acoustics. With its 128 microphones and a sample frequency of 48 kHz per microphone the array developed in the Institute is the biggest and most powerful in the world. In the present article next to the presentation of some of the possible methods of measurement with this device, some examples of musical instruments are discussed.

The contribution of Beurmann and Schneider is about the acoustic investigation of a historic cembalo of Andreas Ruckers. The instrument belongs to the best preserved cembali of the workshop of the Ruckers family and is judged to be exemplary for the art of cembalo building in the Netherlands in the 17th century.

Electronic Musical Instruments are the most frequently used sound generators, especially in the field of popular music. In the paper of Pfeiffle and Bader, a new instrument is proposed, which realizes a whole body Banjo with a FPGA hardware device. This real-time method is a solution proposition in the field of electronic sound generation with real musical instrument geometries in real-time internationally discussed today, which already works in reality.

Music psychology today also deals with the Neurocognition of Music. In the contribution of Bader and von Ruschkowski an EEG-study is presented, which investigates the brain activity in regard to musical meaning. Therefore, subjects were presented parts of mostly well-known musical pieces in their original and in a blurred version. An overall intensified activity for the blurred version could be found, which additionally is of interest in terms of noise research.

In the paper of Schneider, von Ruschkowski and Bader musical roughness is investigated as music psychological counterpoint to musical consonance in terms of an intense review of the literature as well as additional empirical studies. It is shown, that most of the roughness algorithms used today may be working with industrial noise sounds, indeed for most musical cases they fail. Coming most close to musical reality seams to be the algorithm of Setharis, which again comes close to the one of Helmholtz.

In seinem Aufsatz Forensic Music Psychology geht Albrecht Schneider auf ein Teilgebiet der angewandten Musikpsychologie ein, nämlich der Musikpsychologie als Hilfsdisziplin bei der Klärung von Rechtsfällen. In dem Beitrag kommen sowohl strafrechtliche wie zivilrechtliche Aspekte zur Sprache; die deutsche Rechtsprechung wird vor allem bei Plagiatsfällen bis zur Gegenwart berücksichtigt.

In their paper *Forensic Music Psychology*, Albrecht Schneider is dealing with a branch of applied music psychology, namely music psychology as a supporting discipline when settling legal cases. In the paper criminal as well as aspects of civil law are discussed. The German jurisdiction is considered primarily in cases of plagiarism up to the present.

The paper of Bader, Dietz, Elvers, Elias & Tolkien is dealing with a fundamental renewal of music theory, where a theory of musical thought according to the Logic of Aristoteles, the Syllogistic is performed. As alternative to modern music theories which mostly are more like 'handcraft' theories rather than philosophical founded, a method is made applicable for music, which for all fields of music as well as for extra-musical areas, for musical meaning can be made useful. A definition of the cadence, a positive definition of consonance as well as a foundation of tonality as logical reasoning is performed.

The investigation of the blow pipes in Russia is discussed in the paper of Uli Morgenstern using selected examples. The *duda*, which had appeared in the areas of Pskov, Tver', and Smolensk, and is no longer existing is presented using written sources. The author discusses history, tuning and playing technique of the instrument.

The singing voice in popular music is discussed in the article of Martin Pfleiderer. After a detailed description of the different singing genres, the physiology of the human voice and their ways of articulation, Pfleiderer investigates along the piece 'Come back, Baby' of Ray Charles and further pieces of vocal music their vocal techniques using spectral analysis. In a pilot study about emotional association of expression, an assignment of different musical examples to contentious and emotional perception is performed.

Helmut Rösing in his contribution investigates the relevance of music for society. He is pointing to the fact that music has existed in all cultures and all times and finds the social-psychological power of music, to communicate and produce emotions as the main factor in music. Also he discusses the functions of music in societies as proposed by Max Weber and uses popular music as typical example.

Hamburg, 16. 9. 2009

Rolf Bader

Vorwort

Die systematische Erforschung von Musik kennt viele sich ergänzende Herangehensweisen an das Thema, mit deren disziplinär spezifischen Methoden, Sichtweisen und Fragessstellungen. Das umfassende Verstehen von Musik wird durch die Befruchtung, Ergänzung und Beeinflussung dieser unterschiedlichen Disziplinen erst möglich. So kann Musik durch Instrumenten- und Raumakustik, die Physiologie des Ohres und des Gehirns, durch Psychologie der Wahrnehmung, Musikalische Logik und Musiktheorie, aber auch durch Musikalische Bedeutungslehre, Emotions- und Bewegungsforschung, sowie anhand von gesellschaftlichen, ethischen und politischen Aspekten behandelt werden. Erst die Beschäftigung mit all diesen Teildisziplinen ermöglicht ein umfassendes Wissen und Verstehen von Musik, welches daraufhin wiederum auf die Musik zurückwirkt, durch die Erfindung von klangerzeugenden und klangverändernden Algorithmen zur Musikproduktion, zur Erstellung von musikpsychologischen und musiklogischen praktischen Musiktheorien für die Komposition, im Lärmschutz, zur Erfassung und Verbesserung gesellschaftlicher Strukturen und Systeme oder im Kulturaustausch durch Verstehen der Musik und Kultur fremder Ethnien.

Der vorliegende Band stellt einige der z. Z. am Musikwissenschaftlichen Institut der Universität Hamburg im Bereich der Systematischen Musikwissenschaft stattfindenden Forschungen, Erfindungen, Trends und Diskussionen dar. Er ist einer von zwei Teilbänden, wovon der zweite Band in der Historischen Musikwissenschaft angesiedelt ist. Der in den letzten Jahren am Institut erfolge Generationenwechsel ließ nun erneut eine aktuelle Vorstellung der ansässigen Forschung, in Anlehnung an den ersten Band der Reihe ‚Hamburger Jahrbuch für Musikwissenschaft' wünschenswert erscheinen.

So sind sämtliche Autoren des Bandes Professoren, wissenschaftliche Mitarbeiter oder Studierende des Hamburger Instituts. Die internationalen Vernetzungen der verschiedenen hier vorgestellten Forschungen mit Universitäten, Gesellschaften und regelmäßigen Konferenzteilnamen in aller Welt fließen hier also indirekt durch die behandelten Themen, welche sich meist an ‚hot spots' aktueller internationaler Forschung befinden ein. Erwähnt seien hier nur die *International Cooperative in Systematic and Cognitive Musicology* mit ihrer nun seit vielen Jahren etablierten *International Summer School for Systemtatic Musicology*, regelmäßige Organisation von Sessions zur Musikalischen Akustik und Musikpsychologie bei der *American Acoustical Society*, der *European Acoustical Association* und der *Deutschen Gesellschaft für Akustik*, sowie der *European Society for the Cognitive Science of Music*, dem *International Council for Traditional Music*, der *Society for Ethnomusicology* und vielen anderen.

Die verschiedenen Beiträge dieses Bandes beschäftigen sich mit den oben genannten Teilaspekten der Forschung in der Systematischen Musikwissenschaft. Die in den letzten zwei Jahren im Institut entwickelte ‚Akustische Kamera' zur Visualisierung von Schallfeldern, der Abstrahlung von Musikinstrumenten, sowie zur flächigen Messung der Schwingungen dieser Musikinstrumente wird im Beitrag von Bader, Münster, Richter und Timm vorgestellt. Geräte dieser Art sind heute kommerziell sehr erfolgreich in vielen Gebieten der Akustik eingesetzt, wobei diese allerdings qualitativ für die Musikinstrumentenakustik nicht genügen. Mit seinen 128 Mikrophonen und einer Samplefrequenz von 48 kHz pro Mikrophon ist das in der Systematischen Musikwissenschaft entwickelte Gerät das größte und leistungsstärkste weltweit und so werden in dem vorliegenden Beitrag neben der Vorstellung einige wenige der möglichen Meßmethoden mit diesem Gerät anhand einiger Beispiele verschiedener Musikinstrumente diskutiert.

Der Beitrag von Beurmann und Schneider hat die akustische Untersuchung eines historischen Cembalos von Andreas Ruckers zum Gegenstand. Das Instrument gehört zu den am besten erhaltenen Cembali aus der Werkstatt der Familie Ruckers und gilt als exemplarisch für die Kunst des niederländischen Cembalobaus im 17. Jahrhundert.

Elektronische Musikinstrumente sind heute die häufigsten Klangerzeuger, gerade im popularmusikalischen Bereich. Im Beitrag von Pfeiffle und Bader wird ein neues Instrument vorgestellt, welches durch ein Physical Model des gesamten Korpusses eines Banjos in einer FPGA Hardware realisiert wird. Diese echtzeitfähige Methode stellt einen Lösungsvorschlag für das z. Z. international diskutierte Problem der elektronischen Klangerzeugung gesamter Geometrien in Echtzeit dar, welches bereits physikalisch vorliegt.

Die Musikpsychologie beschäftigt sich heute verstärkt mit Musikalischer Neurokognition. Im Beitrag von Bader und von Ruschkowski wird eine EEG-Studie vorgestellt, welche die Aktivitäten des Gehirns in Bezug auf musikalische Bedeutung untersucht. Hierfür wurden Versuchspersonen Ausschnitte aus bekannten Musikstücken in Originalfassung und in einer verzerrten Version, welche Bedeutungsinhalte verwischt, vorgespielt. Es konnte eine durchweg erhöhte Aktivität für die verzerrten Versionen festgestellt werden, was ebenfalls für die Lärmforschung interessant ist.

Im Beitrag Schneider, von Ruschkowski und Bader wird musikalische Rauhigkeit als musikpsychologischer Kontrapunkt zu musikalischer Konsonanz anhand intensiver Literaturrecherche sowie eigener Versuche und mittels eines Algorithmus, welchen die Helmholtzschen Rauhigkeitsdefinition implementiert untersucht. Hierbei wird gezeigt, dass die meisten der heute verwendeten Rauhigkeitsalgorithmen zwar für industrielle Lärmprobleme geeignet sein könnten, für viele musikalische Fälle aber unzutreffende Ergebnisse zeitigt. Der musikalischen Realität scheint der Setharis'sche Algorithmus, welcher dem Helmholtzschen sehr nahe kommt am besten geeignet.

In seinem Aufsatz Forensic Music Psychology geht Albrecht Schneider auf ein Teilgebiet der angewandten Musikpsychologie ein, nämlich der Musikpsychologie als Hilfsdisziplin bei der Klärung von Rechtsfällen. In dem Beitrag kommen sowohl strafrechtliche wie zivilrechtliche Aspekte zur Sprache; die deutsche Rechtsprechung wird vor allem bei Plagiatsfällen bis zur Gegenwart berücksichtigt.

Mit einer grundsätzlichen Neufassung einer Musiktheorie befasst sich der Beitrag von Bader, Dietz, Elvers, Elias, & Tolkien, in dem eine Theorie des Musikalischen Gedankens aufbauend auf der Aristotelischen Logik, der Syllogistik unternommen wird. Als Alternative zu heutigen Musiktheorien, welche eher Handwerkslehren als philosophisch fundiert sind wird hier eine Methode für die Musik anwendbar gemacht, welche für alle Gebiete der Musik, sowie für außermusikalische Bereiche, also musikalische Bedeutungslehre prinzipiell ebenfalls anwendbar wäre. Es wird eine Definition der Kadenz, eine positive Definition von Konsonanz, sowie Tonalität als logische Argumentation hergeleitet.

Die Untersuchung von Sackpfeifen in Russland wird im Beitrag von Uli Morgenstern anhand ausgewählter Beispiele diskutiert. Die *duda*, die in den Gebieten den Gebieten Pskov, Tver' und Smolensk vorkam und dort praktisch ausgestorben ist wird anhand schriftlicher Quellen und Überlieferungen vorgestellt. Der Autor zeichnet Geschichte, Stimmung und Spieltechnik des Instruments nach.

Die Singstimme in populärer Musik ist im Beitrag von Martin Pfleiderer diskutiert. Nach einer grundlegenden Beschreibung verschiedener Gesangsstile, der Physiologie der menschlichen Stimme und deren Ausdrucksmöglichkeiten untersucht Pfleiderer anhand des Ray Charles Stücks ‚Come back, Baby' und weiterer Gesangsstücke mittels Spektralanalysen deren Vokaltechniken. In einer Pilotstudie zur emotionalen Beschreibung des Ausdrucks wird die Zuordnung verschiedener Musikbeispiele zu inhaltlicher und emotionaler Perzeption ermittelt.

Helmut Rösing untersucht in seinem Beitrag die Relevanz von Musik für die Gesellschaft. Er weist darauf hin, daß Musik in allen Kulturen zu allen Zeiten existiert hat und sieht die sozialpsychologische Kraft der Musik, Emotionen zu vermitteln und zu erzeugen als wesentliches Merkmal von Musik. Auch diskutiert er anhand der von Max Weber vorgeschlagenen Funktionen von Musik in Gesellschaften die Popularmusik als typisches Beispiel.

Hamburg, 16. 9. 2009

Rolf Bader

Rolf Bader
Malte Münster
Jan Richter
Heiko Timm

Microphone Array Measurements of Drums and Flutes

Introduction

Microphone array techniques record sound pressure fields radiated by any sound source using many microphones, often arranged as a microphone grid or array. These recordings show the radiation field of the radiating source. Furthermore, the sound data collected by the array can be used to back-propagate the sound field to the radiating source surface. By doing so, the vibration of surfaces and other radiating bodies, but also those of e.g. air columns, can be measured without disturbing the vibration process itself. This unique technique is therefore most suitable for analyzing musical instruments as vibrating structures, giving insight into phenomena which could hardly be measured in any other way.

As musical instruments often radiate high frequencies, and as initial transients of musical instruments are often the most important part of the sound, it is necessary to have a sampling rate covering this frequency range. Moreover, as the vibrating patterns can be very complex, a high spatial distribution of points, meaning many microphones, is necessary to fulfil such a task. So a microphone array is needed which is able to record with many microphones simultaneously and with a high sampling frequency.

Fig. 1: One of the three microphone arrays at the Institute of Musicology in Hamburg consisting of 128 electret microphones.

Such a microphone array was built at the Institute of Musicology in Hamburg for the last two years. It consists of 128 microphones recording simultaneously with a sampling frequency of 48 kHz (see Fig. 1), so it is possible to obtain a high spatial resolution while recording with a high temporal resolution to cover the whole hearing range up to 20 kHz. There are many commercial products on the market using microphone array techniques, but to our knowledge, the array in Hamburg is the largest and fastest available. While in industrial applications often only single frequencies in the low or mid range need to be investigated, with musical instruments the whole frequency range is of interest. This is because musical instruments are constructed for sounding, and therefore show the full range of vibrating possibilities. Hence, a microphone array technique which tries to cover most of this vibration behaviour is needed to meet the criteria mentioned above.

Back-propagation of recorded sound field to the source

The unprocessed pressure field data are of interest with respect to many sound field problems, like the radiation from musical instruments, the detection of shock waves from horns, the reconstruction of binaural properties of musical instruments etc. Still there is a great interest in the source behaviour itself as shown with many examples below. Therefore, after recording the sound field of a tone played by an instrument with the microphone array, the recorded spatial pressure distribution must be back-propagated to the sound source. So a mathematical method is needed which reconstructs the radiation sound pressures at the surface of instruments bodies or at the holes of wind instruments from the recorded pressure field.

Many suggestions have been made how to perform such a reconstruction. Today many commercial products (so-called 'Acoustic Cameras') are on the market, making it one of the hot topics in Acoustics today. One of the first methods used for back-propagation was called Acoustic Holography (Williams et al. 1980, Meynard et al. 1985). Here, the same mathematical framework that was developed for Optical Holography is applied, the Holograms known to produce 3D pictures by directing a laser to an optical device. This method uses an angular spectrum, which decomposes the two-dimensional recorded field into spatial wave vectors k_x and k_y. This spectrum is then propagated in the z-direction to the radiating surface by propagating the phase and amplitude with respect to the k_z wave vector. After propagation, the new angular spectrum is inverse Fourier transformed to end up with the reconstructed sound field at the radiating surface.

This method is straightforward but has certain trade-offs. One is the problem of side bands which occur with Fourier decompositions using just a few points, in our case those of the microphones. Another is the treatment of evanescent waves. Those physically present waves occur with radiating plates or membranes from the cut-off frequency on. For these waves, the cancellation by phase differences of neighbouring radiating points leads to an exponential decay of the amplitude in the radiating direction. So after approximately 10-20 cm these waves are gone (Williams 2000). This is the reason why the Holography method is near-field, meaning that the recording of the radiation is done right in front of the vibrating structure to capture these evanescent waves. As these waves occur physically, all reconstruction methods need to be aware of this problem. Still, the Near-field Acoustic Holography (NAH) is back-propagating these waves inevitable in an exponentially growing manner, as inverse of the exponential decay. So the solution is 'blowing up' very fast leading to unrealistic values. To avoid this, the angular spectrum may be low-pass filtered, which on the other side skips a good part of the radiation. Also, the spacing of the microphone array grid can be made larger. However, then the distance between the array and the radiating body should be increased, which on the other hand might be too much in terms of evanescent wave inclusion. Many versions of NAH have been developed over the years (i.e. see Williams et al. 2003, Prager 2007, Scholte et al. 2005).

A second method proposed is the solution of the Helmholtz equation, the wave equation (Wang & Wu 1995, Wu 1999, Wu et al. 2005, Rayess & Wu 1999). When solving for spherical coordinates, the radiation of a point source can be decomposed by radiators of different orders, e.g. monopole, dipole, quadrupole, or hexapole, including their phases. Indeed, all possible radiations can be decomposed into these radiators with suitable amplitudes. Thus when recording a complex sound field, these amplitudes can be estimated from the recording. Afterwards the sound field at any place is known and calculated simply by inserting new spatial coordinates into the analytical radiator functions and superposing them. However, this Helmholtz-Least-Square Method (HELS) has certain trade-offs, too. In practice, the reconstruction may depend crucially on the virtual source point assumed behind the radiating surface. Also, the reconstruction using superposition of spherical radiators is not easy to handle, as more complex structures need many radiators with sometimes unreasonable large amplitudes. This restricts this method to radiations which are known to come close to one or two of the pole shapes.

An approach taking complex radiating surfaces into account is the Boundary-Element method for reconstruction (see i.e. Bai 1992). The radiating surface is discretized into finite elements and the enclosed air of the structure is taken as radiating source. Then the sound field is calculated taking spatial boundary conditions into consideration. This method is able to define the radiation structure clearly, which is of great advantage. However, the construction of a complete boundary mesh for all structures investigated is very time consuming and also the solution process is difficult.

Yet another way to reconstruct the surface radiation is to extend the point source idea to a multiple point source radiator (Ochmann 1995, 1999), where the optimum amount of radiators and pole orders can be estimated. Taking this a step further, many monopole sources can be used comparable with the Equivalent Source methods (see i.e. Bouchet & Loyau 2000). The method developed for the array used here works in a similar manner. For an overview of the history of Sound field reconstruction methods see Magalhães & Tenenbaum (2004).

Some attempts have been made to study musical instruments using Microphone Arrays. One of the first investigated the radiation from guitar top plates (Strong et al. 1982). As the sampling frequency was quite low in the early times of computation, only the eigenmodes of the plates were reconstructed. Still, it was found that the sound hole is the dominant radiator with low frequencies. The radiation from violins was studied by Wang & Burroughs (2001). They found frequency dependent differences between top and back plate radiation. A very detailed study of Bissinger et al. (2007) investigated the f-holes of violins. It was found that the A1 mode has a nodal line right through the f-holes, and is therefore radiating as a dipole. But as dipole radiation is very low in energy because of phase- and anti-phase-cancellation, this mode is contributing little to the sound.

Minimum energy reconstruction method

The method of reconstructing the vibration at radiating surfaces used here was developed especially for the task of musical instrument research, where complex radiation pattern with very low and very high frequencies need to be back-propagated (Bader 2009a).

The jth pressure p_m^j measured at the microphone array is assumed to be a superposition of i monopole radiating pressures p_g^i (see Fig. 2) like

$$p_m^j = \sum_{i=1}^{N} p_g^i \, R_0^{ij}$$

The radiation matrix R_0^{ij} contains phase relations and amplitude drops between the radiating points i and the microphones j like

$$R_0^{ij} = \frac{1}{\Gamma_0^{ij}} e^{\imath k r^{ij}}$$

$k = 2 \pi f / c$ is the wave vector, where f is frequency and c is the speed of sound in air. r^{ij} is the matrix of all distances. The amplitude drop is

$$\Gamma_0^{ij}(\alpha) = r^{ij}(1 + \alpha(1 - \beta^{ij})),$$

where β determines the direction of radiation with β = 1 being the normal direction and β = 0 being the direction ±90°. The crucial parameter is α which is the directivity of the radiation. For α = 0 the source is a monopole, for α > 0 the radiation is getting more narrow with increasing α.

In Fig. 3 the difference of directivity is shown. The perfect half-circle is the monopole with α = 0. When α is increased, the radiation is getting more and more narrow. So α is an overall directivity value for the radiation. We will find with the tambourine discussed below that higher order modes, which have a more directed radiation, indeed show a higher value for α.

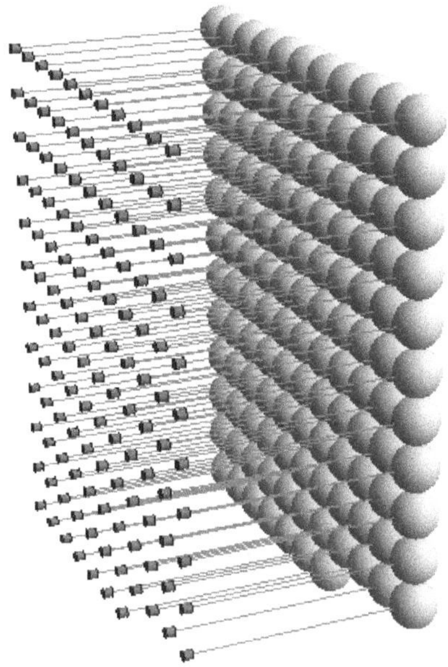

Fig. 2: Schematic view of the relations between the radiating points on the right consisting of 128 monopole radiators and the corresponding microphones on the left. Each radiation point is radiating into all microphones according to a directivity value α.

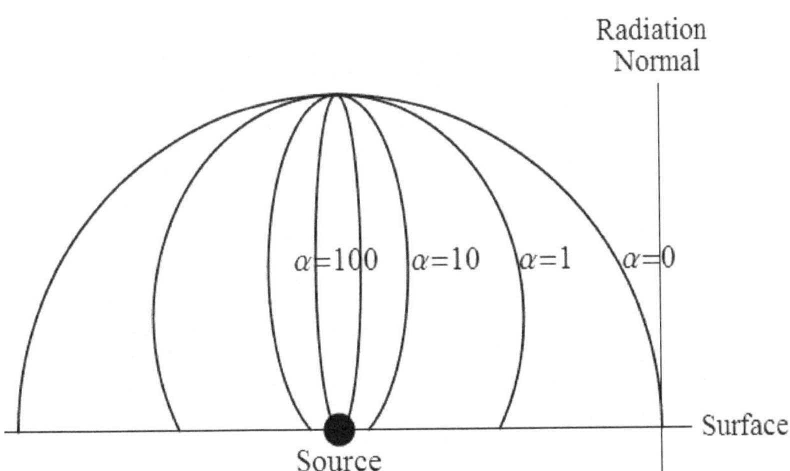

Fig. 3: Radiation directivity according to the value α, where $\alpha = 0$ is a monopole radiator and $\alpha > 0$ is narrowing the radiation with respect to the normal radiation direction, which is the up-axes.

When the matrices and vectors are defined, a linear equation solver reconstructs the pressure p_g on the radiating surface from the measured pressure values p_m. The solver can only do so, if a value for α is assumed. This assumption has to be arbitrary first, as we do not know the directivity. But it appears that the correct value for alpha is the one which minimizes the reconstruction energy. This energy is proportional to the squared reconstruction pressures. If several α values are tried and the reconstructions are performed, a curve α vs. reconstruction energy can be plotted as shown in Fig. 4. This curve has a minimum where indeed the correct reconstruction is achieved.

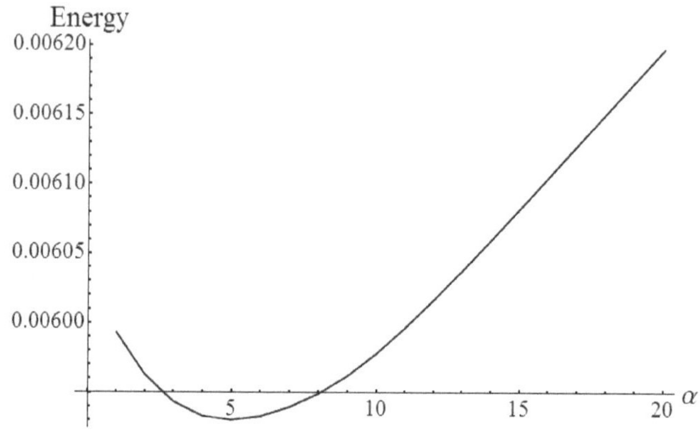

Fig. 4: Reconstruction energy vs. directivity value α. The point of minimum energy is the correct reconstruction value.

This method can be understood intuitively, which makes it very stable and reasonable. If α is very large, it is assumed that each radiating point is only radiating into the microphone opposite to it. Then, the pressure measured is necessarily only produced by this one radiation point in front of the microphone. As in reality, the radiating sources superpose, thus the reconstruction will show an unrealistic high energy. Because of this superposition the recording itself is always a blurred version of the real radiating surface. Decreasing α leads to a sharpening of the reconstruction, that can be followed visually very easy. If the correct value for α is reached, the reconstruction is perfect and the minimum energy is reached. If α is decreased even further, then the influence of neighbouring radiation points onto the recording of one microphone would be overestimated. In such a case the linear equation solver would try to balance and therefore explain the sound field by unreasonable high reconstruction amplitudes. This means, that the reconstruction energy is increasing very fast again. So the minimum of the reconstruction energy is the correct solution to the propagation problem. Additionally, we obtain an estimated overall radiation directivity by this value, indeed a very interesting parameter for musical instrument research.

1. Instruments with membranes

1.1 Degenerated modes of an Usbek tambourine

Circular membranes have complex vibrational pattern. The differential equation governing the vibration can be solved theoretically (Morse & Ingard 1986), resulting in mode shapes with radial and circular nodal lines. Each mode is vibrating with its own frequency and damping rate. If the tambourine is struck, the vibration is a superposition of all these modes fusing to the played tone. The elementary low frequency mode shapes are monopole, dipole, quadrupole, hexapole etc. with increasing frequency. If the membrane is not homogeneous, e.g. if the tension or density of the goat skin used is nonuniform, modes degenerate, resulting in a doubling of the modes. Therefore many mode shapes appear twice on the membrane with slightly different orientation and also with slightly different frequencies. These frequencies differ usually by a few Hertz only and so a beating is produced, i.e. an amplitude modulation which makes the sound very rich and lively. The precise strength and depth of this beating depends on the differences in frequencies, which again depends on the kind of inhomogeneity of the membrane.

The tambourine used in the study shows clearly at which point it was beaten over the years. At one area of the membrane, the goat skin is more transparent and so it can expected to be a bit thinner there. The sound itself is very rich and beating can clearly be heard.

Yet the beating does not last long, as the percussive nature of the sound makes the membrane vibrate for about 200 ms only. Subsequently, the tone is gone as is the case with the sounds of small tambourines. Also, since the higher frequencies are damped more than the fundamental, the beating caused by the higher partials is even shorter. As a consequence of this the beatings appearing with the sound fuse to a tone. This is the very nature of the tambourines' sound. Hence when investigating it, it is necessary to use a method which is able to analyze one strike only. Methods artificially accelerating the membrane with a sinusodial force at the eigenfrequencies of the membrane are able to show the shapes of the membrane clearly. However, this does not correspond with the perception of a tambourine being played. With sounds of such short length the uncertainty principle holds. Thus it is crucial to investigate how well the degenerated modes can be distinguished one from another and in which way they may fuse.

In this study, the tambourine was put in front of the microphone array and struck once. From this single strike the analysis of the first lowest modes was performed using the Minimum Energy Method discussed above. As the analysis is using the phase information, the reconstructed amplitudes also include the phase of the amplitude. This is done by using amplitudes as complex numbers consisting of a real and an imaginary part, where the absolute value is the amplitude strength and the argument of the complex number is the phase. With this information, we can discuss the vibration of the tambourine in three ways: 1.) we can plot the real and the imaginary part of the reconstruction to show the different phases of the vibration; 2.) we can then plot the absolute radiation strength without phase information; 3.) we can plot the phases of the reconstruction points in a circular phase plot ranging from 0 to 2 π. The last plot gives us information about the uncertainty of the modes. If a mode is vibrating for a long time, uncertainty is decreasing and the phase relations between all reconstruction points can only have two discrete values opposite one to another. Only for the fundamental mode all phases have to be the same. The uncertainty is influencing the results considerably if the vibration time is short. Namely in a way, that the two degenerated modes can no longer be separated completely, because they are very close together in terms of their frequencies. So both vibration patterns of the degenerated pair show, besides their own shape, additionally a bit of the other one. This can be seen to some extend in the plots of the real and imaginary parts of the membrane, but it certainly appears clearly in the plot of the phase relations, which are no longer simply two opposite phases, but rather show a wider spread of phase relations.

In Fig. 5 the first seven modes of the tambourine are shown as they appear in the spectrum. It is interesting to see, that the fundamental mode is a single mode and the dipole, quadrupole and hexapole modes are indeed appearing twice as degenerated modes. The frequencies are shown in Tab. 1, the beatings are 10 Hz for the dipole pair, 8 Hz for the quadrupole pair, and 26 Hz for the hexapole pair. 8 Hz and 10 Hz beatings can be perceived clearly as fast beatings, but 26 Hz is too fast to be heard as an amplitude modulation, as the threshold for pitch perception is around 20 Hz. However, because of damping this last pair disappears so fast, that it hardly would be perceived with a beating. For the other two, when listening carefully, one is able to perceive the beating. However, normally beating is not heard with such a short percussive sound. Rather, the amplitude modulation appearing with the fast decaying tone is more part of the overall timbre of this tone, creating its character. So when modifying the inhomogeneity of the membrane, the tone colour is likely to be altered more by a change of the beating between the degenerated pairs rather than by a slightly changing frequency shift of the partials.

When taking a closer look at the mode shapes in Fig. 5 and especially at the phase relations, we get the picture as expected. The fundamental mode, shown in the top row, has the same shape for the real and the imaginary part and the phases are all the same. The situation changes moderately when examining the dipole pair in row two and three. The basic pattern of two phases opposite one to another appears clearly; still they show a bit of a spreading. This is explained by the real and imaginary shapes shown in column one and two. If no uncertainty would be present and both modes would vibrate independently, both plots (real and imaginary) would be the same, where only colours would change from black to white indicating a π phase shift. When comparing the real with the imaginary mode shape of the lower dipole mode in row two (first and second plot) we find, that the imaginary plot is shifted a little bit counter clockwise compared to the real one. A similar situation appears with the higher dipole mode, where the imaginary part also moves a bit counter clockwise compared to the real one. So when investigating one mode, the nodal line is not strictly fixed but moving, because of the influence of the other degenerated mode which has a different mode shape. We could also say, the Q-values of the frequencies are so low and therefore the peaks are so wide, that a certain amount of amplitude of one peak is explained by the influence of the broad peak tail of the other degenerated pair mode.

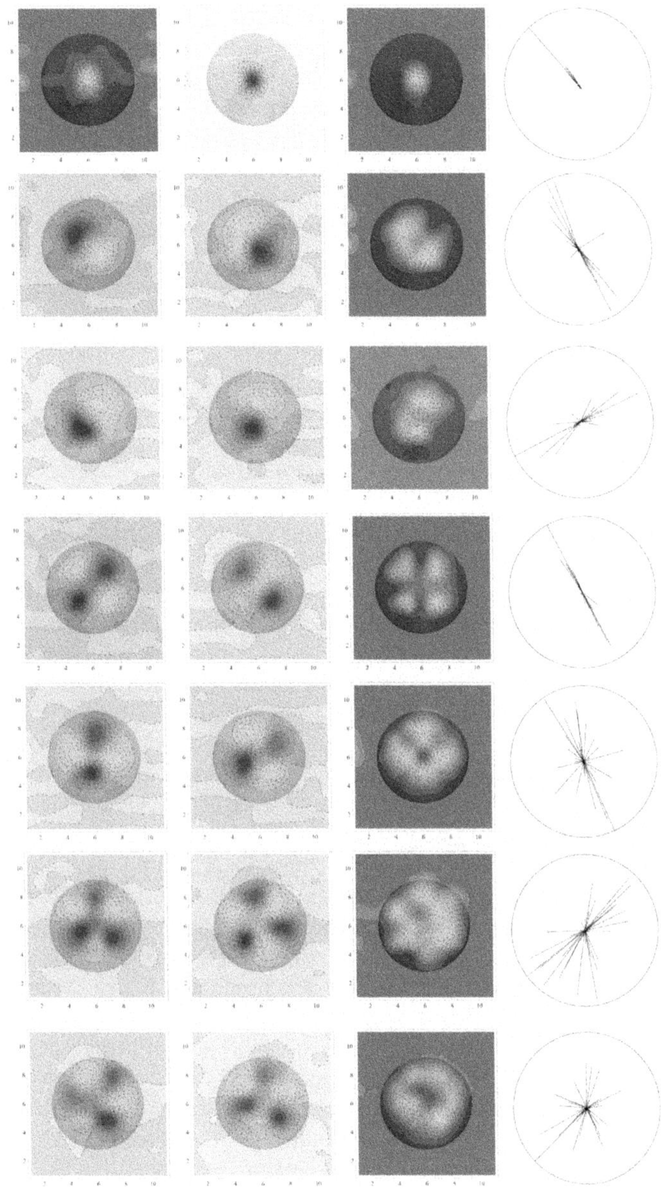

Fig. 5: Modes of an Usbek tambourine. Lines from top to bottom show increasing mode complexities: monopol, dipole, dipole, quadrupole, quadrupole, hexapole, hexapole. From left to right: real part, imaginary part, absolute part, phase plot for 121 reconstruction points. All data from one strike on the membrane. For frequencies see Tab. 1.

The quadrupole mode is a bit different from this situation. We find the lower mode shape to be very stable in its vibration pattern, which is verified by the phase plot in the fourth row. Still, the higher quadrupole shows much more unstable relations. Again, the real and imaginary plots confirm this very clearly. The lower mode shows the real and imaginary shapes to be the same with only black and white exchanged, the π phase shift. The higher mode shape again shows a circling mode, which in its imaginary part is getting closer to the lower mode shape. The reason for this behaviour is the amplitude relation (not shown here). The higher mode has much less amplitude and is therefore influenced by the lower one much more than vice versa. Thus the lower quadrupole mode shows nearly no influence from the higher one.

Finally, when looking at the hexapole modes, the phase relations are very unstable in both cases. Both modes are circling modes as can be seen from the real and imaginary plots in rows six and seven, showing strong influences. This is caused by the fact that these modes are damped very fast and so the uncertainty relation is holding to a much greater extend. Still, the basic mode shapes can be seen clearly.

Three more interesting modes are shown in Fig. 6. The first is 158 Hz and therefore below the fundamental mode, indicating a Helmholtz resonance of the air within the wooden frame, which is all in phase as can be seen in the top row. A second mode at 252 Hz is in-between the monopole and dipole mode. Here a medium strong peak occurs. This mode seems to be a coupling mode between an air mode and the dipole mode of the membrane, where unsymmetrical vibration of the lower and upper membrane occurs. The third mode at 874 Hz, quite strong in the spectrum, is a combination of eigenmodes in this frequency region. This can be seen in the phase plot, where about four distinct phase regions are present indicating a mixture of two membrane modes. This frequency, like the other ones discussed above, is gone that fast, that the uncertainty is mixing two modes into one peak.

So we have seen that the real struck tone of the tambourine can only be understood if we take the uncertainty principle into consideration, explaining the phase interplay between the degenerated modes appearing with the strike. Of course, these modes are radiating in the same complex manner and therefore are producing a rich, ever-changing timbre, which will also be changing with respect to binaural hearing, causing interesting stereo effects. Everybody knows the difference between a 'perfect', nearly homogeneous mylar membrane sound used for modern drumming and the much more inhomogeneous material sound of an animal skin drum head. The main reason are not the changing frequencies; and for such fast decaying sounds nor the perception of a beating between the degenerated modes appearing with such inhomogeneity either. The effects rather are:

- enriched timbre because of mode coupling between degenerated modes caused by the uncertainty principle, and
- enriched binaural or stereo effects of circular modes again caused by the mode coupling due to the uncertainty principle.

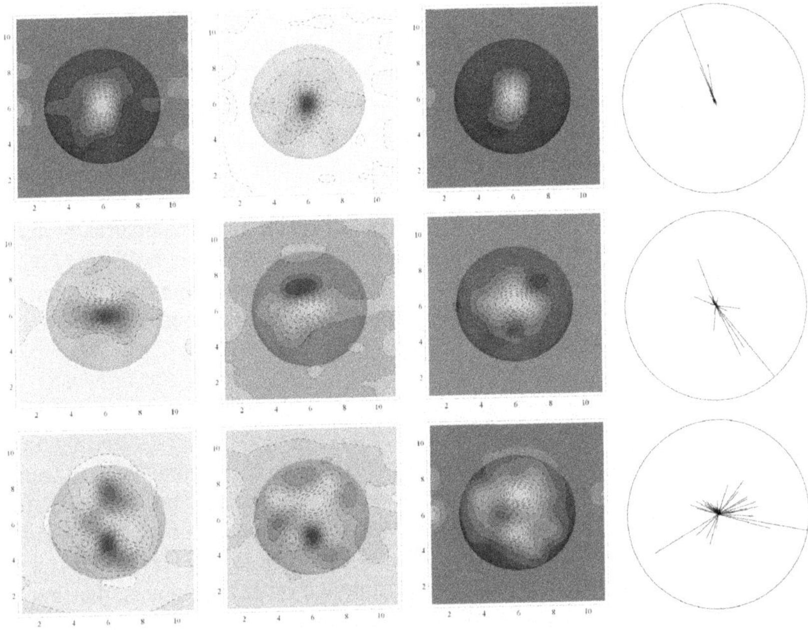

Fig. 6: Additional modes of the tambourine with low amplitude showing complex vibration pattern. Ordering from left to right like Fig. 5. Frequencies from top to bottom: 158 Hz, 252 Hz, and 874 Hz.

Frequency	194 Hz	316 Hz	326 Hz	436 Hz	444 Hz	540 Hz	566 Hz
n-pole	monopole	dipole	dipole	quadrupol	quadrupol	6-pole	6-pole
alpha (radiation characteristic)	0.8	2.2	1.8	4.8	4.8	10.0	9.0

Tab. 1: Directivity values for seven modes of an Usbek tambourine from Fig. 5.

Tab. 1 shows the directivity values α for the first seven modes of an Usbek tambourine. The values are getting larger for higher modes as expected. These modes show a more narrow directivity and so the overall directivity value α is getting larger. As the dipole, the quadrupole and the hexapole modes are all degenerated, the stability of the α-value is clearly appearing, as two modes of the same mode shape have nearly or indeed the same values for α.

1.2 *duff* frame drum

Another frame drum which sounds much longer than the small Usbek tambourine has been analysed. The *duff* frame drum from Pakistan is also made of goat skin. It has a diameter of about 45 cm and is used in Sufi music. Two basic striking positions are used. A rim shot producing a very rich overtone series sounding nearly metallic is the most typical for this instrument. The other one is a low sounding strike in the middle of the membrane. It is not used here because it is not so characteristic for the instrument.

In Fig. 7-9 18 mode shapes up to about 1 kHz are shown, representing the main peaks found with the single rim shot strike analyzed in this study. Only the real parts are shown as phase relations will not be discussed here. Many modes are degenerated, starting from the dipole mode at 239 Hz / 250 Hz, pointing to an inhomogeneous membrane again. Higher mode shapes are also found, e.g. a radial 18-pole mode at 942 Hz can be seen in Fig. 9. Above the plots, the value for alpha is shown, which is increasing with increasing mode complexity as expected. We have already displayed the directivity values for the tambourine in Tab. 1, where the same holds for higher modes. There, in more detail, the differences in directivity between the degenerated modes are shown, too, which are basically the same as expected. So indeed, the radiation of higher modes shows a stronger directivity.

The drum was also investigated by comparing two different humidity conditions, one of normal and one of more than 90 % humidity. When the air is getting moister, the membrane relaxes and thus the frequencies decline. Still from the spectrum one can hardly identify which mode has moved, as only peaks and not mode shapes can be compared. The shift caused by humidity was indeed tremendous with the *duff* (see Fig. 10). The quadrupole mode dropped from 334 Hz down to 186 Hz, which is a relation of 0.56, thus nearly halving the pitch. Concerning the other two modes, the first ring mode drops from 359 Hz to 201 Hz (relation: 0.56) and the octopole mode from 513 Hz to 275 Hz (relation: 0.54). This consistency in the drop indicates an overall drop of the membrane tension. As this tension is appearing to be linear and the frequency appearing quadratic in the eigenvalue equation of the membrane, we would expect the downward shift of the frequencies to be of a square root nature. The fact that at least in this frequency range the drop is linear, points to a much more complex nature of humidity influence. Indeed, as the damping of the membrane is much increased, the drum has much less higher harmonics, too. So a complex interaction between these two variables may cause this effect, which should be studied in further detail.

Another test for the microphone array method was done with the *duff*, too. The octopole radiation pattern measured for the single rim shot strike was compared to a Finite-Element (FEM) solution of a membrane, using the homogeneous and simplest eigenvalue formulation of a two-dimensional circular membrane (see Fig. 11). On the right, the 'perfect' mode shape of the FEM solution shows equally spaced nodes and anti-nodes in a circular manner. The measured octopole mode of the rim shot shows a slightly different behaviour for some details. Nevertheless, the overall shape is pretty much the same, indicating a good fit of expected to measured results. The small differences may be caused by several things. First, as the time window consists of only one strike, the membrane may change its shape during the initial transient phase of the strike. For higher frequencies this is indeed the only phase where they appear. Thus when integrating over the whole range, slight deviations may be caused by this initial struggle. Another potential influence has already been discussed in detail above: the slightly inhomogeneous nature of the membrane causing the mode shapes to differ from an analytically expected shape. To investigate the inhomogeneity quantitatively, the next section is proposing a method which can be used with a microphone array. Furthermore results from a study using this method for the *duff* drum head are presented.

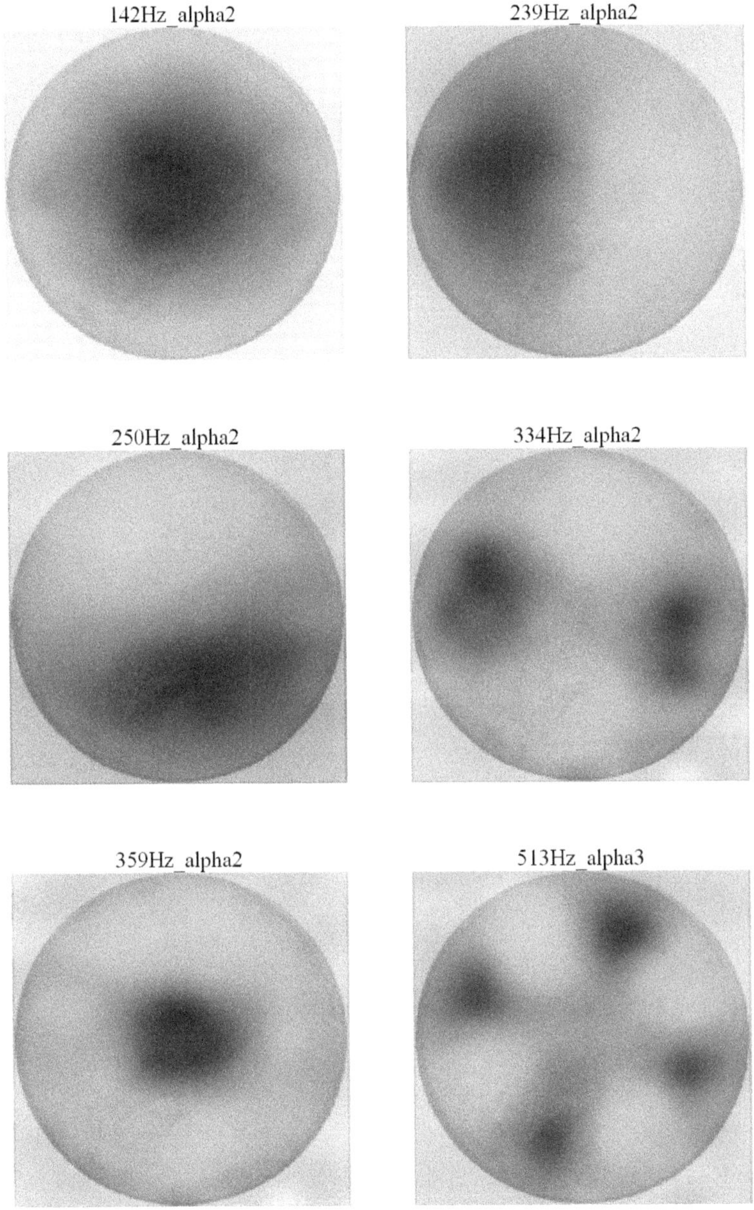

Fig. 7: Modes of the *duff* membrane as measured by the microphone array and back-propagated to the membrane surface up to 513 Hz.

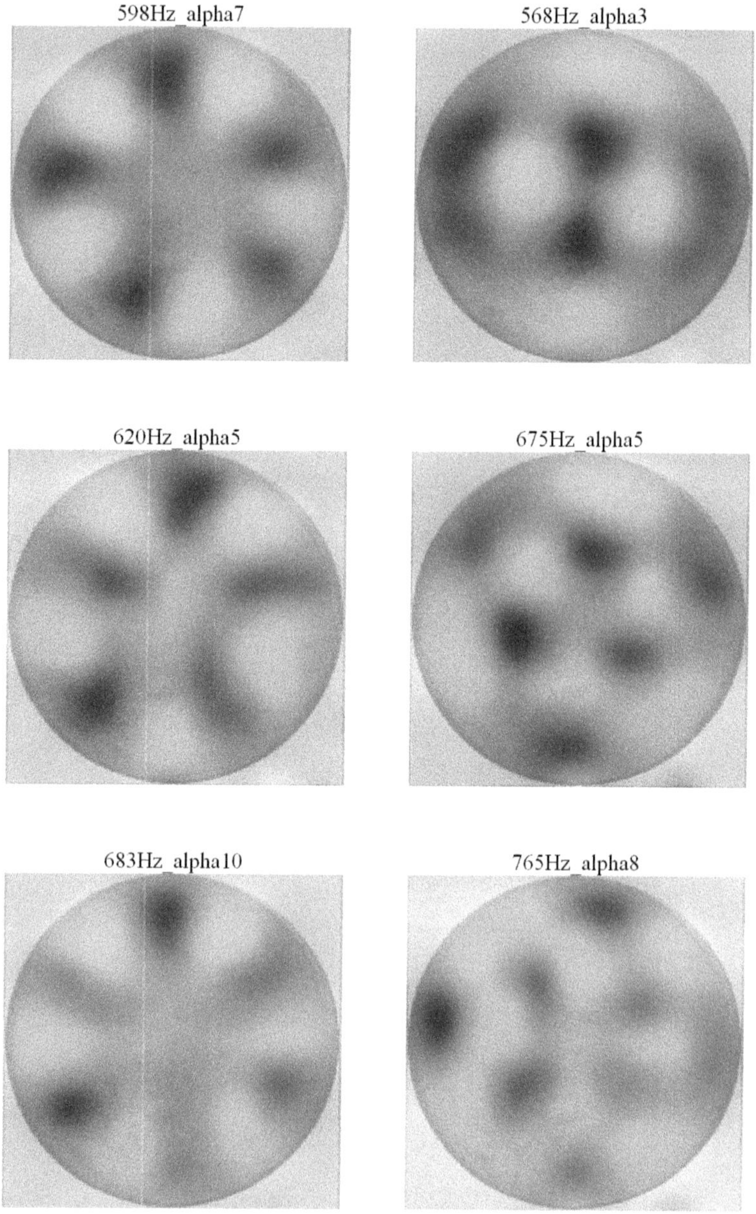

Fig. 8: Modes of the *duff* membrane as measured by the microphone array and back-propagated to the membrane surface up to 765 Hz.

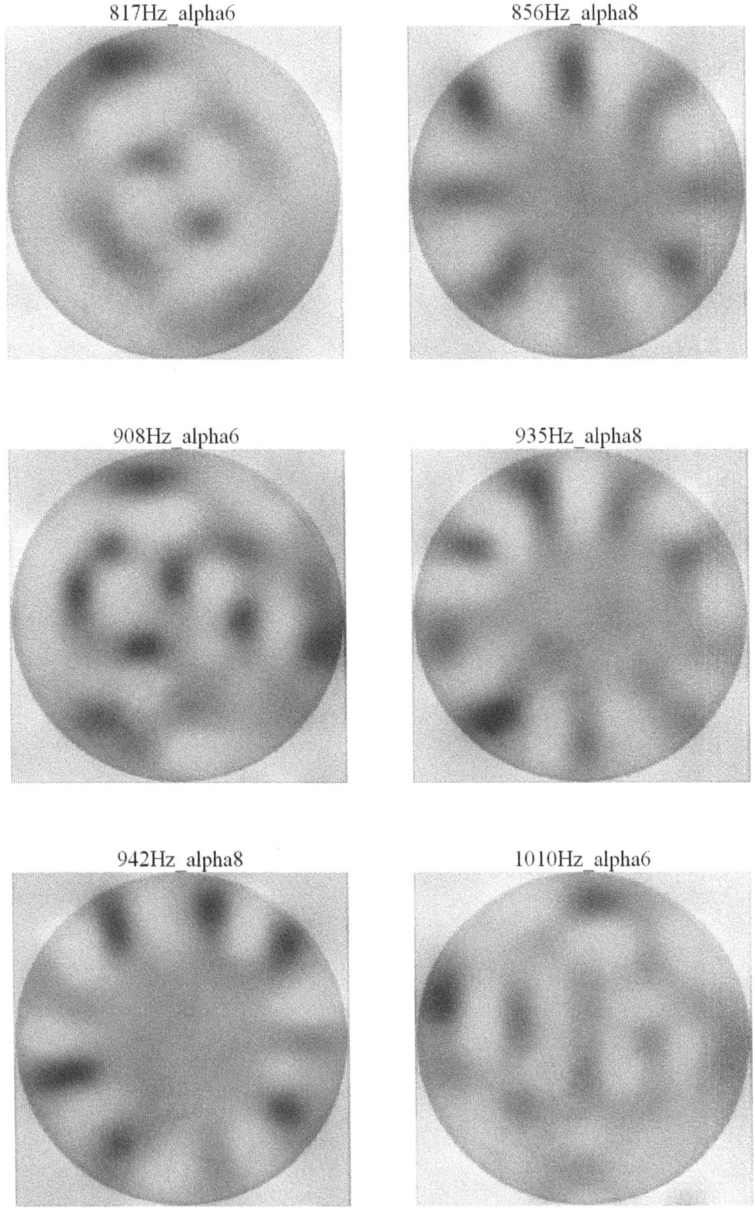

Fig. 9: Modes of the *duff* membrane as measured by the microphone array and back-propagated to the membrane surface up to 1010 Hz.

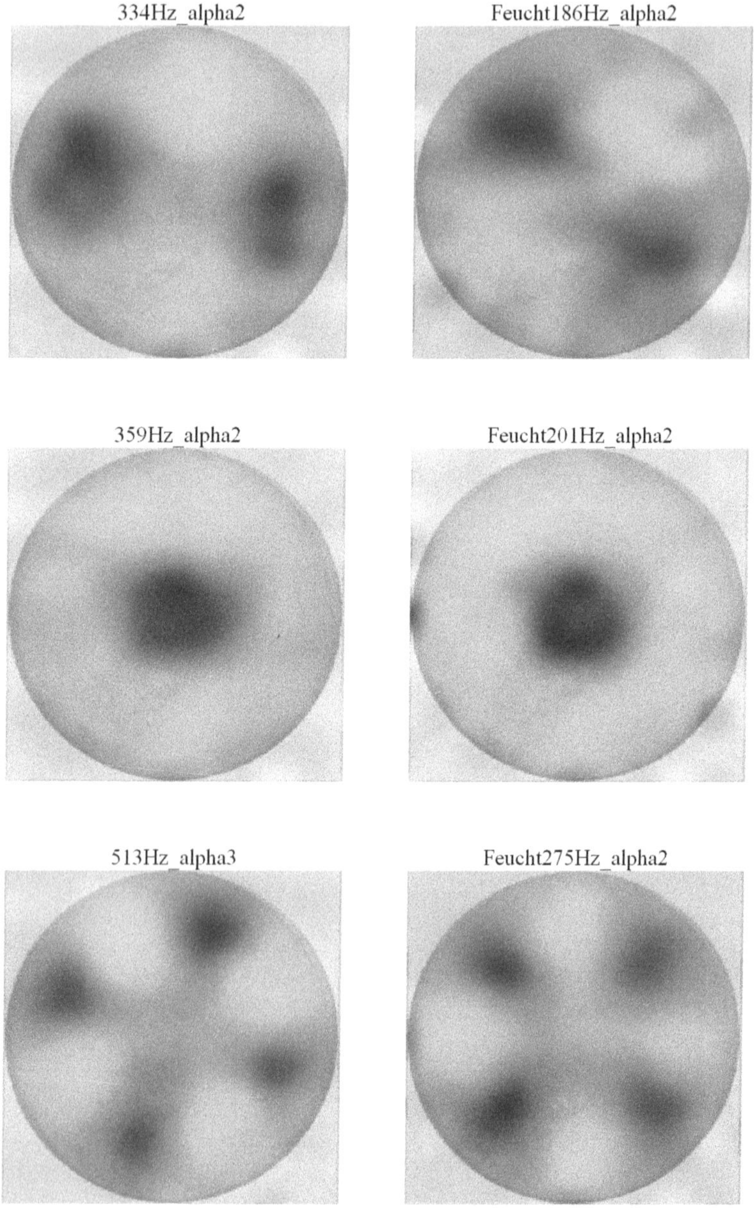

Fig. 10: Comparison between modes of the *duff* at normal humidity (left) and at high humidity (right) for three mode shapes with dramatically changing frequencies.

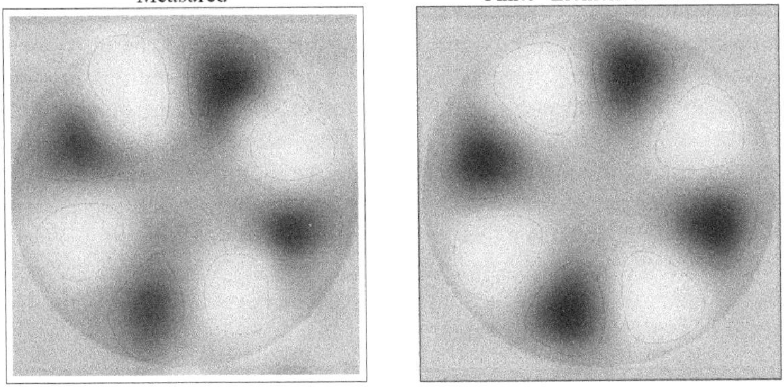

Fig. 11: Comparison between the measured mode shape of the *duff* and a Finite-Element solution for the same octopole mode.

1.3 Membrane Tension

The tension of the *duff* membrane changes along its area mainly for two reasons. The first reason is the material (goat skin) which is not perfectly homogeneous. The thickness of the skin is varying, and also the Young's modulus may change when the skin differs in terms of its molecular content. As skin is a natural material, these changes are arbitrary both along one membrane and between different drums. Additionally, the changes in thickness may also be caused by wear and tear of the drum when played. At the areas where the drum is struck by the hand or the fingers the material is getting thinner by-and-by. This leads to a more systematic change of the thickness, as the points where the membrane is struck are often held constant over the years.

Another reason why the membrane is not perfectly homogeneous is its attachment to the wooden rim. The *duff* investigated here is glued and therefore fixed to the rim of the instrument. Therefore the tension cannot be changed easily by hand. One would need to unglue the membrane, something an instrument builder would only do with a broken instrument. Thus, normally the player would not change the tension of the *duff* by himself. The tension is very hard to keep constant when it is glued to the rim. It can only be achieved if along the whole boundary of the membrane the same force would act in the direction to the midpoint of the drum. This is hard to obtain perfectly. Still it is necessary to get as close as possible to an ideal distribution of tension as otherwise nonlinear effects change the sound of the drum considerably.

The main audible effect on the sound of membranes not attached with equal tension along its rim is a change of pitch with each strike. Immediately after hitting the drum its pitch glides down to a lower note which is easily audible. Often a whole tone or even a minor third pitch glide is found which last for about 50 ms to 100 ms. This pitch glide is unacceptable for normal drums because it is a very prominent effect. A drum sound needs to contribute rhythmic sound rather than a sound associated with pitch, and therefore it should maintain its pitch over time.

The pitch glide is mainly caused by the differences in the tension of the membrane. When tuning a drum with tuning pegs attached to the rim of the instrument the tension may change considerably around the tuning pegs, causing regions of low and of high tension. When the drum is hit, the boundaries of such regions of different tensions act partly as a reflecting boundary for a travelling wave. Therefore, the vibrating area is smaller than the membrane as a whole around the region of the higher tension. After the initial transient phase of the strike, standing waves form which use the membrane as a whole as vibrating area. As this area is larger than the vibrating area of the initial transient phase the pitch is lower. But because these boundaries are never strict but complex it is hard to determine the precise trajectory of the pitch glide. Nevertheless it will occur to a higher or lower extent.

We should add that there is another reason for a pitch glide with membranes caused by a nonlinear effect. If the membrane is struck extremely hard it is displaced strongly. The restoring force of the membrane may not correspond linearly to this displacement, but can rather be larger. In such a case with great displacements the pitch is higher than with smaller ones. And as the membrane is reducing its maximum displacement with time because of damping, the pitch glides from a higher value at the beginning to a lower one at the end of the tone. Nevertheless, we do not find this kind of pitch glide with the duff. First, such heavy strikes are usually not performed. Second, even if they were, the effect would be negligible.

Pitch glides are well known to drummers who tune their instrument. The tuning is made possible by a certain physical behaviour of drums with circular membranes. These membranes show a harmonic overtone series next to the inharmonic partials. The harmonic components are caused by the angle solution part of its differential equation, which is similar to the solution of a harmonic string. The inharmonic partials stem from the radial solution part of this differential equation, which is a Bessel function with lead to inharmonic frequencies when applying circular boundary conditions. Thus the sound of a round membrane is harmonic and inharmonic at the same time. Because of its inharmonicity it may be associated with percussion instruments which normally show inharmonic spectra. On the other hand it may be considered as a tunable instrument suitable for playing melodies due to its harmonic overtone series. This solution of the overtone series is unique for percussion instruments. Other percussive instruments like the steel pan drums or the instruments of the Indonesian *gemelan* orchestra made of bronze can be used for melody playing mainly because the higher harmonics damp out very fast. So even without a harmonic overtone series the instruments will sound harmonic because one fundamental sinusodial will last longer than a few milliseconds. And other percussion instruments like bells or xylophones are built in a way to come close to a harmonic overtone set, at least for the first few partials.

Even if the *duff* drum is struck at the rim it shows many higher partials. Some of these partials form a harmonic overtone series and therefore the fundamental can be heard as the pitch of the drum. Moreover, when striking the drum at different positions higher harmonics can clearly be perceived and identified as additional pitches. The arbitrary character of the *duff* sound is one of the main reasons for its use in mystical *Sufi* music. First of all, different playing techniques change the spectral centroid of the sound, i.e. its brightness. Even more important is the arbitrary nature pitch perception caused by different perceivable pitches within one strike and between strikes at different positions on the drum. This richness of sound with its nearly endless possibilities enhances the feeling of an inner eternity of the *duff* sound, producing a multidimensional perceptual space. The inner eternity is associated with the highest potential of god, which the *Sufi* can 'hear' in the drum as an aesthetic presence.

Each drum has the potential of causing a similar perception. Still, drums differ because every drum has its own homogeneity. So we might say that the character of a drum is caused by the distribution of the density and the tension along its area. Physically, these parameters are force per area, the tension T and a linear density σ, which is mass per area. They are inserted in the differential equation governing the vibration of the membrane as the fraction T/ σ like

$$\frac{T}{\sigma}\left(\frac{\partial^2 u}{\partial x^2}+\frac{\partial^2 u}{\partial y^2}\right)=\frac{\partial^2 u}{\partial t^2}.$$

Here, u is the displacement of the membrane, which is differentiated twice with respect to the spatial dimensions x and y and with respect to time t. The equation describes the equivalence of the temporal change of displacement of each point on the membrane to the spatial curvature at this position and therefore the vibration behaviour of the membrane.

When examining the frequencies of the vibration of the membrane one can assume a sinusodial solution for the time component of u, and by solving the time differential accordingly, arrive at the eigenvalue equation of the membrane

$$\frac{T}{\sigma}\left(\frac{\partial^2 u}{\partial x^2}+\frac{\partial^2 u}{\partial y^2}\right)=\lambda u.$$

Here, λ is the eigenvalue and the frequency f is part of it: $\lambda = (2\pi f)^2$. When measuring the membrane tension with the microphone array, the resulting shapes of the eigenmodes are discrete, consisting of 11 x 11 points. Hence we need to reformulate the eigenvalue equation as a discrete matrix equation like

$$\frac{T}{\sigma}\mathbf{M}\mathbf{u}_0=\lambda_0\mathbf{u}_0.$$

The displacement u has become $\mathbf{u}_0$, now being a vector containing the complex displacements at discrete points on the membrane. The matrix M contains the derivatives of the analytical equation.

In this equation the tension and linear density of the membrane is assumed to be constant everywhere. But the character of the drum is mainly caused by the distribution of these values along the membrane. So we need to define a vector **ε**, containing the changes of the mean T/ σ resulting in a new vector **k**, which is also defined at each discrete membrane point like

$$\mathbf{k} = \frac{T}{\sigma}\epsilon \; .$$

So the new vector k is taking tension and area density into account, whereas **ε** is only carrying the distortion from a mean value. The elements of **ε** are around 1. Values larger than one mean a higher tension or density, while values lower than one mean a decline of these parameters.

As this change of the membrane tension and/or area density needs to be included in the differential matrix **M**, we can modify this material matrix to

$$\mathbf{M}_m = \mathbf{M}\mathbf{k} = \frac{T}{\sigma}\mathbf{M}\epsilon \; .$$

With this new material matrix and the distortion vector **ε** we can reformulate the eigenvalue equation:

$$\frac{T}{\sigma}\mathbf{M}\epsilon\mathbf{u}_m = \lambda_m \mathbf{u}_m \; .$$

Here, $\mathbf{u}_m$ is the mode shape measured by the microphone array, and λ_m is the eigenvalue of the mode shape. In this equation all variables are known except T/ σ **ε** = k. So we can reformulate this eigenvalue equation to become the linear equation system

$$\mathbf{R}\mathbf{k} = \mathbf{r} \; ,$$

where

$$\mathbf{R} = \mathbf{M}\mathbf{u}_m$$

and

$$\mathbf{r} = \lambda_m \mathbf{u}_m \; .$$

This equation system can easily be solved for **k**. If we take the mean value of **k** as the overall value for

$$\frac{T}{\sigma} = \frac{1}{N} \sum_{i=1}^{N} \mathbf{k}_i \,,$$

we can calculate **ε** by solving the linear equation system again, slightly altered to

$$\frac{T}{\sigma} \mathbf{R} \epsilon = \mathbf{r} \,.$$

Now we have calculated both a mean value for the membrane tension holding at each point of the membrane and the vector of alternating tension and area density at all membrane points. We can expect this tension to be associated with the two physical reasons for such variations: the different glue strength at the rim and the material differences within the membrane itself. The method needs to be consistent over different striking positions of the membrane.

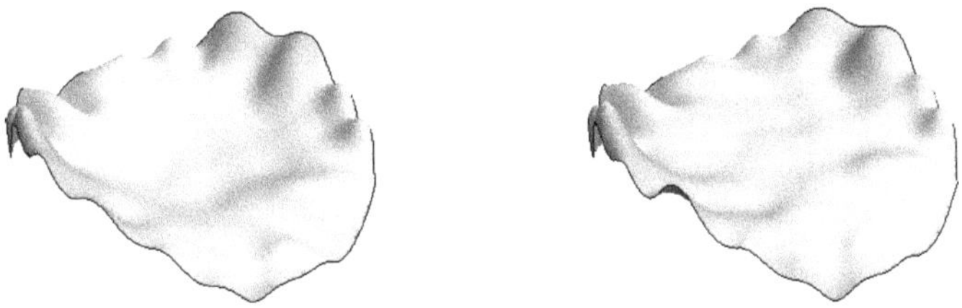

Fig. 12: Tension and area density distortions of the *duff* membrane measured using the method shown in the text for two different strikes. Left: struck in the middle of the membrane. Right: struck at the rim points.

Figure 12 shows two plots of the calculated values of **ε** for two different strikes, using the fundamental mode of the membrane as shown in Fig. 7. On the left, the result for a strike at the middle of the membrane is shown. In case of the figure on the right, the membrane was hit at its rim position. Both plots show the same tendency, but nevertheless differ a bit mainly within the middle region of the membrane.

2. Wind instruments

Investigations of wind instruments using a microphone array are concerned with the radiation from the different holes of the instrument. Several features of wind instruments can be studied in this way:

- Radiation impedance of tone holes or blowing holes of flutes, horns, and bells with reed and brass instruments
- Determining the real length and shape of the pressure wave along the open tone holes and the end of the tube by measuring the radiation from holes beyond the tone hole
- Turbulence sound production with flute instruments
- Phase relations between blowing holes and tone holes predicting end-corrections of the tube lengths
- Phase relations between partials and the related impulse shape of the travelling wave in the tube, as well as mode-locking between partials
- Directivity measurements of sound radiation from wind instruments
- Calculation of Spaciousness and Interaural Cross-Correlations with binaural hearing for different microphone and listening positions.

Based on these parameters several conclusions can be drawn, concerning among others the

- Instrument characteristics determined by the radiation pattern over different blowing regimes
- Articulation of a player in terms of musical phases
- Instrument quality description in terms of leakage detection or reasons for low radiation
- Surround sound reconstruction of a sound field of a performance from a mono or stereo recording
- Sound synthesis, taking movements of a player into consideration.

Three examples for flutes are shown below, investigating a Balinese *suling*, a Japanese *shakuhachi*, and a Chinese *dizi* flute. They are all made of bamboo. The *suling* and the *shakuhachi* are end-blown, while the *dizi* is a transverse flute. The *suling* serves as an example of a soft bamboo flute with simple radiation patterns. The *shakuhachi* sound is very noisy, and therefore turbulent sound production may occur. The *dizi* flute is the most sophisticated one as it has an additional tone hole between the blowing and the sounding tone hole, which is covered by a thin bamboo membrane. The membrane adds a bifurcation scenario to the sound leading to a much more rough tone than one would expect from a bamboo flute. The flutes are investigated in terms of a comparison between the radiation strength of the blowing hole and the tone hole, and the phase relations appearing between them. Also some special features of the flutes will be shown, some of them needing further investigations in the future.

2.1 Balinese suling flute

The *suling* flute exists in several lengths. It is played in ensembles as well as solo instrument. The flute investigated here is 21.5 cm in length and can therefore be considered as a small version of the instrument. It is played with circular breathing, producing ornamented melodies. It has six tone holes of equidistant tuning. The flute under investigation was bought by one of the authors in Bali in 1999, and so the production of the flute was documented. After the first tone hole was drilled, the other five were cut at a constant distance to the first one by using a small bamboo stick, which was also used to measure the distance between the holes. Thus the equidistance of the tuning is explicitly wanted, and the builder was also aware of this tuning when asked. This tuning is not common with the Balinese or Javanese *gamelan* orchestras, which are tuned to *pelog* (7-tone) or *slendro* (5-tone) tuning. However, these tunings differ a lot between different ensembles. Modern Balinese tunings use tone distances which favour some very sharp intervals, sometimes only 70 cent, followed by very large ones. In contrast the intervals in older tunings were distributed more equally over the octave. So the *suling* flute will never fit in perfectly in any *gamelan* orchestra, and for this reason it is mostly played as a solo instrument.

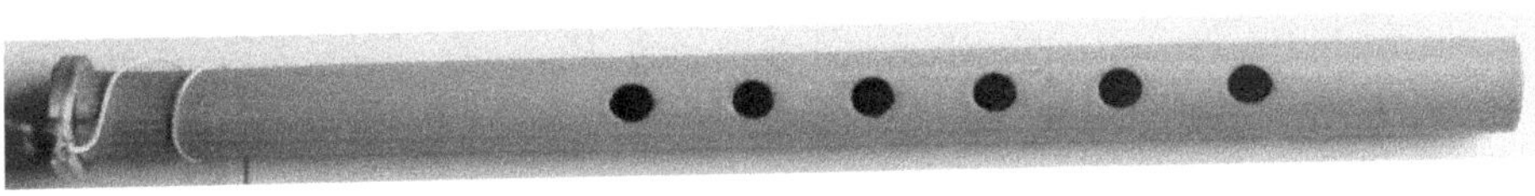

Fig. 13: Top: Balinese *suling*. Bottom: labium is behind the bamboo ring.

The *suling* is easy to play. The cutting edge, where the player is blowing in to produce a self-sustained oscillation is cut at the other side of the tube near the end, where the bamboo stick is covered by a ring of bamboo slides, tied together by a thread. The other side of the thread is wrapped around the tube loosely so if the ring is occasionally sliding from the tube it will not get lost. The ring is used to direct the air blown in by the player towards the tone edge. Above the edge that is located below the ring, the tube is slightly carved to let pass an exiguous and directed air stream. Therefore the flute is easy to play and everyone is able to produce a tone immediately. The *suling* sounds very nice and tender and has a typical bamboo flute sound. The playing style is governed by ornamentation rather than by different blowing styles as appearing with the *shakuhachi* flute discussed below. So when investigating the flute only one tone needs to be played for each fingering.

Both the labium and the tone holes are small. The labium is a rectangle hole with a side length of 5 mm. The tone holes have a diameter of 7 mm. The tube itself has a diameter of 1.4 cm and the bamboo is about 1 mm thick. So we can expect the radiation from this instrument to be similar to a monopole, at least in the front plane which is investigated here. The monopole radiation should result in a directivity value of around zero.

Fig. 14 shows the first four partials of a *suling* tone. The vertical line indicates the boundaries of the tube, which is blown at the very top of the plot. The vertical line shows the position of the first open tone hole. As the *suling* has a very soft sound, higher harmonics appear with low amplitude only. Therefore the tone is represented more or less completely with these four plots. It can be seen that both the blowing hole (at the upper edge of the plots) and the tone hole are the radiating points of this flute. However, the degree of strength varies with the partials. Concerning the low partials, the blowing hole is the stronger radiator. On the other hand, with the fourth partial the tone hole is the most prominent radiator. Additionally, the radiation from the blowing hole is much more complicated than that from the tone hole. This is because the cutting edge is on the opposite side of the tube and therefore, all radiation we get into the front plate shown in the plots is caused by refraction and scattering. Here, the first and the second partial have a broad radiation both at the tube and at its sides as the waves need to travel around it. Looking at the third partial one can see that the sides of the tube are radiating stronger than its middle part. This indicates that in this case the refraction is more directional. The highest partial can be neglected because it has such a low radiation from this point.

Tab. 2 shows the directivity values for the four partials. They range from $\alpha = 0.0$ to $\alpha = 0.1$, both indicating monopole radiation as expected. We will find this with the other flutes, too. Therefore we can say that the radiation from flute tone holes is of pure monopole character in the front plane of the flute.

partial	α
1. (580 Hz)	0.1
2.(1160 Hz)	0.1
3. (1770 Hz)	0.0
4. (2370 Hz)	0.0

Tab. 2: Directivity value α for the four partials of the *suling* tone. All partials radiate like a monopole.

Fig. 15 shows the results of radiation strength quantitatively. For higher partials the tone hole radiates more than the blowing hole. When examining the phase relations between the blowing and the tone hole for these first four partials, we find a reasonable relation for the second, third and fourth partial, where relations of zero or π are found. This can be explained by the normal mode shapes of a tube. However, the fundamental partial shows a relation of $\pi/2$ which is more difficult to explain. One possible explanation could be that the fundamental partial has a different end-correction compared to the others and therefore needs more time to appear in the radiation. But it could also be that the boundary conditions of the flute are different for each partial. Because both explanations can be seen as complementary, further investigations on the *suling* should be done to give insight into the frequency-dependent end-correction of this flute.

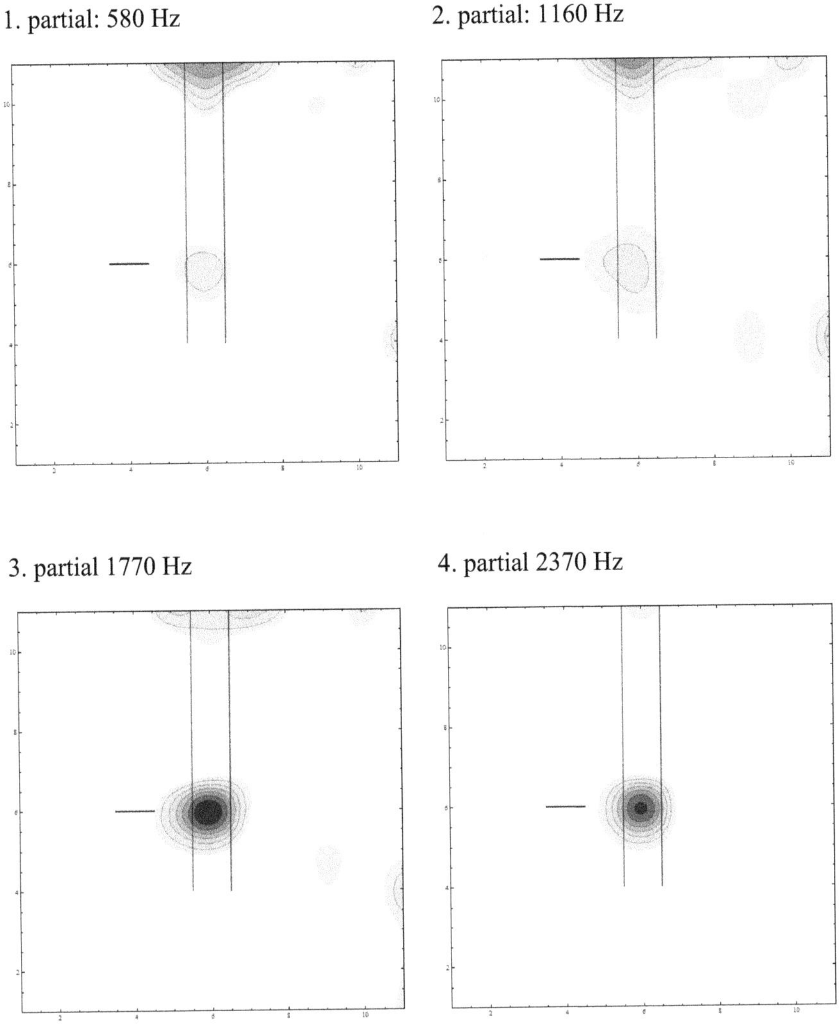

Fig. 14: Radiation of the first four partials of the *suling*. The vertical lines indicate the tube, the horizontal line shows the first open tone hole position.

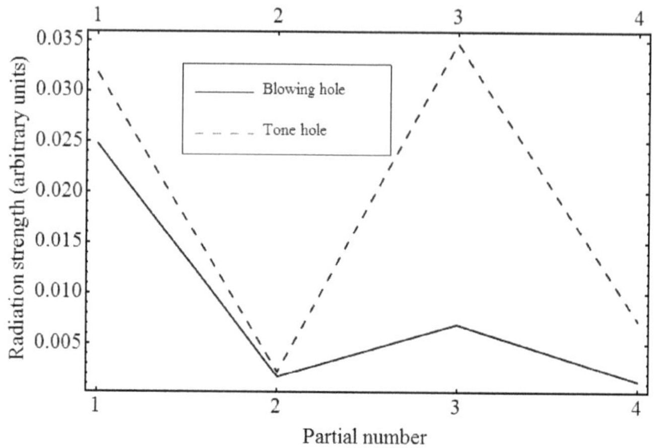

Fig. 15: Radiation strength of blowing hole vs. tone hole for the first four partials of the *suling* tone. The tone hole shows clearly stronger radiation compared to the blowing hole with higher partials.

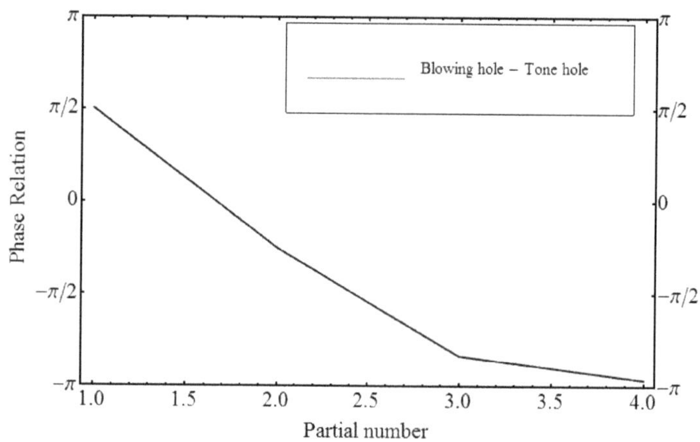

Fig. 16: Phase relations between blowing hole and tone hole for the first four partials of the *suling* tone. Phase differences as expected for the second, third and fourth partial. The first partial shows a $\pi/2$ relation.

2.2 Shakuhachi

Compared to the *suling*, the *shakuhachi* flute is more complicated in terms of radiation. The flute has several playing techniques to alternate the pitch and the amount of noise. Normal fingerings determine the pitch of the instrument, but it can also be overblown to one higher register. Additionally, pitch slides can be achieved either by altering the distance of the lips to the labium or by slowly opening and closing the tone holes. Furthermore the *shakuhachi* can be played with much breath to enhance the noisy sound of the flute. These playing techniques are all common. As the melodies are very slow and tones are held for several seconds, the timbre of the instrument is the most crucial part of the playing rather than the small amount of pitches being used.

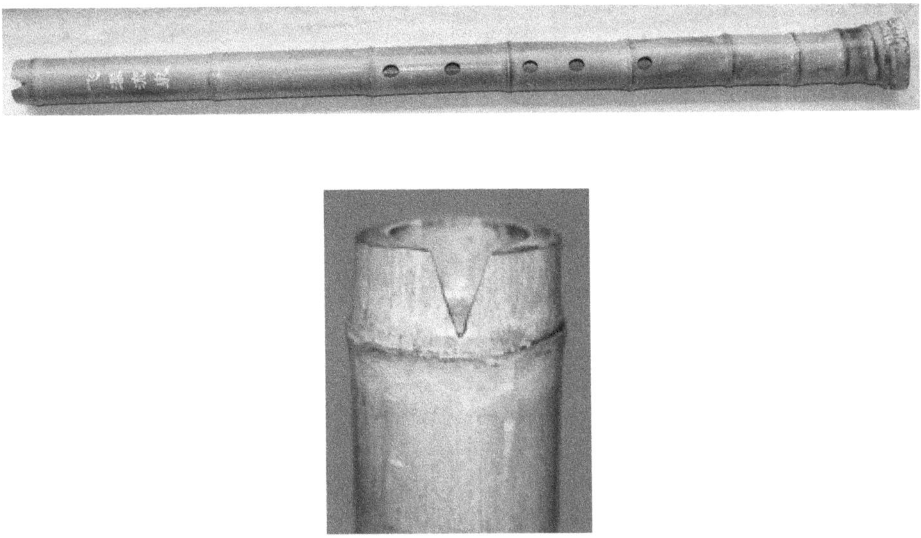

Fig. 17: Top: the *shakuhachi* used in the investigation. Bottom: labium of the flute. The lower lips cover most of the open hole.

So with this instrument we can expect a huge amount of turbulence appearing at the labium which could lead to a brightening of the radiation area around the blowing hole. Also, radiation from the first open tone hole is expected.

Two tones were played and than compared, a normal playing tone and an overblown tone with the same fingering. When overblowing the *shakuhachi* it is easy to create a multiphonic sound, where both the normal tone and the overblown tone occur. This is astonishing at first, because the overtone spectra of both sounds overlap completely. Nevertheless a multiphonic sound is perceived and so it was used as the overblown sound in the investigation. In the analysis the overblown tone starts with the same frequency as the normal tone. It is, although quite low, clearly present in the spectrum, and also clearly perceived.

Fig. 18 shows the radiation patterns of the *shakuhachi* in comparison. It first appears that the radiation from the labium compared to the radiation area at the tone hole is decreasing with higher harmonics. This was also found with the *suling*. However, with the overblown tone this part of the flute is much less radiating compared to the normal tone and so this effect can be neglected here. Even for the first partial of the overblown sound the labium radiation is low. Also, the radiation area around the labium is small. We did expect a much larger radiating field as we expected strong turbulences to appear with the flute tone. But it seems that this turbulent velocity field is not reaching very far and so radiation is indeed focussed to the labium region. This may also be the case because the blowing pressure, at least for the normal tone, is not very high.

Normal tone, 1. partial: 396 Hz

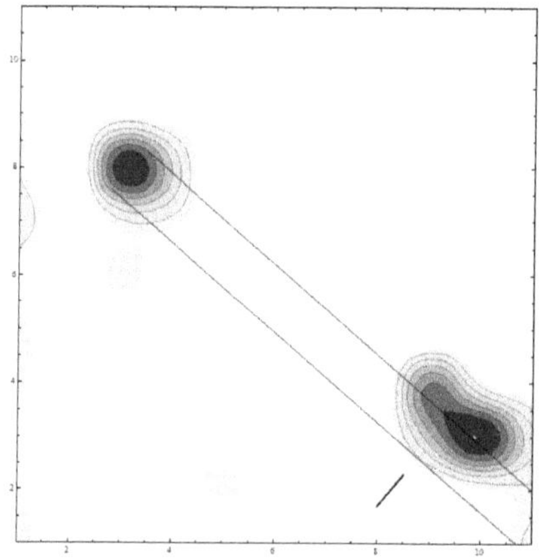

Normal tone, 2. partial: 795 Hz

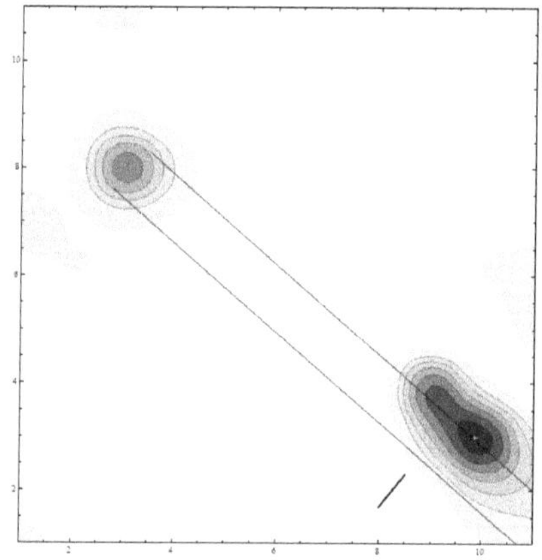

Overblown, 1. partial: 396 Hz

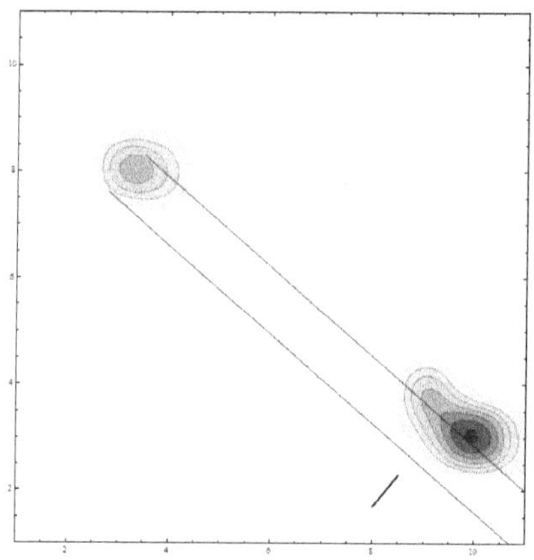

Overblown, 2. partial: 795 Hz

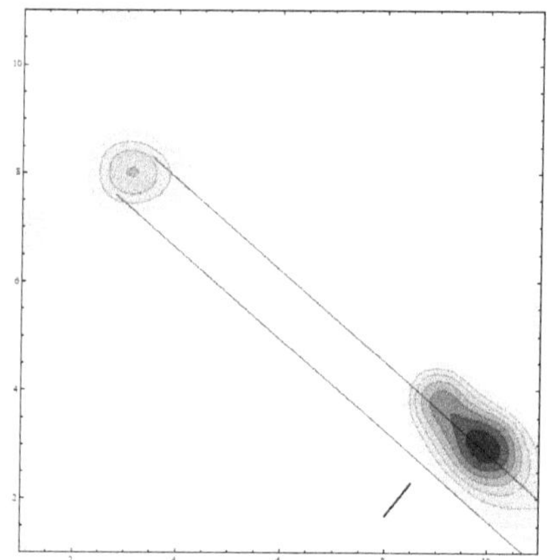

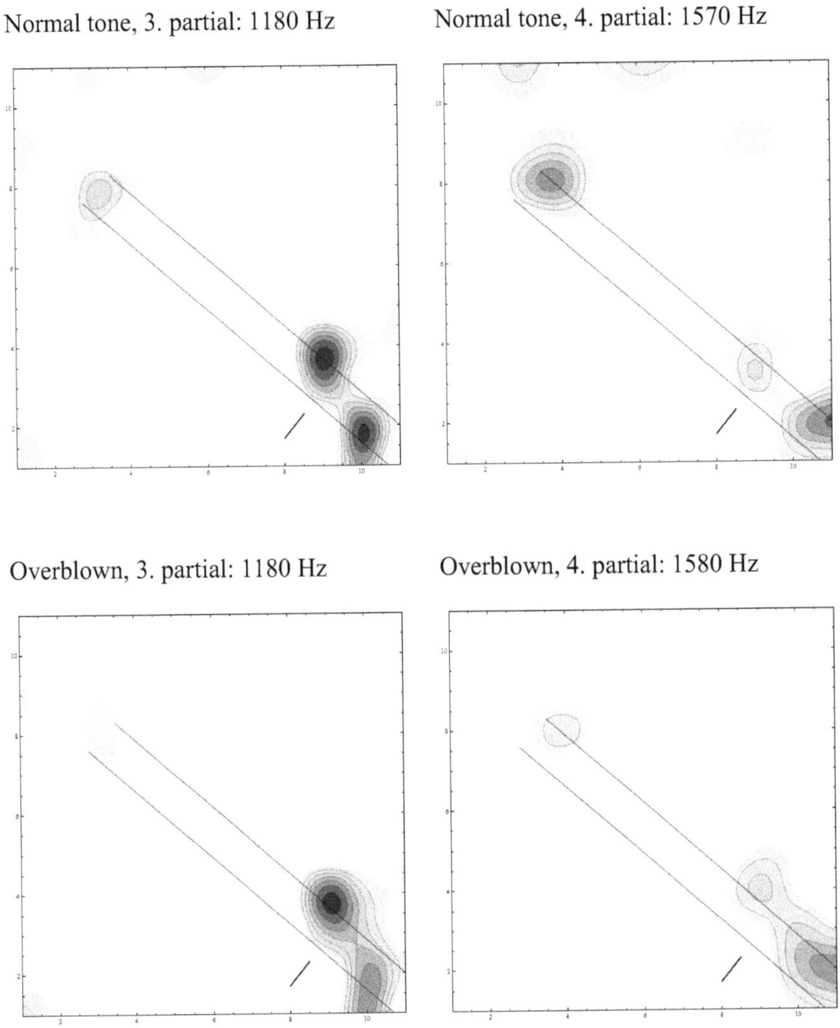

Fig. 18: Comparison between the first four partials of a normal and an overblown tone of the same fingering for the first four partials.

Another interesting aspect is concerning the radiation area around the tone hole, which is indicated at the lower right of the plots by a short solid line. Complex radiation appears around the tone hole for all partials. When comparing the patterns for the normal and the overblown tone for each partial, strong similarities appear. With the first and second partial the radiation area is shifted downwards to the bottom of the flute before splitting up into two separate regions. This behaviour can be associated with the fact that with higher partials, the whole tube is vibrating with the open tone holes. Because the open holes participate in the radiation, a complex radiation pattern is achieved. In the plots shown in Fig. 18 only parts of the flute are shown, as mainly the radiation from the blowing hole was about to be investigated. Further studies need to be done taking the open holes into consideration, too, especially with higher frequencies.

Normal	alpha	Over-blown	alpha
396 Hz	0.2	395 Hz	0.2
795 Hz	0.2	802 Hz	0.2
1180 Hz	0.2	1180 Hz	0.4
1570 Hz	0.2	1580 Hz	0.0

Tab. 3: Directivity values of radiation of the *shakuhachi* in normal and overblown condition. All radiations are monopole radiations.

As one can see in Tab. 3 the radiation directivity is a monopole radiation just as with the *suling* (see above).

When examining the radiation strength and phase differences for the normal and the overblown tones (Fig. 19 and Fig. 20) the overall stronger radiation of the tone hole stands out clearly, at least in Fig. 19. It is not that clear in Fig. 20 because the radiation is distributed over a large area, while in the plot only the radiation point at the tone hole was taken. When integrating over the area, the increased strength will stand out more.

On the other hand the phase differences with the overblown tone are not so easy to interpret. The normal tone shows the expected relationship between the phases of zero phase, π and $-\pi$ relation respectively. In contrast for the overblown tone the relation is nearly zero with all four partials. This cannot be explained completely yet, still the overblown tone has a multiphonic character and therefore the first and the second partial may be treated separately. If so, the second partial has the fourth partial as its first overtone and so the third partial is only associated with the first one. Thus they need to show the same phase relation which they indeed do. Also the second and the fourth partial show the same phase relation, and therefore are perfectly reasonable. However, the second and forth partial are also present in the overtone spectrum of the first partial and therefore the radiation measured would be a superposition of these two phases. To separate them, we would need to associate them to a theoretical framework of impulse and impedance behaviour of this flute, which is beyond the scope of this paper.

So two phase relations are not fully understood yet: the first partial of the *suling* and the phase relations of the overblown tone of the *shakuhachi*.

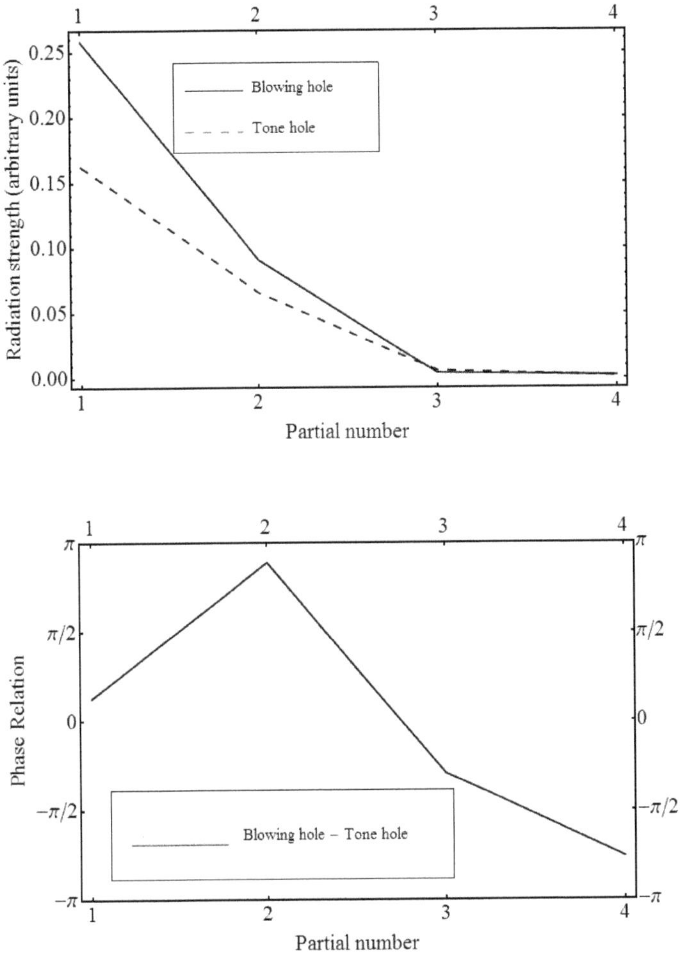

Fig. 19: Radiation strength and phase differences between the blowing hole and the tone hole for the normal *shakuhachi* tone. The tone hole radiation is stronger again. Still, only the radiation right of the tone hole point is taken here, and not the hole radiation area. The phases show a behaviour as expected from such a tube.

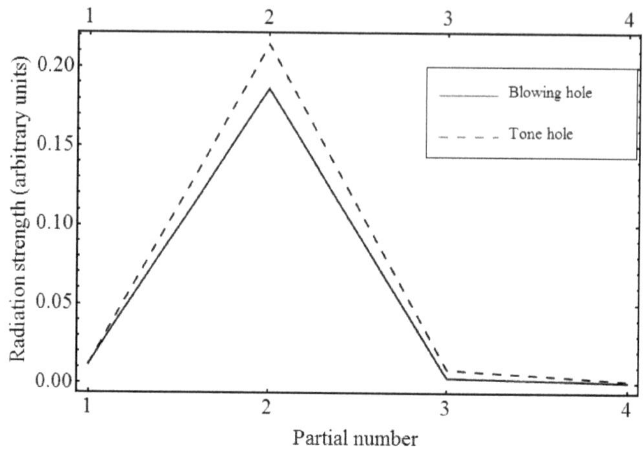

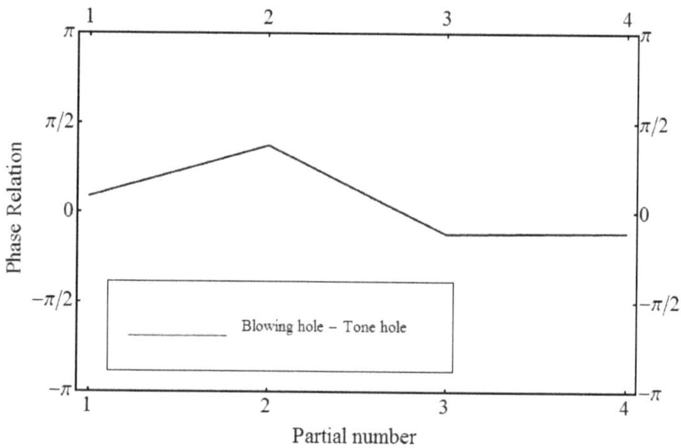

Fig. 20: Radiation strength and phase differences between the blowing hole and the tone hole for the overblown *shakuhachi* tone. The tone hole radiation is stronger again. As above, only the radiation right of the tone hole point is taken here, and not the hole radiation area, which makes the difference much more clear compared to the plots of Fig. 18. The phases show a behaviour of nearly equal phase. This may be caused by the multiphonic character of the tube, making the first and third and the second and fourth partial different tones. Still this aspect is not perfectly clear yet.

2.3 *dizi* flute

The *shakuhachi* showed complex radiation pattern around the tone hole region. This behaviour can be expected from the *dizi* flute, too. Additionally, we can expect some kind of radiation from the hole which is located between the blowing hole and the first open tone hole and covered by a thin bamboo membrane. The type of *dizi* used in the study comes from the northern part of China. It has a rough sound that is caused by a nonlinear coupling between the membrane and the air column. The coupling results in a spectrum of bifurcating partials, which can be described by the classic formula

$$f_{n,m} = n\ f_0 + A_m\ m\ f_b \qquad n = 1, 2, 3... \quad m = 0, \pm 1,\ \pm 2, \pm 3, ...$$

Here, the spectral components of frequency $f_{n,m}$ are composed of a harmonic part f_0 and a deviating part f_b, where f_b is usually small compared to f_0. The amplitudes A_m are decreasing with increasing m. So each spectral component of the harmonic overtone series is surrounded by several additional components, which are equally spaced in the frequency domain around the harmonic one. This feature of the *dizi* sound is well known. The formula is also used for multiphonic sound description.

Fig. 21: Top: *dizi* flute. Bottom: membrane hole covered with a bamboo membrane.

When investigating the lower partials of the tone as shown in Fig. 22 we find a strong radiation from the blowing hole and the tone hole. As with the other flutes the radiation from the blowing hole is strongest with the lowest partial. With the second and third partials, the radiation of the tone holes increase in relation to the blowing hole. In the plots, the blowing hole is at the left and the tone hole at the right, indicated by the vertical lines. It is interesting to see that the blowing hole radiation is appearing at the body of the flute, while the radiation from the tone hole is appearing above it. The flute is played towards the array, and so the blowing is outward the normal direction of the plots while the tone holes are at the top boundary of the flute and therefore in the up direction of the plots. Thus the expected radiation of the flute is above the flute body as indicated in the plots. But this is not true, especially for the blowing hole. Here, the radiation is appearing at the flute body, indicating that the turbulent self-oscillating velocity flow of air around the blowing hole is producing sounds mainly when travelling downwards the flute. This is reasonable, as the flow is only shortly entering the tube and for most of the time stay outside of it. So this periodic motion is mainly appearing when the air inside the tube is driven, which would cause the flow to travel along the outward tube wall causing radiation around the tube area stronger than above the tube.

This picture changes when looking at the higher partials. In the plots of Fig. 23, the membrane hole is indicated by an additional vertical line between the blowing and the tone hole. Again, the radiation from the tone hole is clearly there and above the tube. Also radiation is happening at the blowing hole. But now, additional radiation is taking place at the membrane hole. This is getting stronger with higher frequencies, and with the 7th partial, the membrane hole radiation is indeed stronger than both, the blowing and the tone hole. This indicates, that the membrane hole is driven by higher frequencies which makes sense in a way, that the membrane itself is very small - indeed the same diameter as the one of the tone hole - and so vibrates with a very high fundamental frequency.

What is also interesting to see is an interaction between the blowing hole and the membrane region outside the tube. This is strong for the 4th up to the 6th partial. With the 5th partial, an additional separate region betweeen and a bit above the two other regions can be found. We can conclude, that here, coupling of the air outside the *dizi* tube is happening, which need to be taken into consideration when understanding the sound production of this flute. Also, as the radiating region is now tremendously enlarged to the upper air field of the tube, we may find additional radiation caused by turbulent air motion in this area. So we find the membrane hole not only to radiate strongest with higher partial but also to couple to the blowing hole in a region around 3 kHz. The tone investigated here was played with two tone holes closed. Further investigations could look for couplings between the membrane hole and the tone hole, when the flute is played with all tone holes open, because then the first tone hole is so near to the membrane hole that coupling could occur between these two, too.

Another aspect of the *dizi* radiation is similar to the findings of the *shakuhachi* flute, where radiation from additional tone holes beyond the open hole existed. This is also true for the *dizi* flute, and clearly appears with partials 4, 6 and 7. Again, it is not clear from the plots which holes contribute to which extend as the flute continues on the left and only the part of the flute from blowing hole to tone hole was investigated here. If additional energy is coming from any side of the field, where the algorithm was told to expect radiation, the resulting radiation plot would show some energy coming from the side as a kind of sideway radiation influence. This is happening with the plots of Fig. 23. Further studies would again need to take a closer look at this region.

As mentioned above, the sound of the *dizi* flute shows a bifurcation scenario with additional partials surrounding the harmonic overtones. Still, when comparing the radiation from the different partials in the region around one harmonic overtone, the radiation does not change considerably, sometimes it is hard to see any differences. The radiation seem to depend on the frequency region solely. This also means, that the bifurcating frequencies, generated by the nonlinear behaviour of the air column – membrane interaction all vibrate the same way. Physically this means, that both kinds of vibrations are travelling all along the tube and so radiate through all the holes. As radiation depends upon the radiation impedance of the tone holes and blowing holes, and as this impedance is frequency dependent, the radiation pattern of nearby frequencies is expected to be very similar. This is indeed found with the bifurcating spectral components. Still, the travelling of the different waves through the tube need to be investigated in more detail here, possibly with a physical model.

Fig. 24 sums up the radiation strength of the holes quantitatively. The tendencies seen in the plots are clearly there, the blowing hole radiates most with lower frequencies and much less with higher ones. The membrane hole has practically no radiation with the 1st partial and is most loud from the 5th partial on. The tone hole stays in the lower middle region of radiation being least from the 4th partial on. Tab. 4 shows the directivity of the partials, again all are monopoles as expected.

The phase relations shown in Fig. 25 are more complicated. The phase difference between blowing and tone hole starts with zero phase as expected. The second and third partial have a $\pi/2$ and $-\pi/2$ relation respectively. So with the second partial, the blowing hole is ahead of the tone hole while with the third partial it is behind. A simple tube would show a π difference for the 1st partial and an in-phase radiation with the 3rd. The occurance of a $\pi/2$ relation points to a complex pressure behaviour in the tube. Even more strangely, the blowing hole and the membrane hole nearly exactly show the same relations for these first three partials. This means, that here, the membrane hole is in phase with the tone hole, which is verified by the third line in Fig. 25. This in-phase relation between tone hole and membrane hole could point to a separation of the flute at the membrane hole. The part between membrane and tone hole would act as one vibrating air column on its own. This would be the case for the first three partials only, as then the phase relations change again. Then, from the 3rd to the 7th partial, the relation between the blowing hole and the membrane hole are constant around $-\pi/2$.

Again subject to further investigations, we still may say, that for the different frequency regions, the overall behaviour of the flute changes fundamentally. The precise relations and influences within such a tube need to be studied in further detail. Of course with such a strong nonlinearity added to a tube, we would expect such a complex phaenomenon, as is the bifurcation scenario of the spectrum, a clear indicator of multiple coupled relations.

1. Partial 571 Hz

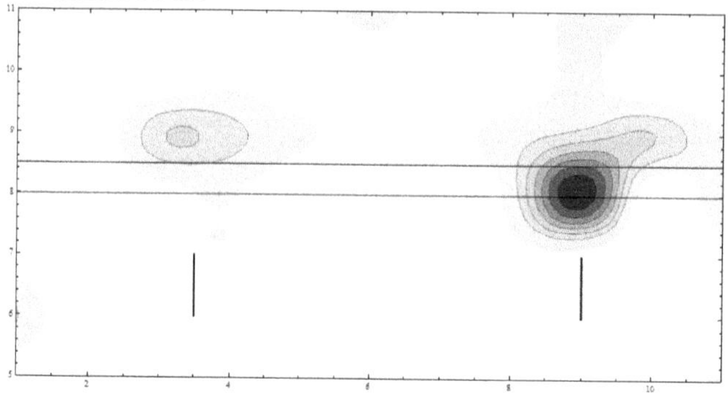

2. Partial 1140 Hz

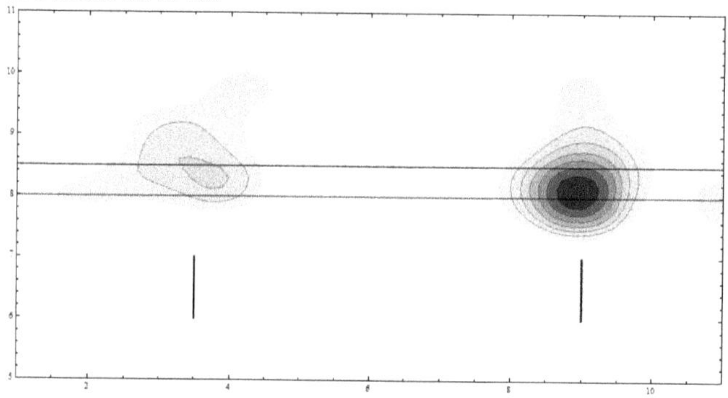

3. Partial 1720 Hz

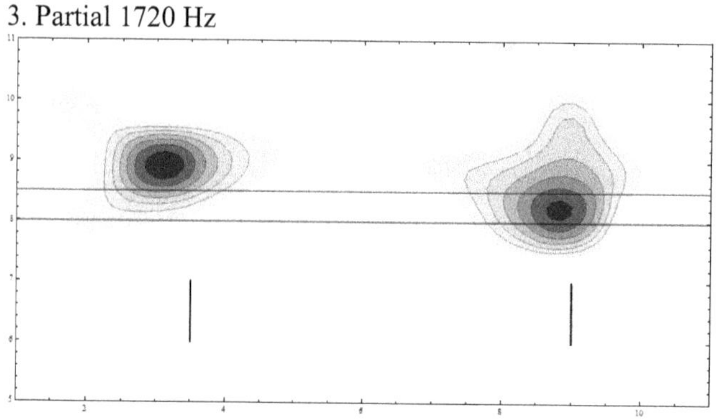

Fig. 22: First three partials of the *dizi* tone. Flute indicated by the horizontal lines, blowing hole at the right and tone hole at the left indicated by vertical lines.

4. Partial 2290 Hz

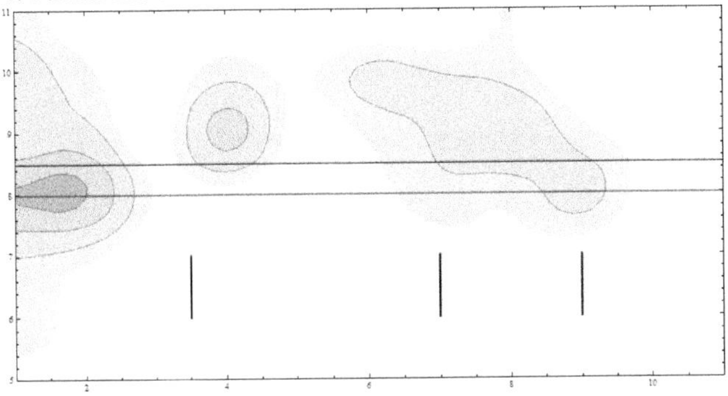

5. Partial 2860 Hz

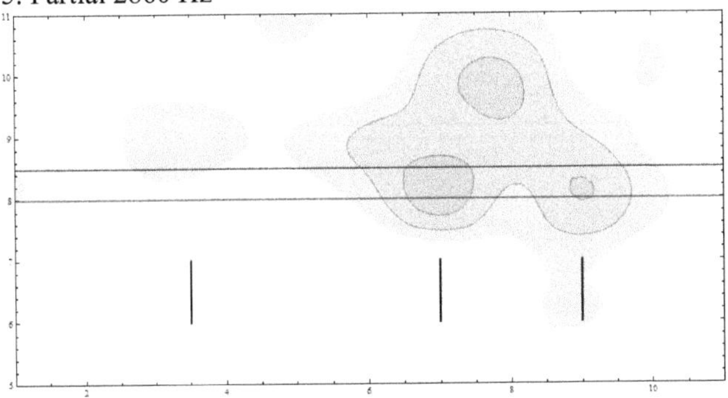

6. Partial 3435 Hz

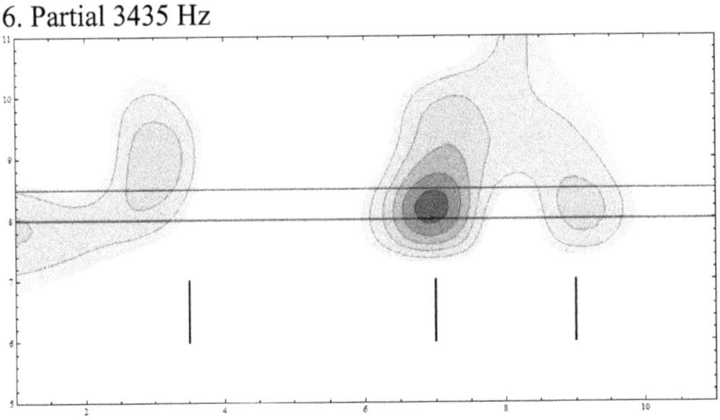

7. Partial 4010 Hz

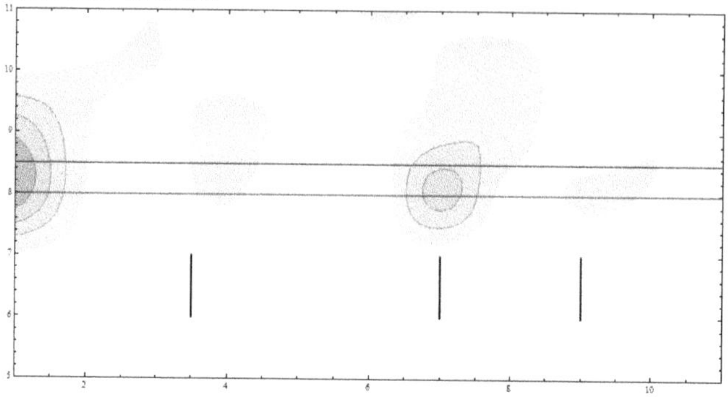

Fig. 23: Forth to seventh partial of the *dizi* tone. Additional hole covered by the membrane indicated by third vertical line between blowing and tone hole.

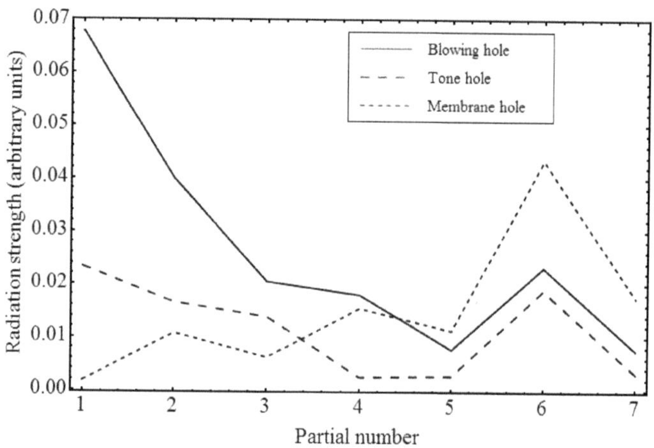

Fig. 24: Radiation strength vs. partial number for blowing hole, tone hole and membrane hole. The blowing hole dominates with the low frequencies, while with high frequencies, the membrane hole is the loudest radiating point.

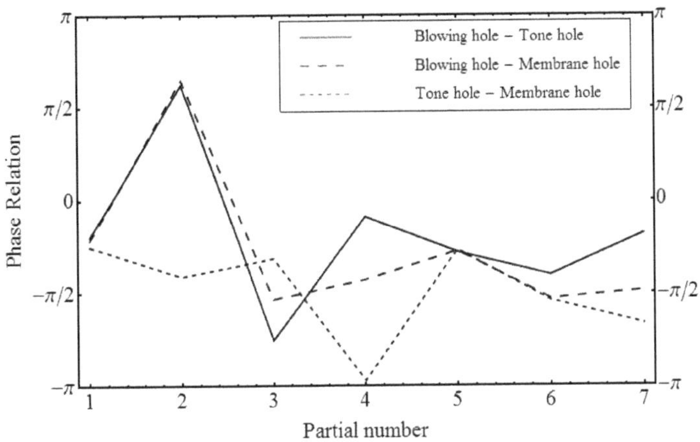

Fig. 25: Phase relations of partials between the three different holes of the *dizi* flute.

Partial	alpha
571 Hz	0.0
1140 Hz	0.6
1720 Hz	0.0
2290 Hz	0.0
2860 Hz	0.0
3435 Hz	0.0
4010 Hz	0.3

Tab. 4: Directivity of the *dizi* flute. Again, all partials radiate as a monopol source.

Conclusions

Five instruments were investigated using Microphone Array techniques, an Usbek tambourine, a *duff* frame drum, a Balinese bamboo flute, the *suling*, the Japanese *shakuhachi* bamboo flute and the Chinese *dizi* flute. In the case of the tambourine, the directivity of the first seven partials could clearly be measured.

With the *duff*, an algorithm was capable of calculating the changing tension and area density distribution over the drum area. Both examples gave an insight into basic properties of the instrument, including mode shapes and the existence of degenerated modes. Also the character of the instrument could be calculated by looking at the characteristic deviations of the membrane from a homogeneous one. Many further investigations are needed here to show the transient behaviour of the instruments at the initial transient phase. This important part of the sound lead to very complex radiation pattern and to binaural and stereo effects needed for a good instrument. Also comparison between instruments in terms of their changing area tension and therefore their character is needed.

With the flue instruments, only some new insights into the radiation of these instruments and their phase and amplitude behaviour was shown. Even the simplest bamboo flute, the *suling*, showed some properties which could not be explained straightforward. The *shakuhachi* was even more complex in terms of the radiation of different tone holes. It is crucial to investigate the phase relations between these radiating points to again get an idea of the binaural perception of this instrument. As it is nearly only played for timbre modulations rather than for ornamented melody playing, the radiation of the turbulent noise parts also need to be considered further. Here, for the partials of the played note, the radiation of turbulent jets only appeared very narrow around the blowing hole. The most difficult instrument showed up to be the *dizi* flute due to the added membrane. Here, the radiation is complex and shows a strong influence of the membrane hole in overall radiation. Also strong radiation of the additional tone holes beyond the tone hole could be found. The phase relations point to a complex travelling impulse behaviour of the flute to be studied in more detail in further investigations.

Although only a small portion of the possibilities of measurements of a Microphone Array could be presented, still with the results presented here, new insights into the instruments could be obtained. As Microphone Arrays are able to measure and back-propagate radiating sources of both, vibrating structures and vibrating air columns, it makes the method suitable for musical instrument acoustics with many applications waiting for in the future.

References

Arfken, G. B., H. J. Weber (2005), *Mathematical Methods for Physicists*, Amsterdam: Elsevier Academic Press.

Bader, R. (2009a), Reconstruction of Radiating Sound Fields using Minimum Energy Method, *submitted J. Acoust. Soc. Am.*.

Bader, R. (2009b), Vibration patterns of the classical guitar top plate with strings plucked using microphone array minimum energy method, *Submitted J. Acoust. Soc. Am.*.

Bai, M.R. (1992), Application of BEM (boundary element method)-based acoustic holography to radiation analysis of sound sources with arbitrarily shaped geometries, *J. Acoust. Soc. Am.*, 92 (1), 533-549.

Bissinger, G., E. G. Williams, N. Valdivia (2007), Violin f-hole contribution to far-field radiation via patch near-field acoustical holography, *J. Acoust. Soc. Am.*, 121 (6), 3899-3906.

Bouchet, L., Th. Loyau (2000), Calculation of acoustic radiation using equivalent-sphere methods, *J. Acoust. Soc. Am.*, 170 (5), 2387-2397.

Magalhães, M. B. S., R. A. Tenenbaum (2004), Sound Source Reconstruction Techniques: A Review of Their Evolution and New Trends, *Acta Acustica united with Acustica*, 90, 199-220.

Meynard, J. D., E. G. Williams, Y. Lee (1985), Nearfield acoustic holography: I. Theory of generalized holography and the development of NAH, *J. Acoust. Soc. Am.*, 78 (4), 1395-1413.

Morse, M. Ph., U. K. Ingard (1986), *Theoretical Acoustics*, Princeton University Press.

Ochmann, M. (1995), The source simulation technique for acoustic radiation problems, *Acustica*, 81, 512-527.

Ochmann, M. (1999), The full-field equations for acoustic radiation and scattering, *J. Acoust. Soc. Am.*, 105 (5), 2574-2584.

Prager, J. (2007), Approximate reconstruction of sound fields close to the source surface using spherical nearfield acoustical holography, *J. Acoust. Soc. Am.,* 122(4), 2067-2073.

Rayess, N., S. F. Wu (1999), Experimental validations of the HELS method for reconstructin gacoustic radiation from a complex vibrating structure, *J. Acoust. Soc. Am.*, 107 (6), 2955-2964.

Scholte, R., N. B. Roozen, I. Arteaga, I. Lopez (2005), Regularization in PNAH by means of L-curve, *Proceedings of the Forum Acusticum Budapest,* 2579-2583.

Strong, W. Y. Jr., T. B. Beyer, D. J. Bowen, E. G. Williams, J. D. Maynard (1982), Studying a Guitar's Radiation Properties with Nearfield Holography, *J. of Guitar Acoustics*, 6, 50-59.

Veronesi, W.A., J. D. Maynard (1988), Digital holographic reconstruction of sources with arbitrarily shaped surfaces, *J. Acoust. Soc. Am.*, 85 (2), 588-598.

Wang, Z., S. F. Wu (1995), Helmholtz equation-least-squares method for reconstructing the acoustic pressure field, *J. Acoust. Soc. Am.*, 102(4), 2020-2032.

Wang, L.M., C. B. Burroughs (2001), Acoustic radiation from bowed violins, *J. Acoust. Soc. Am.,* 110 (1), 543-555.

Williams, E. G., B. H. Houston, P. C. Herdic (2003), Fast Fourier transform and singular value decomposition formulations for patch nearfield acoustical holography, *J. Acoust. Soc. Am.*, 144 (3), 1322-1333.

Williams, E. G., J. D. Maynard, E. Skurdzyk (1980), Sound sourcereconstructions using a microphone array, *J. Acoust. Soc. Am.*, 6 (1), 341-344.

Wu, S. F. (1999), On reconstruction of acoustic pressure fields using the Helmholtz equation least squares method, *J. Acoust. Soc. Am.*, 107(5), 2511-2522.

Wu, S. F., H. Lu, M. S. (2005), Bajwa, Reconstruction of transient acoustic radiation from a sphere, *J. Acoust. Soc. Am.*, 114 (4), 2065-2077.

Andreas Beurmann
Albrecht Schneider

Acoustics and sound of the harpsichord: another case study

1. Introduction

Recently, we reported on acoustic measurements on a French harpsichord built by Nicolas Pigalle of Dijon from 1771 (Beurmann & Schneider 2008). In this paper, we present data from measurements we performed on a harpsichord housed in the collection of the first author. It is a Flemish harpsichord built by the famous Andreas Ruckers the elder in 1628 (Beurmann no. 24; see Beurmann 2000, p. 80-90). This instrument had been studied previously in regard of its acoustics (Elfrath 1990). Several methods including modal analysis and impedance measurement had been employed. In our investigation, special emphasis was put on the generation and the onset of sounds since here many details can be studied relevant for the acoustics of the harpsichord as well as for perceptual aspects. The plucking of a string in a harpsichord is a fairly complex process which yields several types of vibration (see below). Since the plucking can be regarded as an impulse-like excitation first of the string, and then (via the bridge) of the soundboard, a transient part of vibration results before a stable and periodic pattern is established. In regard of perception, the transient part at the onset of harpsichord sounds is particularly rich in information that is contained in variable temporal and spectral features.

2. Scaling of the Ruckers 1628 harpsichord

The instruments which have been built by the Ruckers family are highly esteemed for their excellent craftmanship and sonic quality (O'Brien 1990); in fact, they have become the model for many modern makers who have studied these harpsichords which then have been emulated. Harpsichords and virginals built by members of the Ruckers family are at the core of the Flemish tradition which, in the 17th and 18th century, became influential throughout Western Europe.

The specimen which we investigated was built by Andreas (or Andries) Ruckers the elder, in 1628. It is a double-manual harpsichord that originally had only two sets of strings, one 4', and one 8'. The keys of the two manuals formed a transposition by a fourth in relation to each other (cf. Beurmann 2000, p. 88). In 1728, the instrument was modified so as to make both manuals equal, and to offer a second 8'. Consequently, it now has 8' + 4' and 8' in the upper and lower manual, respectively. Also, the compass of the instrument has been expanded from the original C/E ←→ f^3 to G,/B, ←→ c^3, and recently (1960) the notes A_1, Bb_1, $c\#^3$ and d^3 have been added[1].

The scale and the plucking points (p.p.) are given in table 1.

---

[1] In this article, octave designations are used according to Helmholtz which relate to the U.S. Standard in the following manner: G, (Helmholtz) = G_1 (U.S.), C = C2, c° = C3, c^1 = C4, etc.

Table 1: Scale and plucking points (8'), Andreas Ruckers harpsichord 1628.

Tone/String	length L (mm)	p.p. l (mm)	ratio L/l
G,	1695	187	9.064
C	1672	182	9.187
F	1547	172	8.994
Bb	1344	159	8.453
c°	1267	156	8.122
f°	1039	146	7.116
c^1	732	132	5.545
f^1	546	125	4.368
c^2	361	114	3.166
f^2	273	108	2.528
c^3	166	98	1.694

From the data in table 1, and with the c^2 serving as a reference point, the scaling which shows the actual reduction towards lower octaves can be derived as follows:

Note	c^2(C5)	c^1(C4)	c°(C3)	C(C2)	G.(G1)
Theory	1	2	4	8	10.666
Measured	1	2.028	3.51	4.632	4.695

To avoid inappropriate string tension which might cause structural damage to the historic instrument, the diameters of the strings chosen are rather small (G, = 0.65 mm, A, = 0,62 mm, B, = 0.5 mm, C = 0.47 mm, C# - F = 0.45 mm, F# - G = 0.4 mm, G# - A = 0.38 mm, A# - c° = 0.35 mm, c#° - f° = 0.33 mm, f#° - c#' = 0.3 mm, d' - f^2 = 0.27 mm, $f\#^2$ - c^3 = 0.25 mm (G, - A = brass, A# - c^3 = iron). By the time of our measurements, the instrument was tuned to a' = 372 Hz which implies a fundamental frequency of 186 Hz for the a°(A3), of 93 Hz for the A (A2), and of 46.5 Hz for the A, (A1). In the meantone temperament the Ruckers was tuned to, the low C then comes close to 52 Hz. A long-term average spectrum (LTAS; see figure 1) of the sound of the C string recorded with a condenser microphone (Neumann TLM 170) shows that the amplitudes of partials no. 9, 18, 28, 37, 46, 55, 65 (etc.) are weak or close to zero due to the L/l-ratio of, in this case, 9.187, which leads to cancellation of partials at $k \cdot L/l$ (k = 1, 2, 3,..)[2]. If the d'Alembert approach to wave propagation on an elastic string is applied (Morse 1981, ch. 3), initially there are two waves travelling into opposite directions from the place of excitation. After reflection at the boundaries of the string fixed at both ends, superposition leads to a standing wave pattern, which implies that harmonics corresponding to waves with a node at the point of plucking will be absent. Hence the spectrum calculated from the quasi-stationary vibration of the C-string is quasi-periodic with characteristic maxima and minima.

---

[2] The amplitudes (dB) are calculated relative to 0 dBfs as a reference point. In an undistorted digital signal, amplitudes of spectral components must be ≤ 0 dBfs.

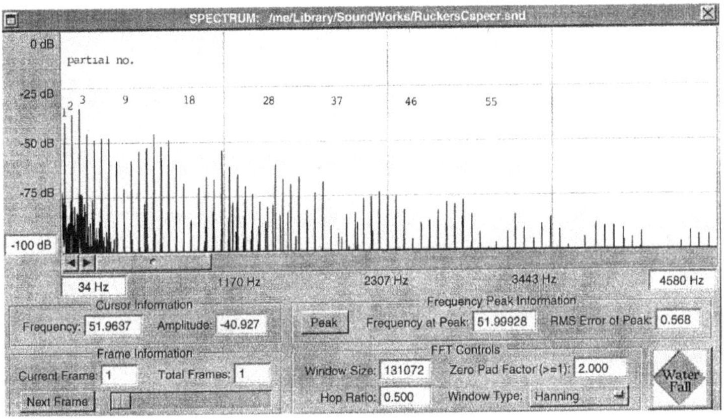

Figure 1: Ruckers 1628: Spectrum (LTAS, 131072 pts) of tone C ($f_1 \approx 52$ Hz).

3. Plucking and excitation mechanism

The plucking mechanism of the harpsichord string has been studied in detail (Cuesta & Valette 1988, 1989, Elfrath 1990). Basically, one can observe that the string is lifted by the plectrum as the jack is raised by pressing the respective key. At the moment the plectrum touches the string (which is under a tension, $s = \Delta F/\Delta A$, F = force, A = area), the string exerts a normal force, F_N, on the plectrum, which is the restoring force, F of the string times cos(α): $F_N = F \cdot \cos(\alpha)$. As the jack is raised further, the flexible plectrum (usually made of Delrin or a plastic of similar quality) lifts the string and begins to bend. The string is still 'glued' to the plectrum due to an adhesive frictional force, F_R, which is the product of F_N and an adhesive friction coefficient, μ0 (see figure 2a, adapted from Elfrath 1990). Hence

$$F_R = m0 \cdot F_N = \mu 0 \cdot F \cdot \cos(\alpha)$$

As the bending of the plectrum increases, another force, F_H is acting on the string in a direction opposite to F_R (figure 2a). F_H brings the string to slip from the plectrum as soon as $F_H > F_R$. Before the string slips from the spectrum, it performs a twisting movement on the plectrum (whereby F_R and F_H act as opposite forces tangential to the string; figure 2b, adapted from Elfrath 1990).

Figure 2a

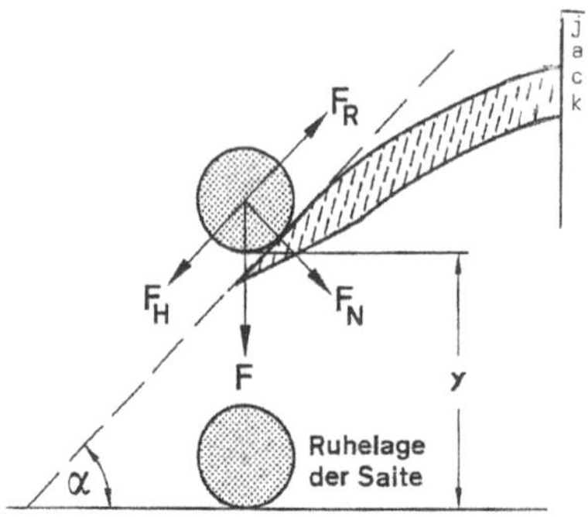

Figure 2b

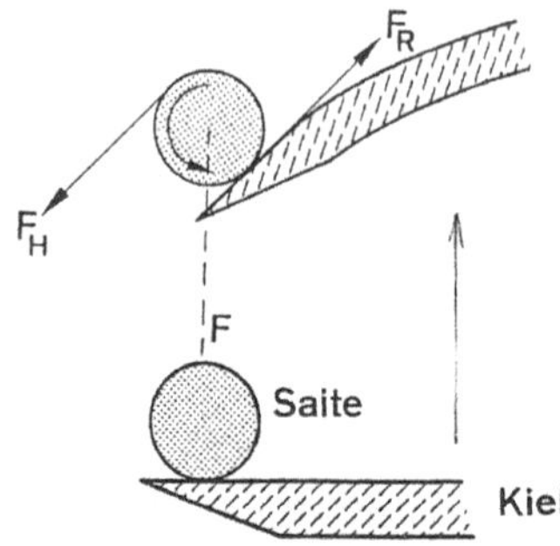

Figure 2a: Forces that act on the string and on the plectrum when the string (cross section) is lifted upwards (y [mm]); adapted from Elfrath 1990.
Figure 2b: As the elongation of the string increases, so does its tension, *s*. Increasing restoring force F_N causes the plectrum to bend. The string undergoes some twisting movement before it slips from the plectrum.

The tangential forces F_R and F_H which lead to the twisting movement of the string give rise to torsional vibration as soon as the string becomes free of the plectrum. Note that the torsional modulus, G, for a thin elastic body typically is much smaller than the modulus of elasticity, E, relevant for stretching, i.e. $E/2 > G > E/3$ if Poisson's ratio m is $0 < m < 0.5$. Consequently, the string performs longitudinal, torsional, and transversal vibration, the latter both perpendicular and parallel to the soundboard.

It can be assumed that the bending of the plectrum increases in proportion to the distance the jack moves upwards. Also, the elongation of the string can be regarded as being proportional to the distance y (mm) the jack (plus the plectrum) travel. Since the distance y is at least several mm, the elongation ξ of the string is not infinitesimal small. In thin homogeneous brass or iron strings with diameters $0.65 \geq d \geq 0.25$ mm as have been used for the Ruckers (see above), bending stiffness will be small, and will be of little if any relevance in regard of the long strings in the low register. Therefore, it is reasonable to assume that the raise of the jack and plectrum, and the elongation ξ of the string at the plucking point, l, indeed would lead to a triangular shape of the string (viewed from the side) in case the plucking process would be performed as slowly as is possible (what we did to study the different stages of excitation). There are strong indications we obtained from measurements of transients in a Kirckman harpsichord (cf. Beurmann & Schneider 2003) that the elongation of the string in fact increases the tension s of the string to $\sigma + \Delta\sigma$, whereby the fundamental frequency of the string goes up by more than 50 Cents. In a situation of real playing, excitation of course usually is achieved in a fast manner, to the effect that the excitation at point l perhaps is closer to an impulse than to the triangular elongation of the string viewed in „slow motion". This view, however, is useful for understanding the plucking mechanism step by step.

Indications that the plucking rather must be impulse-like in effect come from the actual pattern of vibration of strings as well as from the multitude of partials which appear in the spectrum at the onset. The velocity of the string can be recorded by a magnetic device which consists of a small yet strong magnet plus two pole shoes (see Cuesta & Valette 1989, Gätjen 1995). With the string moving up and down in the magnetic field between the shoes, it induces a current which can be picked up at the two ends of the strings. The current and the voltage are very faint though, and need amplification. We have used such a device in another study (Beurmann & Schneider 2008). In the investigation of the Ruckers, vibrations of individual strings were recorded by means of a magnetic pickup and a condenser microphone, both placed above the string, the pickup a few centimeter behind the jack, the microphone more to the middle of the soundboard. All other strings were damped with textile materials to prevent sympathetic vibration. The magnetic pickup was a single coil model[3] that has the advantage of inducing a fairly high voltage that could be recorded on DAT (48kHz/16 bit) with no further amplification for analysis. The voltage is proportional to the velocity of the string, and proportional to the strength of the magnetic field in which the string vibrates. The disadvantage – compared to other methods of measurement (cf. Gätjen 1995, ch. 3) – is that the magnetic field of the pickup may not be homogeneous. However, the signal obtained from the pickup output can be used to study the temporal development of the string vibration apart from its interaction with the bridge and soundboard. Also, the pickup is sensitive practically only to transversal vibrations perpendicular to the soundboard, whereas it does neither cover longitudinal vibration nor transversal vibration parallel to the soundboard. The pickup signal for the note B is shown in figure 3.

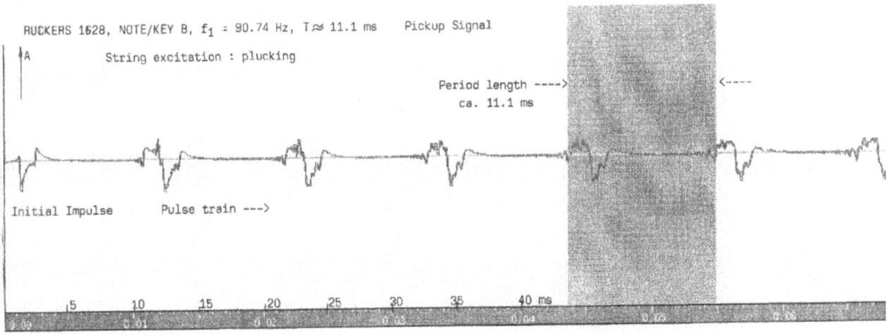

Figure 3: Pickup signal, note/string B, shows pulse train with period T ≈ 11.1 ms ≈ 90.74 Hz.

At the very left, the initial impulse can be seen, followed by a sequence of pulses which have a period of T ~ 11.2 ms corresponding to the fundamental frequency of the note B which, in the tuning chosen, is ca. 90 Hz. The spectrum of the pickup signal taken immediately after the excitation of the string (FFT 2048 pts, triangular [Bartlett] weighting function; see DeFatta et al. 1988) is shown in figure 4. Evidently, there is spectral energy up to 8400 Hz, enough to drive the bridge and the soundboard (see below). The impulse-like excitation can be inferred not only from the time function of the signal (figure 3) yet also from the 3D-spectrum as shown in figure 5 that was calculated by

[3] Seymour Duncan SJAG-3N Quarter Pound; DC resistance of the coil 12.15 kOhm, inductance 6 Henry.

means of a phase vocoder algorithm (sndan, Beauchamp 1995) with a short FFT (1024 pts, Hamming) to allow for appropriate temporal resolution. The microphone signal of the same string is much more complex (figure 6) since at least some of the soundboard vibration is contained in the signal that includes a transient part at the beginning. The period of the microphone signal of course is identical with the pickup signal (T ~ 11 ms as indicated in figure 6). The corresponding spectrum (figure 7) also is periodic with characteristic minima at partials no. 9, 18, 27, etc.

4. Wave propagation and soundboard response

The case of the Ruckers harpsichord (including the bottom below the soundboard) is made from poplar, in a special construction (cf. Beurmann 2000, p. 88/89) which ensures stability yet does not require thick walls. The soundboard is made of spruce and is a relatively thin plate (thickness ca. 3 mm) so that this dimension is very small compared to the length and width of the plate. This leads to considering the soundboard as a structure halfway between a (in this case, approx. triangular) plate and a membrane. A plate of a certain thickness, *h*, because of structural stiffness performs bending waves which undergo frequency dispersion (that is, the wave speed c_B for bending waves on a flat plate of unlimited size is a function of the frequency according to $c_B \sim \sqrt{f}$; see Schneider 1997, 201-216). A thin membrane is a much simpler structure that can be treated, basically, as an extension of a thin one-dimensional string into a second dimension (*x*, *y*). Since a membrane adapts easier to an external periodic driving force than does a wooden plate (where, in addition to the bending stiffness also the grain structure of wood is of importance), instrument builders always seek to thin out plates such as soundboards of harpsichords or top plates of violins, guitars, etc. in order to make them resonant with a periodic excitation provided by, in most cases, transversal vibrations of a string. The number of eigenmodes calculated for quadratic membranes amounts to more than 40 for the first three octaves (Kalähne 1913, p. 156, Tab. 13). The density of eigenmodes we measured (see table 2) seems to be significantly higher than the results obtained from a previous modal analysis of the same instrument (Elfrath 1990).

The impedance (Z = force applied/velocity gained) of the bridge has been measured in detail (Elfrath 1990, pp. 48). Even with a standard impulse of given force, the impedance varies considerably as a function of the place of excitation. Similarly, the impulse response of the soundboard varies as a function of place since the stiffness of the soundboard is higher close to the bridge and hitchpin rail than it is for those parts which have no structural reinforcement or load. We measured the impulse response of the soundboard by applying an impulse on the bridge close to the pin of note C by means of a light impulse hammer (Kistler 9722A500) with a tip that can produce an impulse with an upper frequency limit of well over 2 kHz. Recording of the impulse response took place between the 8' and the 4' bridge, and was achieved with a small accelerometer (Kistler 8614A500, mass = 0.7 g) that does hardly influence the vibration of the soundboard as a load.

Figure 8 shows the time function of an impulse and of the corresponding impulse response (recorded at 48kHz/16bit). The spectrum obtained from three of the impulse responses contains a number of distinct peaks indicating resonances. The frequencies and relative amplitudes (reference: 0 dBFS) of 24 spectral peaks in the range of 27-375 Hz are listed in table 2.

Table 2: Impulse response of soundboard, spectral peaks.

Peak no.	Freq. (Hz)	rel. Ampl. (dB)	Peak no.	Freq. (Hz)	rel. Ampl. (dB)
1	27.71	-65.9	13	218.43	-62.5
2	61.54	-64.7	14	233.46	-60.0
3	90.05	-59.7	15	244.52	-58.7
4	104.14	-63.5	16	258.54	-57.0
5	116.62	-57.2	17	270.55	-57.5
6	129.03	-52.1	18	283.74	-59.4
7	141.67	-54.4	19	297.88	-71.9
8	154.43	-62.4	20	309.49	-70.5
9	169.08	-68.7	21	321.98	-64.5
10	180.45	-67.1	22	336.95	-66.5
11	194.04	-66.8	23	360.65	-65.9
12	206.41	-68.1	24	374.83	-75.6

The results are in good agreement with the modal analysis (Elfrath 1990) which did yield 63, 74, 126, 143 Hz for the first four modes. The modal analysis – due to technical limitations – did not find a basic mode where the soundboard (which is clamped at its edges like a membrane yet not stretched with a tension, *s*) would move up and down as a whole. It is possible that the spectral component no. 1 at 28 Hz represents such a mode which can be expected to be weaker in amplitude than the next mode which divides the soundboard by one nodal line so that the two parts of the soundboard move into opposite directions, acting as a dipole source.

In harpsichord as well as in piano strings, longitudinal waves are generated which are preceding transversal waves, and can be measured in frequency and amplitude. In a fairly large French harpsichord, we recorded the longitudinal wave as it spread on the soundboard (cf. Beurmann & Schneider 2008, figs. 11, 12). Playing the C-string of the Ruckers 8', and recording the resulting vibration of the soundboard with an accelerometer, again yields the periodic longitudinal vibration (of small amplitude) which lasts for only a few milliseconds, and is followed by the transversal vibration of large amplitude. The frequency of the sinusoidal longitudinal vibration is ca. 620 Hz which is well in line with earlier results that gave a ratio of 12:1 for the fundamental frequencies of the longitudinal and the transversal vibration (here we have 620 : 51 $\approx$ 12.15). Since we recorded also the sound radiated from the instrument with a condenser microphone close to the C-string above the soundboard, the two time functions, recorded simultaneously, can be compared (figure 9). Also, the two spectra (FFT: 8192 pts, Hanning, zero-pad-factor: 2) corresponding to the respective time functions are of interest. The spectral energy distribution in the signal recorded from the soundboard (figure 10) shows the characteristic pattern of periodic minima, that is, the spectrum of the transversal string vibration perpendicular to the soundboard (figure 4) which carries most of the energy has forced the soundboard to vibrate according to its temporal and spectral pattern.

Consequently, the sound radiated from the string and soundboard again must be similar in spectral composition (figure 11); the difference in spectral energy distribution can be attributed to the fact that the recording with the accelerometer covers the vibration at a confined area of the soundboard whereas the microphone picks up the sound as radiated from the string (as a weak direct signal) and the soundboard. The difference of the two signals is clearly reflected in their respective autocorrelation functions (ACF; see figures 12a, 12b). The ACF of the microphone signal for the note C reflects its period with strong peaks, the correlation coefficient being at $r_{xx'} \approx 0.75$ at the beginning of each period. In contrast, the ACF of the accelerometer output, though showing the same period length (in samples or in ms), does not yield similarly strong peaks per period, and the correlation coefficient which in the microphone signal shifts regularly between ca. ± 0.5, in the accelerometer signal within each period for the most time is close to only ± 0.25, and also does not exhibit the same regular structure found in the ACF of the microphone signal.

The air modes found in a harpsichord built in 1976 after Flemish models have been calculated and measured by Savage et al. (1992). To obtain such measurements, these researchers had to damp the soundboard by putting huge quantities of sand on top of it. In regard of the fragile historic instrument under investigation, such a manipulation was not possible. However, we chose to investigate if, and to which extent, soundboard modes could be picked up within the cavity of the Ruckers 1628 which has a length of ca. 160 cm, a maximum width of 76.3 cm, and a maximum height of ca. 20 cm yielding a volume of, roughly, 130 litres. Viewed as a Helmholtz box resonator, the resonance frequency calculated from the actual geometry is ≈ 110 Hz with the rose removed, and below 100 Hz with the rose in place. To record either a signal from a sine wave generator with variable frequency, or from one of the strings played from the upper 8′ keyboard, we put two microphones into the cavity via the rose hole which has a diameter of ca. 74 mm. The rose made of an alloy (lead and tin) was removed (what can be done with but little effort); its weight was found to be 99 grams what means it is a considerable load in regard of soundboard vibration when in place. In addition to a very small electret condenser microphone (of dimensions 9 x 5 mm) we used a small Sennheiser MKH 405 condenser microphone which has the advantage of providing a fairly smooth frequency response down to 40 or even to 30 Hz. The microphones were placed inside the cavity hanging from their cables through the open rose hole. It is clear that this operation is likely to change the air flow through the rose hole.

Driving the soundboard and the air in the cavity by means of a sine wave generator fed into a HiFi-amplifier with a compact 2-way loudspeaker cabinet connected that was put flat on the 8' bridge and soundboard, the sine sweep did yield clear resonance maxima at ≈ 93 Hz, 118-120 Hz, 195 Hz, and 312 Hz, respectively, and maxima of lesser amplitude also at ≈518 Hz and 720 Hz. Most of these resonances correspond fairly well to spectral peaks listed in table 2 (see above) though there are some deviations which can be explained by taking the air volume in the cavity into account as an additional load coupled to the soundboard. Sounds produced by string and soundboard excitation that were recorded inside the cavity up to 1 kHz basically showed the same spectral energy pattern as the signal recorded directly from the string. The air enclosed in the cavity vibrates according to the excitation effected by the soundboard. In our measurements, frequencies as low as the G, ($G_1 \approx 41$ Hz) were represented in the spectrum with no identifiable decrease in amplitude.

5. Richness of radiated sound, formant-like quality

The sound radiated from many historic harpsichords is particularly rich due to the impulse-like excitation of the string which leads to transients preceding quasi-stationary vibration, the relatively low impedance of the bridge which ensures fast and sufficient energy transport into the bridge and soundboard, and the frequency response of the soundboard which offers a dense yet distinct pattern of eigenmodes serving as resonances. Regarding these qualities, it is no wonder that the actual sound produced and radiated shows a very high number of partials per note played. For example, the note/key B played with the 8' and recorded with a microphone (TLM 170), at the onset contains more than 200 partials. The dynamic evolution of the sound as analyzed with a phase vocoder (Beauchamp 1995) is shown in figure 13 for partials no. 1 ($f = 90.15$ Hz) to no. 200 ($f \approx 18030$ Hz). Section A of the 3D-plot covers the short time the string rests on the plectrum before it slips, and gives rise to a plosive attack which is at the beginning of section B.

At times, the question has been raised if the harmonically rich sounds of historic harpsichords might exhibit a formant-like structure when analyzed over the whole compass of an instrument (Gätjen 1995). Formant-like zones of concentrated spectral energy in theory could occur due to resonance maxima of the vibrating system in total, or due to characteristics of sound radiation of the harpsichord in general, or of a peculiar instrument such as the Ruckers 1628. Since the spectra of the sounds of the individual keys/strings are periodic with minima and maxima of energy (figures 1, 4, 7, 10, 11), one could overlay the spectra of all 54 keys of the instrument under investigation to see if, and probably to which extent, zones of concentrated spectral energy might appear. Such a method is feasible (see Schneider et al. 2006, fig. 4 for a similar investigation of an organ mixture stop), it is not very elegant to perform for 54 keys/sounds though. Instead, we chose to record an actual piece of music (the folk song „*Ach, du feiner Reiter*", in the four-part setting of Samuel Scheidt) with a single microphone (Neumann U 67 + TAB V76 preamp) placed ca. 1 m away from the bentside of the Ruckers. The digital sound file (mono) was analyzed with an algorithm developed for the detection of formants in speech signals available in the Praat environment (Boersma & Weenink 2008). In the method chosen, a Gaussian-like window is applied to each analysis window (frame) for which the LPC coefficients are calculated according to the algorithm proposed by Burg (see Marple 1987, ch. 8). With the formant tool available in Praat, a maximum of five formants can be detected within the frequency range of ca. 50-5500 Hz. The results for „*Ach, du feiner Reiter*" are shown in figure 14 which indeed indicate that the musical piece as played on the Ruckers yields a sound structure which seems to contain four „formant-like" zones of spectral energy located at ca. 500 Hz, 1.6-1.8 kHz, 2.8-3 kHz, and ca. 4 kHz, respectively. This interpretation is based on actual measurement, yet should be taken with care since at the end of the piece (a chord played as arpeggio) the analysis shows that the traces of concentrated spectral energy are not stable in regard of the frequency bands that had been established before. Hence it is seems reasonable to speak of „formant-like" zones of spectral energy which appear in the radiated sound for the most part of the piece that requires both hands playing full chords simultaneously. In order to check whether the "formant-like" effect was due to the pattern of energy radiated from this peculiar instrument, or to the notes prevalent in certain keys, some more pieces of music (from various composers, available on CDs of the first author) played on the Ruckers were investigated with the same algorithm. The results indicate that such "formant-like" concentrations of spectral energy in part indeed seem to be attributable to a prevalence of certain notes in the respective keys (e.g., C-major), however, there are some energy concentrations which are not confined to individual works of music, and which seem to occur in a number of keys.

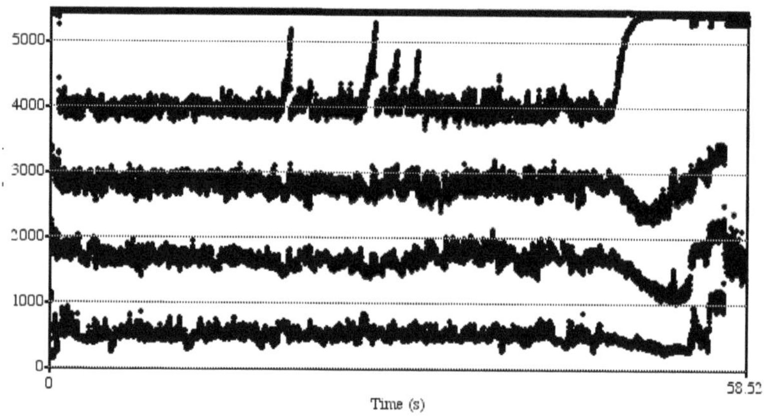

Fig. 14: Ruckers 1628: *Ach, du feiner Reiter*, formant analysis.

6. Conclusion

The Ruckers 1628 which we investigated with the aim to complement and check data published in a previous study (Elfrath 1990) is probably one of the finest extant harpsichords of the Ruckers family. Its rich sound and outstanding sonorous quality has been confirmed also in tests with 30 subjects (among them professional harpsichord players) making judgements based on music played on the Ruckers. The instrument has been used for CD recordings which demonstrate its distinct sound[4].

References

Beauchamp, James (1995), New methods for computer analysis and synthesis of musical sounds. [Proceedings] *32nd Czech conference on acoustics*, Prague, 7-15.

Beurmann, Andreas (2000), *Historische Tasteninstrumente. Cembali, Spinette, Virginale, Clavichorde.* München, London, New York: Prestel.

[4] For examples, listen to the CDs Historische Cembali, Spinette und Virginale. Andreas Beurmann im Barocksaal zu Hasselburg (BMG) and Historische Cembali, Spinette und Virginale. Werke von Bach, Händel, Mozart und Meistern des 16.-18. Jahrhunderts, Andreas Beurmann (Körting-Musikverlag, Hamburg).

Beurmann, Andreas, Albrecht Schneider (2003), Sonological analysis of harpsichord sounds, In R. Bresin (ed.). *Proc. of the SMAC03. Stockholm Music Acoustics Conferencee 2003*, Stockholm: KTH, Vol. I, 167-170.

Beurmann, Andreas, Albrecht Schneider (2008), Acoustics of the harpsichord: a case study, *Hamburger Jahrbuch für Musikwissenschaft* 24, 241-263.

Boersma, Paul, David Weenink (2008), *Praat: doing phonetics by computer*, Version 5.038 (Amsterdam: Institute of Phonetic Sciences, Univ. of Amsterdam).

Cuesta, C., C. Valette (1988), Evolution temporelle de la vibration des cordes de clavecin, *Acustica* 66, 37-45.

Cuesta, C., C. Valette (1989), Le transitoire d'attaque des cordes de clavecin, *Acustica* 68, 112-122.

DeFatta, David, Joseph Lucas, William Hodgkiss (1988), *Digital Signal Processing: a system design approach*, New York etc.: Wiley.

Elfrath, Thomas (1990), *Physikalische Untersuchungen über die qualitätsbestimmenden Einflußgrößen bei Cembali*, Braunschweig: PTB.

Gätjen, Bram (1995), *Der Klang des Cembalos. Historische, akustische und instrumentenkundliche Untersuchungen*, Kassel: Bosse.

Kalähne, Alfred (1913), *Grundzüge der mathematisch-physikalischen Akustik*, T. II. Leipzig, Berlin: Teubner.

Marple, S.R. jr. (1987), *Digital spectral analysis,* Englewood Cliffs, N.J. : Prentice-Hall.

Morse, Philip (1981), *Vibration and sound*, Reprint, [New York:] American Institute of Physics/Acoust. Soc. of America.

O'Brien, Grant (1990), *Ruckers: a harpsichord and virginal building tradition*, Cambridge, Cambridge Univ. Pr.

Savage, William, Edward Kottick, Thomas Hendrickson, Kenneth Marshall (1992), Air and structural modes of a harpsichord, *Journal of the Acoust. Soc. of America* 91, 2180-2189.

Schneider, Albrecht (1997), *Tonhöhe – Skala – Klang*, Bonn: Orpheus-Verlag.

Schneider, Albrecht, Richard von Busch, Dorothea Schröder, Lüder Schmidt (2006), Zur klanglichen Dokumentation von Denkmalsorgeln, *Acta Organologica* 29, 405-426.

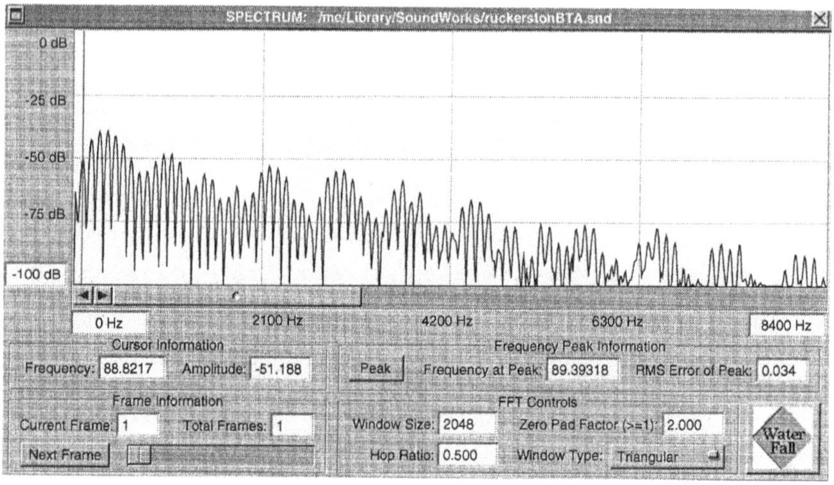

Figure 4: Spectrum of B string immediately after excitation, pickup signal (FFT: 2048 pts, triangular [Bartlett] window).

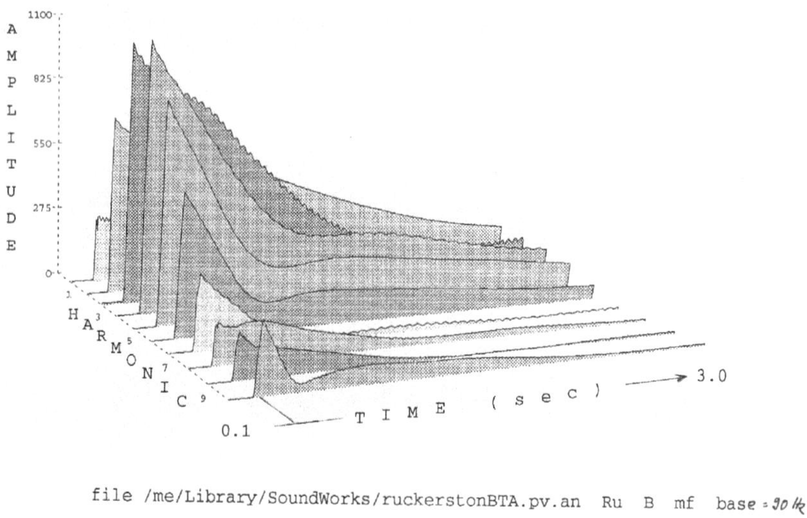

Figure 5: 3D-spectrum (short time FFT: 1024 pts, Hamming), B string, pickup signal, temporal evolution of harmonics 1-10, scaling of amplitudes linear.

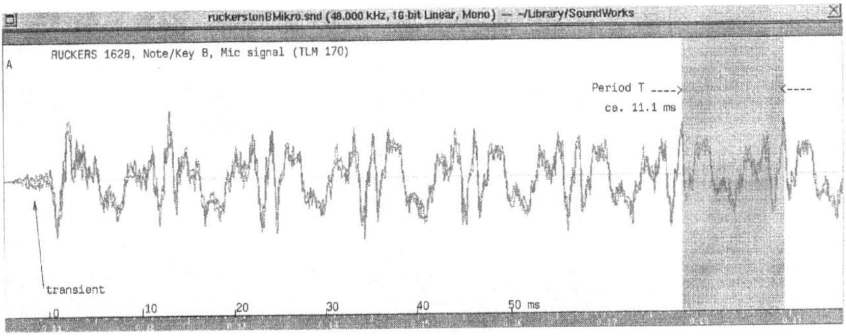

Figure 6: Note/Tone B, time function, microphone signal of radiated sound that starts with a transient part, later to establish a quite stable periodicity at T ≈ 11.1 ms.

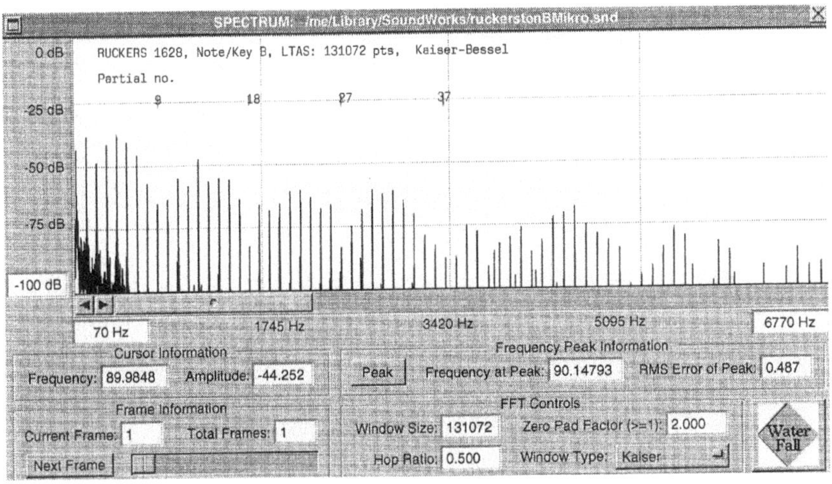

Figure 7: LTAS (131072 pts, Kaiser-Bessel) corresponding to microphone signal of note/string B.

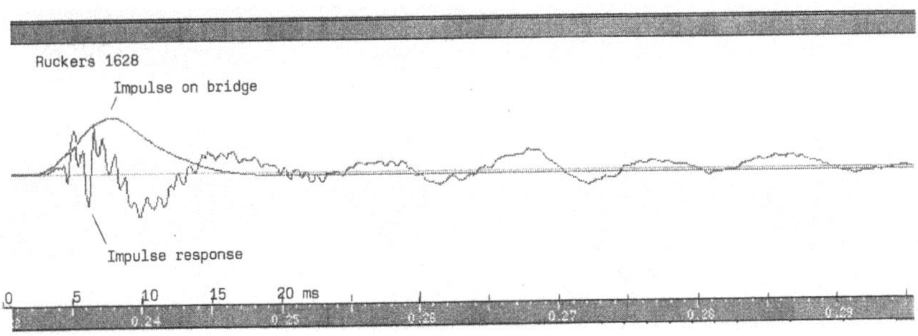

Figure 8: Asymmetric shape of impulse on bridge; impulse response of soundboard.

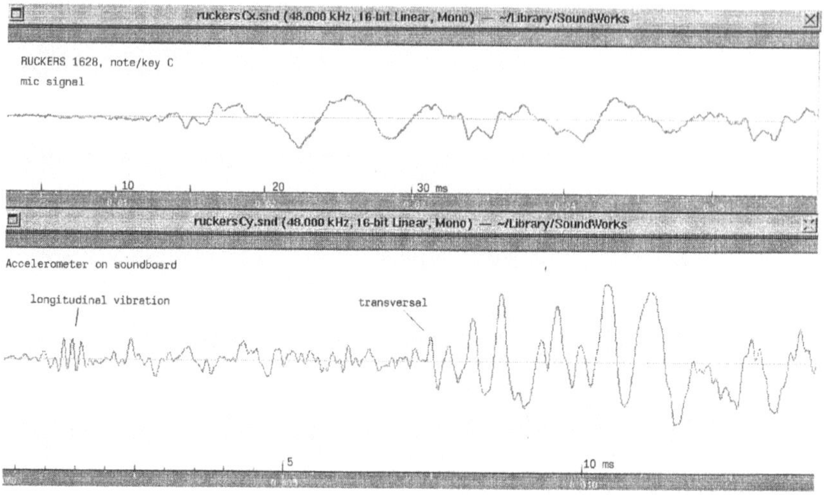

Figure 9: Ruckers 1628, note/key C: time function of microphone signal (upper part of graphic) and of the accelerometer signal of the soundboard (lower part) showing longitudinal vibration at the onset.

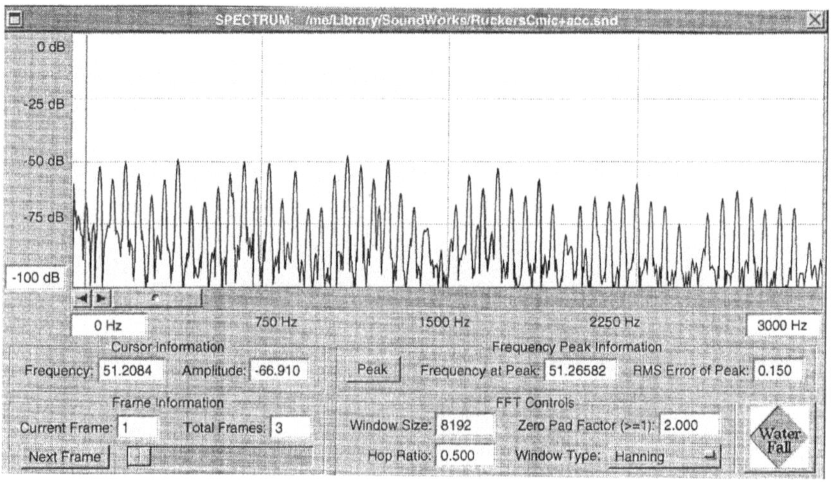

Figure 10: Spectrum, note/key C, accelerometer signal from soundboard, FFT 8192 pts, Hanning.

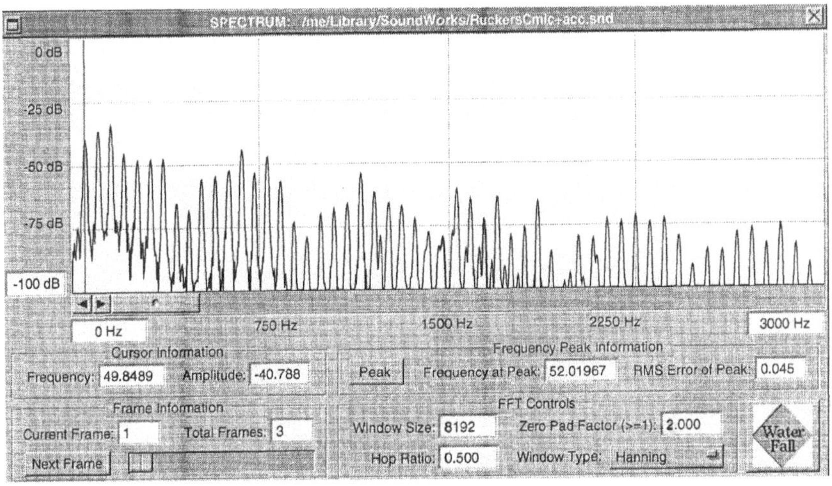

Fig. 11: Spectrum, note/key C, microphone signal of radiated sound, FFT 8192, Hanning.

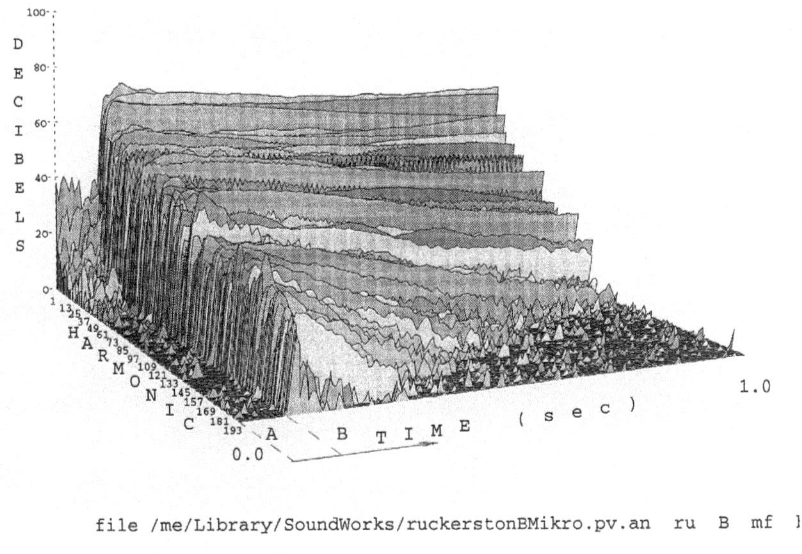

Figure 13: Ruckers 1628, note/key B, microphone signal, transient onset of the sound and temporal evolution of the harmonics 1 – 200 (phase vocoder analysis).

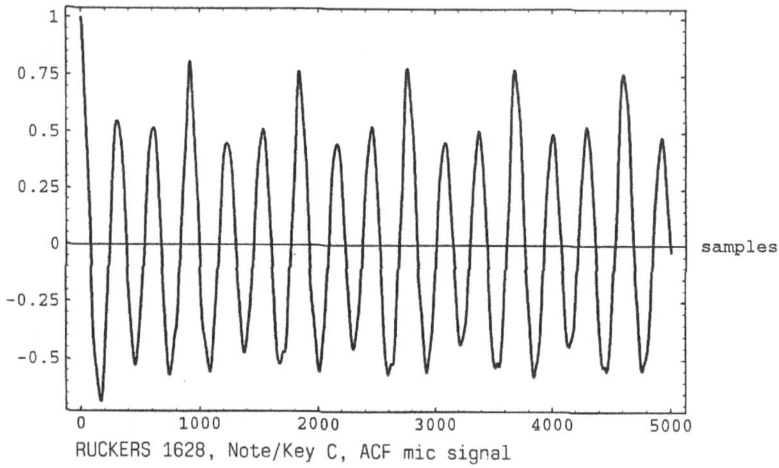

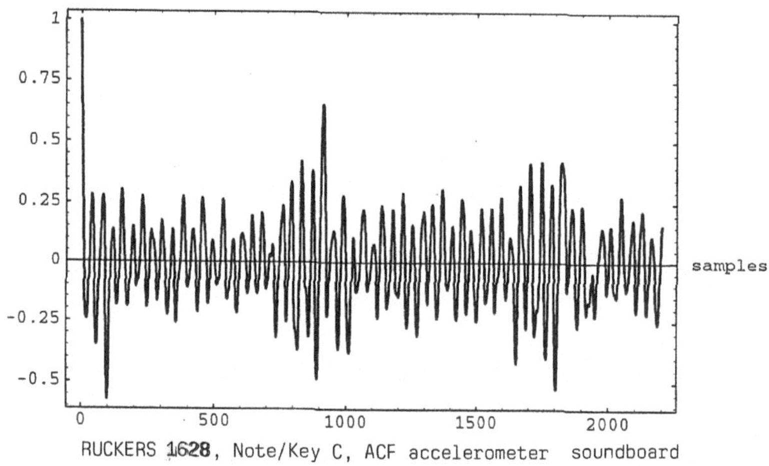

Figure 12a/b: Ruckers 1628, note/key C, autocorrelation function (ACF) of microphone signal (upper) and accelerometer signal from soundboard (lower).

Florian Pfeiffle and Rolf Bader

Real-Time Physical Modelling of a complete Banjo geometry using FPGA hardware technology

Physical Modeling of Musical Instruments

In Musical Acoustics, realizing real-time solutions for Physical Models is a hot topic. The sound quality and variability of musical instrument sounds calculated by whole-body formulations will be pushing this field strongly towards working solutions in the near future. Many attempts are discussed here, yet none of them is working with satisfaction in real-time (Smith 2008). Reducing the number of Degrees of Freedom (DOF) in the models does affect the higher partials of sounds considerably. Changing the mode shapes of musical instruments in terms of spatial distortion is too time consuming to work in real-time. Reducing the complexity of instruments in terms of geometrical simplifications clearly leads to simplified solutions. Also, these models can scarcely be varied in terms of the geometrical fine structure of the instruments and therefore in modelling the parts crucial for musical expression and articulation.

Physical Modelling of Musical Instruments has become one of the major fields in research concerned with properties important for musical performance or instrument building (Bader 2008a). It has been applied to many musical instruments. It could be found that with guitars the coupling of bending and in-plane waves in the guitar body plays a crucial part in the sound (Bader 2005a, Bader 2005b). Also the need to build the guitar back plate under tension to enhance the brightness of the sound could be shown. When systematic changes are applied to the thickness of the top plate, back plates, rims, fan bracing and ribs, the change in brightness of the sound radiated from the different parts do sometimes show a linear but mostly a nonlinear behaviour, caused by the complex coupling of the parts. So e.g. if the fan bracing is added to the top plate and so reduces the flexibility of this plate (resulting in reduced radiation energy of the higher harmonics) this stiffer top plate is able to transport the higher modes to the neck much stronger. That again leads to an enhanced radiation of the high frequencies with increasing fan bracing thickness (Bader 2008c). Other investigations show the possibility to model the guitar in terms of modal synthesis according to the top plate only (Bécache et al. 2005). An example concerning the precision of such a modelling is shown in (Elejabarrieta 2002), where a Finite-Element analysis of a guitar top plate is compared to a real top plate through different stages of construction.

Other instruments have been investigated in a similar way. The pianos' sound board has been modelled in (Giordano 2006), where the fan bracing and the stiffness of the soundboard play crucial roles in the overall sound behaviour. The kantele, a finish dulcimer is investigated in (Erkut et al. 2002). Also a labium is studied in such a way by taking flow-dynamics and turbulence into consideration. It could be shown that the turbulent damping of the flute is crucial for its impedance. Only because of the damping caused by air vortices in the flute at the embochure hole the energy of the flow into the tube is only about 3 % of the whole blowing energy (Bader 2005f). Also fine structures of reeds used with single reed instruments could be investigated. With saxophones, the reed thickness distribution along the area of the reed could be associated quantitatively with the sound expected from different reeds of a commercial reed producer (Bader 2008d). Other investigations concerning the reed show a very similar air vortex distribution in the mouth piece (de Silva 2007).

Also different percussion instruments have been studied using Physical Modelling techniques. Bells show a difficult behaviour within their transient phase as well as in terms of damping or radiation directivity (Schoofs 2000, Lau et al. 2004, 2009). Optimization algorithms have been added to a Finite-Element calculation for bells to find perfect shapes for the bell (Özakča 2004). Other percussion instruments have been investigated, too. Essl and Cook (2000) discussed travelling waves along the rim of a round drum to be important for the transient sound behaviour. In a model/measurement comparison of the complete Bass-Drum, coupling the drums membranes to the wooden box and the inclosed air, it could be shown, that the higher frequencies are radiated from the wooden shell, showing its importance in sound production (Bader 2006c). An investigation of a Balinese percussion instrument (the gender dasa played in a gamelan *wayang* used for puppet theatre) showed, that the trapezoid shape of the plates is producing additional modes through scattering of waves during the initial transient phase of the sound (Bader 2009, 2004). This effect is so strong that it must be considered as the main mechanism producing the overall sound quality of the instrument.

Violins have been studied using whole body Finite-Difference Methods, also implementing the string-bow interactions with changing bowing pressure and velocity (Bader 2005d). Here, also the changes in thickness of different body parts show a complex interaction structure, meaning that linear changes in the violin body normally mean nonlinear changes in the radiated sound. This is also confirmed by investigations of instrument families, maybe most prominent with the violin octet (Bissinger 2003).

Summarizing these investigations, musical instruments do show properties in their sound design which can only be treated by examining the fine structure of the instrument geometry. This fine structure can be the precise shape of the instrument body, the coupling between different body parts, the thickness distribution of plates like violin top plates, the shape of fan bracing appearing with guitars, the distribution of turbulence during blowing, or the thickness distribution along reeds with saxophones or clarinets. All these parameters do influence the sound in a way that they are responsible for the slight changes in timbre which tell a master piece apart from another. Also, these fine structures are responsible for musical expression and articulation, i.e. they are the parts with musical instruments where "the music is happening".

The investigation of these fine structures is often only possible when a geometrical model of the instrument is built, which can show up nearly all the details of the instrument (Bader 2006d, Elejabarrieta 2004, Bader 2009, Bader 2007b, Bader 2007c). These details are responsible for the fine tuning of the sound under different playing conditions. I.e. the quality of a violin may only be judged after using it under many different conditions, in a chamber orchestra, in large concert halls, with folk music, as a solo instrument etc. Under all these conditions, the violin has to fit the needs of presence, loudness, or timbre flexibility. These changes are very hard to detect with methods only concentrating on basic characteristics of the instruments.

The methods used in the field of modelling of whole instrument geometries to date are basically Finite-Element Methods (FEM) (Barthe 2002, Knothe & Wessels 1999), Finite-Difference Methods (FDM) (Bader 2005a), methods of flow dynamics like Lattice-Boltzmann method (LBM), or simplified methods like Waveguides, Delay Lines or Wave Front Methods. All these methods use the basic differential equations for bending, in-plane movement or flow dynamics with appropriate boundary conditions well known from the literature (Leissa 1969a, 1969b, Wagner 1947, Hutchings 1981, Fletcher & Rossing 2000, Flügge 1962). The advantages of FEM or FDM are that geometries of any complexity can be perfectly resolved and any kind of differential equation found to govern the system can be used to underlay the spatial discretization. The disadvantage is the use of many nodal points to resolve the geometry, resulting in high demands on both memory and computation time. So a method which allows the same spatial resolution but on the other side is able to speed up calculation time a lot would be of interest.

Advantages of real-time implementations of Physical Models

For many reasons real-time implementations are of great advantage:

1. Musical instrument shapes are so complex that many alternations are possible. Some sound improvements can only be found by trial and error. Until now, there is no theoretical framework enabling one at first to design a desired sound and subsequently use an algorithm which calculates the adequate instrument shape for that sound. So as the possibilities of instrument shapes are innumerable, a fast real-time tool is needed to try out different possibilities while judging the differences aurally at once.

2. Instrument builders are very seldom familiar with Physical Modelling and computer implementations. They need a soft- or hardware tool to try out different instrument shapes and listen to the results immediately. Although instrument builders might accept several minutes of waiting time for a result, a calculation time of several hours for one sound is too much for a feasible use in musical instrument building environments. Here a fast, if not real-time implementation is needed, which would mean a substantial progress in instrument building practice.

3. Modern music production is based on real-time sound producing algorithms, but no sound designer or musician would wait for hours for a sound to be generated. What they need is a real-time sound synthesis tool. With Physical Modelling one is able to change sounds within musically reasonable regions and therefore the creative possibilities are enlarged to a great extent. Many attempts have been made by commercial soft- and hardware musical instrument building companies (Roland, Yamaha, Apple, Steinberg, etc.) to incorporate Physical Modelling into

their instruments, and as a result hundreds of 'Virtual Musical Instruments' are on the market. Most of them are just samples of real instruments which are filtered and manipulated using modern musical signal processing techniques. Those who try to get as close to a Physical Modelling as possible use too many simplifications which results in sounds that are not convincing, i.e. they do not sound real (e.g. Roland VG 88 modelling guitars where e.g. the body height can be changed).

So the aim is to produce a real-time Physical Modelling hardware implementation using FPGA technology that keeps the full complexity of instruments while performing the sound synthesis in real-time. In the next section the method is explained, and the development status at the Institute of Musicology in Hamburg is presented.

FPGA-Hardware

A Field Programmable Gate Array (FPGA) is a special form of a Very Large Scale Integrated Circuit (VLSI) consisting of matrix-like ordered Logic Blocks, Input/Output-Blocks, and Routing Channels (interconnection network), and therefore similar to an Application Specific Integrated Circuit (ASIC). The Logic Blocks are build on different kinds of logical units, mainly Look-Up-Tables (LUT) with AND- and OR-Gates and other vendor specific logic such as FLIP-FLOPs or MUXs. So FPGAs are similar to single Algorithmic Logical Unit (ALU) microprocessors as logical operations and calculations can be performed, but additionally, they are like memory chips in terms of the amount of logical units performing calculations in parallel. Input and output are realized using the IO-Blocks of the FPGA, equipped with FLIP-FLOPs, which are attached to the physical boundary of the device.

Unlike an ASIC, all elements of the FPGA are freely programmable by the user and can be connected with the interconnection network to realise any logical problem, ranging from a simple adder to a Finite State Machine (FSM) or a complex arithmetic unit. This freely routable matrix-like structure yields the main advantage of an FPGA over other forms of VLSIs.

These capabilities of free programmability and of parallel rather than sequential processing are the great advantages of FPGAs compared to CPUs, DSPs or other Micro Controllers of VLSI architecture. It is therefore possible to compute massive numbers of instructions in parallel within one clock cycle on an FPGA. To benefit from this advantage it is crucial to implement an algorithm that processes as many parallel instructions as possible. It has been shown that an optimized parallel FPGA algorithm is always superior in means of calculation time and maximum clock rates compared to a similar sequential algorithm (Brassail et al. 2007, Subasri et al 2006, Lis et al. 2008, Zouet al. 2006, Inoguchi 2004). This massive parallel computation is implemented on a single small board, and so it is much smaller and much cheaper than a multi-core server system, where MPI can be used to parallel the algorithm (Karniadakis and Kirby 2003). These systems are very expensive even if only about 64 knots are used. Furthermore, more knots do not result in a linear speed-up. The FPGAs on the other hand are capable of processing e.g. 10 000 calculations in parallel on one single small board. An FPGA board can also be implemented easily in a commercial solution.

Another great advantage of an FPGA chip is that it can be programmed from a Personal Computer via a Vendor Specific Flash Tool. The model architecture which is to be implemented on the FPGA device is written in a special Hardware Description Language such as VHDL or Verilog. These Programming Languages contain many High Level constructs like *if-*, *for-*, or *while*-loops and some Low Level aspects like bitwise declaration of signals, variables and constants. The main difference in comparison to other programming languages is the concurrent processing of instructions. This means that unlike e.g. with the C programming language, where sequential code is needed, the code is processed in parallel and needs to be written accordingly. Although the FPGA structure is parallel by nature, sequential statements can also be realised using a Finite State Machine (FSM) as mentioned above and discussed in more detail below.

To avoid time consuming error corrections in codes and to reduce the flashing processing time, a simulation tool is used. Here, similar to a debugging tool, the FPGA code can be simulated using different simulation tools. These simulations can be performed with all kinds of different constraints, like clock speed and changing temperature specifications in the model. Only if the simulated model behaves as expected, one needs to compile and build the source code into a bit-file which then can be flashed onto the hardware-device.

The Build-Process includes different steps like the translation to hardware language, mapping of the IO-Ports and finally Placing&Routing the design so that all preliminaries for the hardware are met. With those advanced programming techniques the engineer is capable of modelling and simulating complex digital hardware systems completely in software before transforming them on a real hardware device. This allows a very fast and stable development of hardware designs without the need of a physical Integrated Circuit (IC) development.

FPGA - Implementations

Because of their high speed processing capability and flexibility FPGA-devices are used in many fields of digital signal processing.

In the area of sound analysis and sound source localization, implementations have been realized for e.g. Real-Time Noise Source Identification (Veggeberg et. al 2008), High Speed Direction of Arrival Algorithms (Hao et. al 2000), Delay-Sum Beamforming (Chen et al. 2008) and countless other applications. All of these works either show a real-time realization of the problem for the first time or a tremendous speed-up using an FPGA with its parallel processing capabilities.

Typical signal processing applications for FPGAs are implementations of IIR- or FIR-filter designs (Meyer-Baese 2004). In Madanayake et al. (2004) 2D/3D Plane-Wave filters are realised using IIR/FIR-filters. Shuang et al. (2008) focus on converting analog to digital controllers using a filter-design with FPGA. Maslenikow et al. (2006) and

Brich et al. (2006) discuss a method how DSP filter-designs (IIR/FIR) can easily be implemented on an FPGA chip.

In the field of sound production, several publications have been published in which an FPGA is used as a function generator for simple signals like a sine wave or a rectangle signal (Meyer-Baese 2004, Reichardt et al. 2006, Kilts 2007). Most of these works focus on high frequency signals like an AM-demodulation chain for a digital radio receiver (Meyer-Baese 2004).

Solving differential equations with FPGAs using finite differences have been reported before, too. Suzuki et al. (2005) discuss the use of an FPGA to simulate an electric field solving Maxwell's equations using a FDTD (Finite Differences in the Time Domain) algorithm. This algorithm is widely used for the analysis of electromagnetic problems (Shlager et al. 1995). Other approaches to solve differential equations on an FPGA implicitly are e.g. the implementation of the conjugate gradient method (Strzodka et al. 2006) or the Euler Method (Jayalakshmi et al. 2006). For parallel algorithms solving linear equation systems see also Karniadakis and Kirby (2003).

Papers focusing mostly on the real-time aspect of FPGA implementations are e.g. Particle Track Recognition (Liu et al. 2008), Direction of Arrival Algorithms (Hao et al. 2002), High Speed Cross Correlation (Von Herzen 1998), Noise Source Identification (Veggeberg et al. 2009), or Digital Beamforming (Wang et al. 2005).

Only one paper recently discussed the use of FPGA to synthesize industrial sounds (Martins et al. 2008). This paper focuses on a low-cost method to process sound on an FPGA without the need for additional electronics and omits musical aspects.

Physical Model of the Banjo

When building a Physical Model of an instrument it is crucial to know a great deal about the instruments' structure and its vibrational behaviour. The banjo's most important parts in the generation of its sound and timbre are the five metal strings and the plastic membrane, made of mylar (see Fig. 1). It has been shown that the wooden/metal body and the peg play a minor roll in the sound production of a banjo, as they are very stiff and therefore mainly fixed. They act as boundary conditions for the membrane and the strings (Dickey 2003). It can be assumed that the bridge serves as a low impedance transfer function element between the strings and the membrane. Only its position on the membrane has an influence on the sound (Stephey et al. 2008). The air beneath the membrane dampens this membrane and lowers the deepest modes of it, similar to the behaviour of a wooden snare drum (Fletcher and Rossing 2000). Although a great deal is known about the single parts of the banjo and their dynamic behaviour, very little research has been done on the coupled system, which is the membrane/bridge/string interaction (Stephey et al. 2008).

Formulating a more detailed model of the banjo as proposed here will lead to many insights, not only into the banjos sound producing behaviour, but also into those of other coupled systems and their dynamic behaviour similar to that of the banjo.

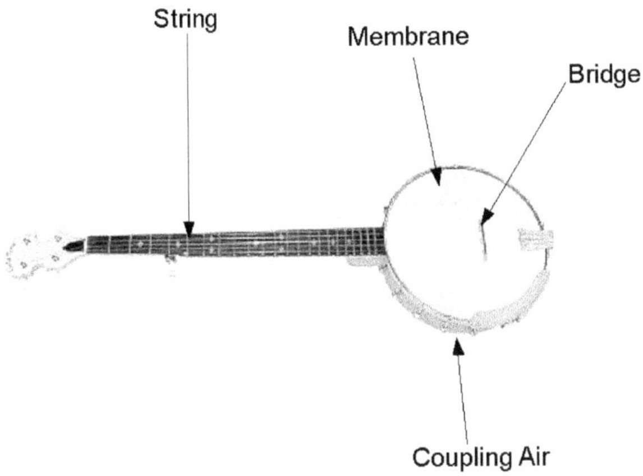

Fig. 1: The Banjo used in the model.

The dynamic system of the modelled Banjo is presented in Fig. 2. The model consists of string, membrane, bridge, and the coupling air beneath the membrane. As mentioned above this model is reduced to the core elements of the banjos sound producing parts. The bridge is the most sensitive part of it. Stephey et al. (2008) have pointed out that the position and the small weight of the bridge are of critical importance for the sound of the banjo whereas the shape is negligible. Because we found that the shape of the bridge is of importance for its transfer characteristics, we have implemented the bridge to be able to vary its shape and mass and to investigate the effects on the resulting sound. There may be a similar case as for the violin, where the two feet of the bridge are crucial to the sound but any carving of it is not, so it could be replaced by a so-called plate bridge (Jansson 2003). An earlier model of the banjo has shown that the bridge can be approximated by a linear factor without a big difference in the resulting sound compared to the sound produced by the final model. Still, in the literature there are many works in instrument acoustics focusing on the bridge, for instance with the violin (Wienreich 1977, Rossing et al.). The impedance function of the bridge may follow the violin, too, where the bridge-hill is crucial for the sound (Woodhouse 2000). Because there are many details that have not been fully understood in the role of the bridges as a transducer we decided to take the bridge into account.

First, each of the banjo parts is modelled separately in MATLAB and C Programming Language with the Finite-Difference Method (FD). The Physical Model of the string is a 100 Node 1D-FD. The membrane has 900 nodes using a 2-dimensional FD and the bridge is a 3-dimensional bending wave PDE using 750 nodes. With the air, 400 nodes are used. All parts take damping into account.

After successful single-system stability tests, the four systems are coupled linearly, yielding the dynamic system of a realistic full-scale banjo. To simulate the behaviour of the complex system, the model can now be excited, where appropriate initial values can be set on any node. To simulate a realistic banjo it is obvious that one has to deflect the string with a triangular shape at time point t = 0. Still with this time-dependent model all kinds of deflections can be realised at any node of the model and so one could also knock on the membrane, the bridge or the string, simulate additional damping anywhere or model unisotropic stiffness of all parts.

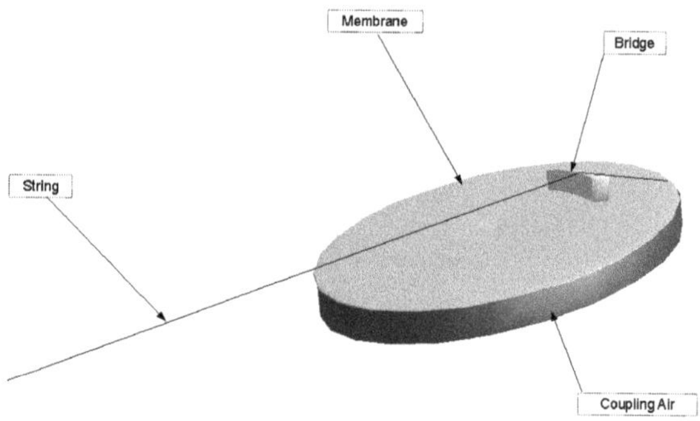

Fig. 2: Geometrical representation of the banjo model

The resulting time series of a triangular deflected string is shown in Fig. 3 in comparison to the recorded and the simulated sound. The sound produced by the model is almost indistinguishable compared to the sound from a real banjo, even for a musical trained ear.

The drawback of this model is its tremendous computational costs. E.g., even with a speed-optimized C-Code the shown time series takes almost 30 seconds to compute on a 2,4 GHz Intel Quad Core Processor. With the normal processing capacities of a standard computer an implementation of the presented banjo model in real-time is not possible. A full-scale model with five strings instead of one string like the presented model can not be realized with the common computer architectures of a personal computer in real time.

FPGA Implementation of the Banjo

To implement the proposed model in real-time with high accuracy the C source-code was transferred into VHDL-programming language. All the k nodes of the model are build in parallel forming a 1-dimensional-array of interconnected parallel pipelined computations. In Fig. 4 a systematic view of the implemented string algorithm is shown. As one can see, the individual nodes are depending on their adjacent nodes, the clock and the FSM signal. Because of the interconnected node structure it is possible to compute one calculation step for all the k nodes in parallel within one clock cycle. Therefore the algorithm is extremely fast and effective.

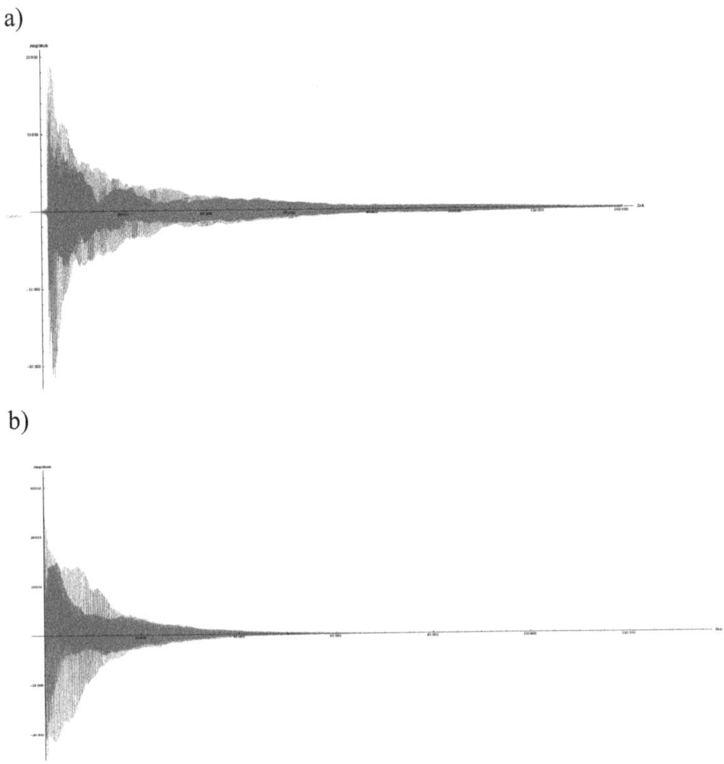

Fig. 3: A physical model of the banjo a) implemented in C++ compared to b) a real recorded sound.

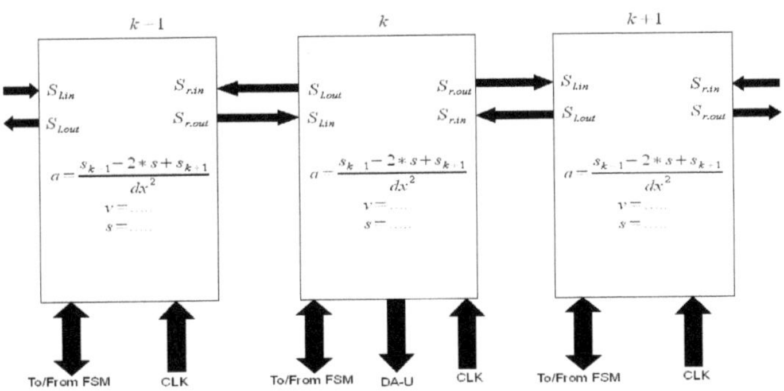

Fig. 4: Finite-Difference Model of the Banjo used for the sound in Fig. 3a.

However, the inherent dependencies of the FD algorithm had to be implemented sequentially. As mentioned above, sequential statements can be realized in a Finite State Machine (FSM) on an FPGA. In Fig. 5 a block diagram of the State Machine and its single states is depicted. As the name suggests the state machine consists of several (in our case six) finite states which are active when the precedent case has finished its task. Every state of the FSM forms one step of the FD algorithm. To use all the parallel processing capabilities of the FPGA, the pipelined FSM-FD algorithm is divided into different subroutines.

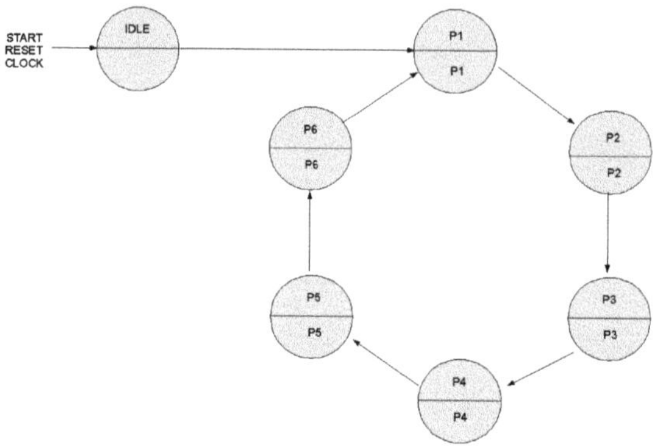

Fig. 5: Finite-state machine of the model

The possible number of nodes which can be computed within one clock cycle depends upon the used hardware device. The hardware described above is able to compute about 50 nodes in 10 clock cycles. The maximum system clock speed is 100 MHz. This means a maximum number of 10,400 nodes for a sampling frequency of 48 kHz.

The banjo model consists of 100 string nodes, 900 membrane nodes, 750 bridge nodes and 400 air nodes. Thus a full-scale model of the banjo with five strings can be modelled with 2,650 nodes, i.e. it can be realized in real time without any restrictions.

The hardware used is a Xinlinx XUP Virtex-2 Pro Development System sponsored by Xilinx company (see Fig. 6 and Fig. 7). The core of this system is a Virtex-2 FPGA Chip with 30,000 logic blocks, 2,448 kB internal Block RAM and 136 18x18 bit multipliers. Besides other features the board is equipped with an AC97 audio codec providing 2 inputs and 2 outputs, a video XSGA output and a 100 MHz system clock. The software used is the latest version of Xilinx ISE Design Suite 10.2 which includes a VHDL/VERILOG editor, a build-, place- and route-tool and a flash-tool. The software used to simulate the model is MentorGraphics Modelsim 6.3f.

A fully functional hardware device has been built that can be attached to the FPGA as a user friendly input-output device. It consists of four voltage controlled (VC) faders, two continuous potentiometers, an alphanumeric LCD display HS286 and a maxim Dallas

8051 Micro Controller. In the current state of development the hardware device communicates with other hardware via the serial RS232-Port using a MIDI protocol.

The aim is to be able to connect a variety of controllers directly to the FPGA Board to influence crucial parameters in real time. Those parameters could be the stiffness of the membrane, the thickness of the banjo string or even the geometry of the bridge, which could then be changed during playing.

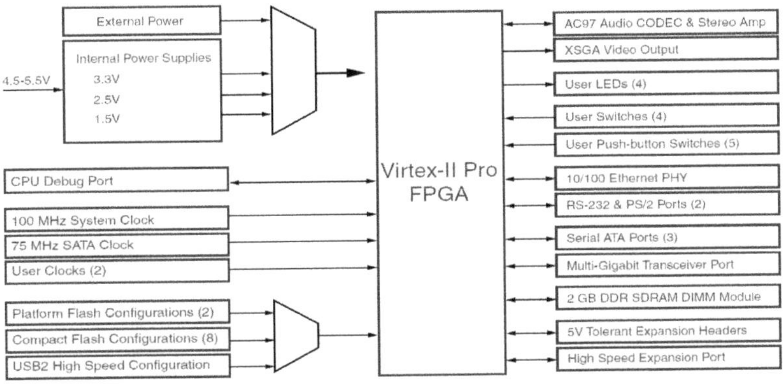

Fig. 6: FPGA board structure

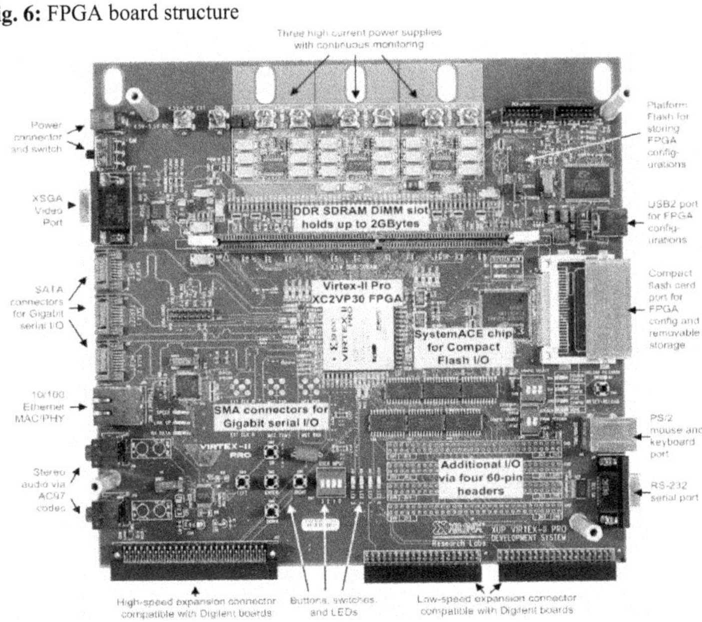

Fig. 7: Xilinx Virtex 2 Pro FPGA Board

Moreover one may even be able to vary the coupling strength between the different parts of the banjo while the banjo-sound is playing. This could deliver many insights into instrument specific parameters and their role in the overall sound. To control the parameters one may use faders, a MIDI keyboard, and any kind of sensors, transducers etc. By using a string-to-MIDI controller, it furthermore would be possible to actually play a real banjo with damped strings which triggers the FPGA model, and so being able to play while changing the instrument geometry in real time.

Conclusion

FPGA technology makes a real-time solution of a complete body model of a musical instrument possible. As via I/O controllers parameters like e.g. geometry, string tension, playing style and articulation can be changed, this synthesis tool is combining the advantages of a sound device of maximum flexibility with the physical properties of a musical instrument. It is therefore the perfect solution to understand the constructional reasons for sound features relevant for music and musicians. Additionally, it might be a helpful tool for instrument builders, because they would be able to listen to different instrumental designs before building them. Concerning the commercial musical production processes this new technology will provide musicians, producers and sound designers with a new gear for sound production. Further studies will include collaborations with musicians and instrument builders to learn about the possibilities and limits of the technique.

References

Bader, R. (2009), Additional modes in a Balinese gender dasa plate due to its trapezoid shape, In: *Concepts, Experiments, and Fieldwork: Studies in Systematic Musicology and Ethnomusicology*, R. Bader, Ch. Neuhaus, U. Morgenstern (Eds.), Hamburg: Peter Lang Verlag, 93-110,

Bader, R., U. Hansen (2008a), Acoustical Analysis and Modeling of Musical Instruments Using Modern Signal Processing Methods, In: *Handbook of Signal Processing in Acoustics*, D. Havelock, M. Vorländer, S. Kuwano (ed.), Berlin, New York: Springer, 219-247.

Bader, R. (2008b), Efficient Auditory Coding of a Xylophone Bar, In: *Hamburger Jahrbuch für Musikwissenschaft*, Hamburg: Peter Lang Verlag, 197-212.

Bader, R. (2008c), Fine tuning of guitar sounds with changed top plate, back plate and rim geometry using a whole body 3D Finite-Difference model, In: *Forum Acusticum joined with American Acoustical Society Paris 08*, 5039-5044.

Bader, R. (2008d), Individual reed characteristics due to changed damping using coupled flow-structure and time-dependent geometry changing Finite-Element calculation, In: *Forum Acusticum joined with American Acoustical Society Paris 08*, 3405-3410.

Bader, R. (2007a), Physical Modeling of the self-sustained oscillation of a Labium using Molecular Dynamic Simulation Techniques, *Proceedings ISMA (International Symposium of Musical Acoustics), Barcelona..*

Bader, R. (2007b), Spatial cognitive timbre dimensions of physical modelling sounds using Multi-Dimensional Scaling Techniques (MDS), *Proceesings ISMA (International Symposium of Musical Acoustics),* Barcelona 2007.

Bader, R. (2007c), Spatial cognitive timbre dimensions of physical modelling sounds using Multi-Dimensional Scaling Techniques (MDS), *J. Acoust. Soc. Am.* 121, 3184.

Bader, R. (2006a), Finite-element transient calculation of a bell struck by a clapper, with Björn-Andre Lau, Peter Wriggers, Albrecht Schneider, *J. Acoust. Soc. Am.* 119, 3290.

Bader, R. (2006b), Amplitude fluctuations within musical instrument sounds because of impulse reflections, *J. Acoust. Soc. Am.* 119, 3382.

Bader. R. (2006c), Finite-element calculation of a bass drum., *J. Acoust. Soc. Am.* 119, 3290.

Bader, R. (2006d), Characterization of Guitars through Fractal Correclation Dimensions of Initial Transients, *Journal of New Music Research*, Vol. 35, 4, 323-332.

Bader, R (2005a), *Computational Mechanics of the Classical Guitar*, Berlin, New York: Springer.

Bader, R. (2005b), Complete Geometric Computer Simulation of a Classical Guitar, *Lay-Language paper of the American Acoustical Society 05*, http://www.aip.org/149th/bader_Guitar.htm.

Bader, R. (2005c), Identifying Musical Instruments through their Microrhythms, *Lay-Language paper of the American Acoustical Society 05*, http://www.aip.org/149th/bader_microrhythm.htm.

Bader, R. (2005d), Whole geometry Finite-Difference modelling of the violin, *Proceedings of the Forum Acusticum 2005*, 629-634.

Bader, R. (2005e), Nonlinearities in the sound production of the classical guitar, *Proceedings of the Forum Acusticum 2005*, 685-689.

Bader, R. (2005f), Turbulent k-e model of flute-like musical instrument sound production, In: *Fractals in Engineering. New trends in theory and applications*, E. Lutton and J. Lévy-Véhel (Eds.), Berlin, New York: Springer, 109-122.

Bader, R. (2004), Additional modes in transients of a Balinese Gender dasa plate, *J. Acoust. Soc. Am.* 116, 2621.

Bader, R. (2003), Physical model of a complete classical guitar body, In: *Proceedings of the Stockholm Music Acoustics Conference 2003*, R. Bresin (Ed.), Vol. 1, 21-124.

Bader, R. (2002), *Fraktale Dimensionen, Infomationsstrukturen und Mikrorhythmik der Einschwingvorgänge von Musikinstrumenten*, Doktorarbeit der Universität Hamburg, (http://www.sub.uni-hamburg.de/volltexte/2002/598).

Bissinger, J. (2003), Modal Analysis of a Violin Octet, J. *Acoust. Soc. Am.* 113 (4), 2105-2113.

Bécache, E., A. Chaigne, G. Derveaux, P. Joly (2005), Numerical simulation of a guitar, *Computers and Structures*, 83 (2-3), 107-126.

Barthe, K.J. (2002), *Finite-Element Methoden*, Berlin, New York: Springer.

FPGA Parallel Implementation of CMAC Type Neural Network with on Chip Learning, *IEEE 2007*, 111 – 115.

Brich, T., K. Novacek, A. Khateb (2006), The Digital Signal Processing Using FPGA, *IEEE 2006*, 322 – 324.

Chen, P., X. Tian, Y. Chen, X. Yang (2008), Delay-sum Beamforming on FPGA, *ICSP2008 Proceedings*, 2542-2545.

Da Silva, A, G. Scavone, M. van Waltijen (2007), Numerical simulations of fluid-structure interactions in single-reed mouthpieces, *J. Acoust. Soc. Am.* 122 (3), 1798-1809.

Elejabarrieta, M.J., A. Ezcurra, C. Santamaria (2004), Vibrational behaviour of the guitar soundboard analysed by means of finite element analysis, *Acustica united with Acta Acustica*, 87, 128-136.

Elejabarrieta, M.J., A. Ezcurra, C. Santamaria (2002), Coupled modes of the resonance box of the guitar, J. *Acoust. Soc. Am.*, 111, 2283-2292.

Essl, G., P. Cook (2000), Measurements and efficient simulation of bowed bars, *J. Acoust. Soc. Am.*, 108 (1), 379-388.

Erkut, C., M. Karjalainen, P. Huang, V. Välimäki (2002), Acoustical Analysis and model-based sound synthesis of the kantele, *J. Acoust. Soc. Am.* 112 (4), 1681-91.

Fletcher, N., Th. Rossing (2000), *Physics of Musical Instruments,* Berlin, New York: Springer.

Flügge, Wilhelm (1962), *Statik und Dynamik der Schalen*, Berlin, New York: Springer.

Giordano, N. (2006), Finite-difference modeling of the piano, *J. Acoust. Soc. Am.* 119, 3291.

Hai, C; W. Ping (2002), The High Speed Implementation of Direction-of-Arrival Estimation Algorithm, *IEEE 2002*, 922-925.

Hutchings, C.M. (1981), Klang und Akustik der Geige. *Spektrum der Wissenschaft*, Dezember 1981, 112-122 (original: Scientific American, October 1981).

Jansson, E.V. (2003), The BH-Hill and tonal quality of the violin, *Proc. Stockholm Music Acoust. Conf. (SMAC 03)*, Vol. 1, 71-74.

Karniadakis, George Em, Robert M. Kirby II (2003), *Parallel Scientific Computing in C++ and MPI*, Cambridge: Cambridge University Press.

Knothe, K., H. Wessels (1999), *Finite Elemente. Eine Einführung für Ingenieure*, 3. Auflage, Berlin, New York: Springer.

Lau, A., R. Bader, A. Schneider, P. Wriggers (2009), Finite-Element transient calculation of a bell struck by its clapper. In: *Concepts, Experiments, and Fieldwork: Studies in Systematic Musicology and Ethnomusicology*, R. Bader, Ch. Neuhaus, U. Morgenstern (Eds.), Hamburg: Peter Lang Verlag, 135-154.

Lau, A., P. Spiess, P. Wriggers, A. Schneider, R. Bader (2004), Analysis of bell vibrations. In: *2nd Conference on Advances and Application of GiD*. Barcelona, Feb. 2004, Preprints. Intern. Center for numerical methods in engineering.

Leissa, A.W. (1969a), *Vibration of Plates*, Publication of the American Acoustical Society (1993).

Leissa, A.W. (1969b), *Vibration of Shells*, Publication of the American Acoustical Society (1993).

Lis, J. D., C.T. Kowalski (2008), Parallel Fixed Point FPGA Implementation of Sensorless Induction Motor Torque Control, *IEEE 2008*, 1359 – 1364.

Liu, M., W. Kuehn, Z. Lu, A. Jantsch (2008), System-on-an-FPGA Design for Real-Time Particle Track Recognition in Physics Experiments, 11th EUROMICRO CONFERENCE on DIGITAL SYSTEM DESIGN Architectures, Methods and Tools, *IEEE 2008*, 599 – 605.

Madanayake, A., L. Bruton, F. Comis, C. Comis (2004), FPGA Architectures for Real-Time 2D/3D FIR/IIR Plane Wave Filters, *IEEE 2004*, III 613 – III 616.

Martins, G., M. Barata, L. Gomes (2008), Low Cost Method to Reproduce Sound with FPGA, *IEEE 2008*, 1932 – 1936.

Maslennikow, O., A. Sergiyenko (2006), Mapping DSP Algorithms into FPGA, *IEEE 2006.*

Meyer-Baese, U. (2004): *Digital Signal Processing with Field Programmibalbe Gate Arrays*. Springer.

Özakča, M.T.G. (2004), Structural Analysis and optimization of bells using finite elements, *Journal of New Music Research*, 33, 61-69.

Subasri, V., K. Lavanya, B. Umamaheswari (2006), Implementation of Digital PID controller in Field Programmable Gate Array (FPGA), *Proceedings of India International Conference on Power Electronics 2006*, 172 – 176.

Schneider, A., R. Bader (2003), Akustische Grundlagen Musikalischer Klänge, *Mitteilungen der mathematischen Gesellschaft Hamburg*, Bd XXII, 27-44.

Schoofs, A. (2002), Computer-aided bell design and optimization in the quality of bells, Eurocarillon 2002, Brugge, Sept. 6th, 2002 (=*Proc. of the 16th Meeting of the FWO Research Soc. on Foundations of Music Research*), Ghent: IPEM, Ghent University 2002.

Shlager, K. L., B. Schneider (1995), A Selective Survey of the Finite-Difference Time-Domain Literature, IEEE Antennas and Propagation Magazine, Vol. 37, *IEEE 1995*, 39-56.

Kai, S., X. Yankai, J. Shan, Z. Hongwu (2008), Converting Analog Controllers to Digital Controllers with FPGA, *ICSP2008 Proceedings*, 486-489.

Smith, J. III, Digital Waveguide Modelling and Signal Processing, http://ccrma.stanford.edu/~jos/wg.html.

Smith, J. III (2008), Musical Acoustics: Virtual Instruments I & II', Session at the Joined Meeting of the American Acoustical Society with the European Acoustical Society, Paris 2008, *J. Acoust. Soc. Am.*, 123 (5, 2), 3521-23, 3664-66.

Strzdoka, R., D. Göddeke (2006), Pipelined Mixed Precision Algorithms on FPGAs for Fast and Accurate PDE Solvers from Low Precision Components, *IEEE 2006.*

Veggeberg, K., A. Zheng (2009), Real-time noise source identification using programmable gate array (FPGA) technology, *Proceedings of Meetings on Acoustics, Vol 5., Acoustical Society of America*, 1 – 10.

Von Herzen, B. (1998), Signal Processing at 250 MHz Using High-Performance FPGA's, *IEEE TRANSACTIONS ON VERY LARGE SCALE INTEGRATION (VLSI) SYSTEMS*, VOL. 6, NO. 2, 238 – 246.

Wagner, K.W. (1947), *Einführung in die Lehre von den Schwingungen und Wellen*, Wiesbaden.

Wang, Z., R. Jin, J. Geng, Y. Fan (2005), FPGA implementation of Downlink DBF Calibration, *IEEE 2005.*

Woodhouse, J. (2005), *O*n the „Bridge Hill" of the Violin, *Acta Acustica united with Acustica,* 91, 155-65.

Zou, Z., W. Hongyuan, W., Y. Guowen (2006), An Improved MUSIC Algorithm Implemented with High -speed Parallel Optimization for FPGA, *IEEE 2006.*

Rolf Bader
Arne von Ruschkowski

Differences in brain activity between blurred and unblurred versions of musical pieces

Introduction

Musical meaning is extracted from features of musical pieces like melody, rhythm, articulation, sound or form. From these features associations to outer-musical meaningful units can be made. Karbusicky (1986) derives his musical semantics from the categories of Pierce (1958). Pierce distinguishes between the Ikon, an onomatopoetic sound (like the "Kuckucksruf" of Beethoven's Pastorale, blues harp mimicking train whistle, ect.), an Index, an emotional association indexed by a musical event ("Klagesekunde", tritone as index for the devil, ect.), and the symbol (national anthm, wedding march, ect.). Pierce derived these notions from the possible combinations of subject - object relations as discussed by Kant[1]. Karbusicky added a fourth category, the material, the musical notes and sounds, where the semantics is derived from.

Other notions of musical meaning, like mimesis as described by Aristoteles and Platon was also often used for music. Platon in his dialog Nomoi (1991) made his famous complaint, that often during sacrifices in temples the musicians do not play the appropriate melodies but rather improvise what come to their minds. These musicians were no longer mimicking the spirit of the god and therefore did play wrong. Also for Th. Adorno (1949) the notion of mimesis ('Mimetisches Moment') was one of his main tools for analysing musical meaning.

Other directions of meaning, like the structuralism of L. Strauss (1964, 1975) searching for meaning in the structure of the myth of Southern American native tails and the structure of family relations comes close to the thinking of Karbusicky as the latter came from the Chech structuralist movement[2]. The semiotics of U. Eco (2002) on the other side does deny universals and treats meaning in terms of communication in society, mainly by speech.

When it comes to brain research, the prefrontal cortex seems to be the focus for music and meaning. The perception and anticipation of long melodies or phrases was studied among others by Leaver et al. (2009), Halpern and Zatorre (1999), or Zatorre et al. (1996). It could be shown, that the prefrontal cortex (PFC) is highly involved in the imagery of known pieces as expected. Furthermore, wider networks were found between

[1] Kant distinguished between ‚practical' (praktisch) and ‚pragmatic' (pragmatisch). This focus was on practical issues where his third critic the 'Critic of practical reason' (Kritik der praktischen Vernunft) discussed ethical principles, most famous the categorical imperative: "Always act as if your doing could be the foundation of a general law." („Handle so, daß Dein Tun stets Maxime einer allgemeinen Gesetzgebung sein könnte.")

[2] The Mythological of L. Strauss is ordered like a Sonata in terms of the organisaiton of the chapters.

the frontal and parietal regions. The role of the medial prefrontal cortex (MPFC) in autobiographical memories of songs familiar with through the personal past was investigated by Janata (2009) for 30 seconds excerpts and also for short time intervals. It was found a parametrical respond of this region to the autobiographical salience of the musical piece. The same author found previously (Janata et al. 2002) this brain region to track the movement of melodies in tonal space. Blood et al.(1999) showed, that this region responded more strongly to consonant than to dissonant musical pieces. Plailly et al. (2007) found, that familiar musical stimuli responded stronger in MPFC than unfamiliar. Janata (2009) concludes, that the MPFC is the region, where music, memories and emotions are associated one to another.

So we would expect the frontal region to be the focus of attention, anticipation, familiarity, emotion, meaning in general. Still, the connection to the parietal region found by Zatorre et al. (1996) is of interest, too. This study is looking with a first overview on the subject if changes in the meaning of music leads to changed brain activity in the different brain regions. It is assumed, that a change in meaning will most likely be processed in the frontal brain or parietal region, where the temporal region of the brain may react, too, but most likely because of a change in the sound rather than with associations and meaning.

All these movements of semantics, semiotics, mimesis or structuralism derive musical meaning starting by the material. As higher ordered musical content like pitch, rhythm, form ect. is all derived from the sound we hear, changing this sound will change some of these features and so influence the meaning of the piece. This investigation is taking a first look into this field by distorting musical pieces using a simple signal processing algorithm and comparing the brain activity of subjects when listening to the undistorted and the distorted piece. As this is only a primarily study in this field, we do not clearly define, which musical meaning is lost in the distorted pieces. So using the same algorithm for all pieces, in a piece with clear melody, this melody is lost and so meaning in terms of 'beauty', 'tension' or 'tenderness' may be lost, while in another piece without clear pitches the timbre distortion is more prominent and sounds associated to musical instruments, genres or emotions are lost.

Distorting the pieces

The pieces used in the study are

1. Classical Music: Wolfang Amadeus Mozart – Eine kleine Nachtmusik
2. Beat Music: The Beatles – With a little help from my friends
3. Child song: Iris Gruttmann – Schnappi, das kleine Krokodil
4. Jazz: Dave Brubeck – Take Five
5. Hip-Hop: Eminem – Without me
6. Heavy Metal: Slipknot – People = Shit

The songs were chosen to represent certain musical styles mostly different one from another. 30 seconds of each song was used in the study. These 30 seconds were presented in the original and in a blurred version. The blurring algorithm randomly interchanged the amplitudes of all frequencies present in one critical band. As theories of loudness perception calculate loudness as a weighted sum of the energies present within each critical band, the algorithm, conserving this energy tried to keep the perceived loudness constant. Nevertheless by exchanging the partials within a critical band, the sound was changed considerably.

An FFT was performed for a time window of 100 ms like

$$F(\omega) = \frac{1}{dt} \sum_{i=1}^{dt} f(t+i)\, e^{i \omega i / s}$$

Starting from sample t of the time series f(t) and using the sample frequency s, for each ω fitting the frame size dt in samples a Fourier transform is performed. 24 critical bands were defined as third-octave band filters. Within each band, the amplitudes were exchanged using a random matrix associating each frequency within one band randomly with another frequency therein. The altered spectrum was then Inverse Fourier transformed to obtain the new time series.

As a time window of 100 ms was used, the lower bands have only a few frequency components. This causes the sound to change mostly in the mid and high frequency range conserving most of the rhythm and the overall form of course. So i.e. the rhythm of the Mozart piece was perfectly the same in the blurred version compared to the original one, still the melody was very different. It was interesting to see, that most people recognized the piece right away only because of the rhythm.

So overall, melodies were not recognizable any more as no harmonic overtone series were present any longer. Vocals could still be recognized as such but could only be understood if they were known in advance, so still the rhythm and part of the articulation left no doubt about how far the piece had proceeded.

When blurring a piece to an extreme, one would only need to replace it by white noise, no matter which piece we have. This seemed not appropriate, as then all pieces would only be compared to a white noise sound. To keep some musical features while changing others was chosen to obtain a reduction but not a total erase of the material for construction of meaning.

Experiment

Five subjects were informed about the investigation to a full extend and all 12 musical pieces (6 original, 6 blurred) were presented to them beforehand. Then their EEG data were recorded while listening three times to all the pieces. They were instructed to concentrate on the music throughout the whole experiment. Still some reported they were drifting away mentally or even falling asleep. The subjects were asked after the experiment about their listening strategy and experience and a qualitative agreement with most of the responses to the EEG data could be found in terms of active or passive listening.

As five subjects listened three times to the 12 sounds, an overall of 180 cases were recorded, 90 original and 90 blurred cases. A filter was used to automatically reject artifacts caused by eye motion or swallowing of 50 μV. As each subject showed an individual pattern of brain activation which was very consistent over all pieces and trials, an average for each individual and each electrode was calculated and substracted from all data of this subject. The remaining activity was averaged over all subjects and tries and a mean activation for each piece of music was calculated for all electrodes. This mean was used as a reference mean for all pieces and each piece was referenced to this mean showing more or less activation in comparison to all other pieces, namely the blurred/unblurred partners.

Results

When comparing all blurred to all unblurred cases, the activity of all brain regions recorded was higher with the blurred versions (Fig. 1). This holds more for the ventral and more for the right hemisphere. Still also the frontal lobe shows a higher activation with blurred sounds.

When taking a close look at the different pieces, for most of them a clear pattern occurs with the original pieces (Fig. 2). Mozart, Beatles, Schnappi and Eminem all have decisive differences at the left frontal lobe, where Mozart shows more activation while the other three show less. The clear unisono rhythm of the intro to 'Eine kleine Nachtmusik' seems to invoke an active association process, while the other three seem to work right in the other direction, more activating the right frontal region and the other brain regions. In this sense, also the Slipknot piece has this pattern. Only the Jazz tune shows a general activation of the frontal lobe while less active in the parietal region.

The patterns for the blurred pieces are much less clear. The activations seem more random, in line with the randomization of the sounds of these pieces. Some of the patterns change only partly, but this is so random, that it cannot be stated significant.

When looking at plots comparing the different pieces at the different brain regions, first the clear dominance of the left frontal side for the Mozart piece appears. This is very consistent throughout the subjects and tries and rules out any other effects. The opposite is true for the right side with the blurred Slipknot piece, still the amplitudes are much lower there. The trash nature of the Slipknot piece is even enhanced with the blurred version and associations more emotional may be the cause for the right side activation.

It is also interesting to have a look at single brain areas shown in Fig. 3-8 for the frontal, temporal and parietal regions.The F3, F4, F7 and F8 plots show the basic trend already mentioned with the overall blurred-unblurred relation: the blurred pieces show more activation than the unblurred ones. Only the Mozart piece is still more for all plots than the average, no matter if blurred or unblurred. Also 'Schnappi' is less than average for all frontal regions. This child song was mostly perceived as a funny, emotional song not suitable for reflections and associations. Eminem is mostly average no matter if blurred or not. This music strongly depends on the lyrics which for a non-native English speaker

– which all subjects were – is very hard to understand and therefore may not provoke as much as it is supposed to do.

A clear difference for the blurred- unblurred case is the Brubeck piece. Except for F8, the activation is slightly less than average for the original piece and strongly positive for the blurred version. The very smooth character of the melody seem to under-evoke associations, while the blurred version was in irritating contrast to this. This also holds for the Beatles piece, which is low for the original version and strong for the blurred one. A similar interpretation may be drawn here.

The picture for the temporal and parietal regions is very clear in terms of the blurred-unblurred relation, especially for P4, Pz, T4, and T6. This is again consistent with the overall relation over all pieces. Still, among the regions, the reaction is much less consistent. Mozart is strong at T3 and T5 and weak for T4 and T6 for the original and more like average for the blurred versions. Schnappi is less for all parietal and temporal with the original, but strongly changing for the blurred version. Beatles are very strong for the blurred version but change very much for the original here. Eminem and Slipknot change for both versions very strongly. Schnappi is always less with the original but changes a lot with the blurred one. We would need to go more into the details of the sounds themselves to explain the differing activations for this brain region.

Discussion

The increase of brain activation due to blurring of musical pieces in the temporal region seems to be caused by the lower consistency of the sounds being much less harmonic and therefore more noisy. The increased activation in the frontal region seems to be due to the search of the brain for structures and orientation. Here the parietal region is indeed involved, too, quantitatively even more than the frontal region. The time window used for analysis was taken only 15 seconds after the start of the piece not to include the brain action used for guessing which piece was the blurred one. In most cases only after a few seconds the right piece was identified and so this guess is not included in the analysis.

Still, the difference between the Mozart piece on the one hand side, activating more the left frontal region and Beatles, Eminem, Schnappi and maybe Slipknot on the other side activating this region less can clearly be associated to the main content and meaning of the pieces. This is also enhanced by the interpretation of the frontal electrode behaviour, which is much more consistent with the content of the music than is the case with the activations of the temporal or parietal regions.

Maybe the first, overall result is the most striking. If meaning is reduced, the brain is more active in all its regions. This may also be interesting in terms of research of noise. As the blurred sounds are less harmonic, they come closer to noisy sounds. This study shows, that more noise means a higher activation of all brain regions. In musical terms it is not surprising, that noise music is much more demanding than harmonic pieces. This demanding effect is clearly present in the results of this investigation.

Fig. 1: Grand average of brain activation for normal musical pieces minus activation for blurred musical pieces. Over the whole brain, the blurred musical pieces showed more activation than the unblurred ones, ventral more than dorsal and right more than left.

Fig. 2: Brain activity of the unblurred (left column) and blurred (right column) version for the six songs used (see text).

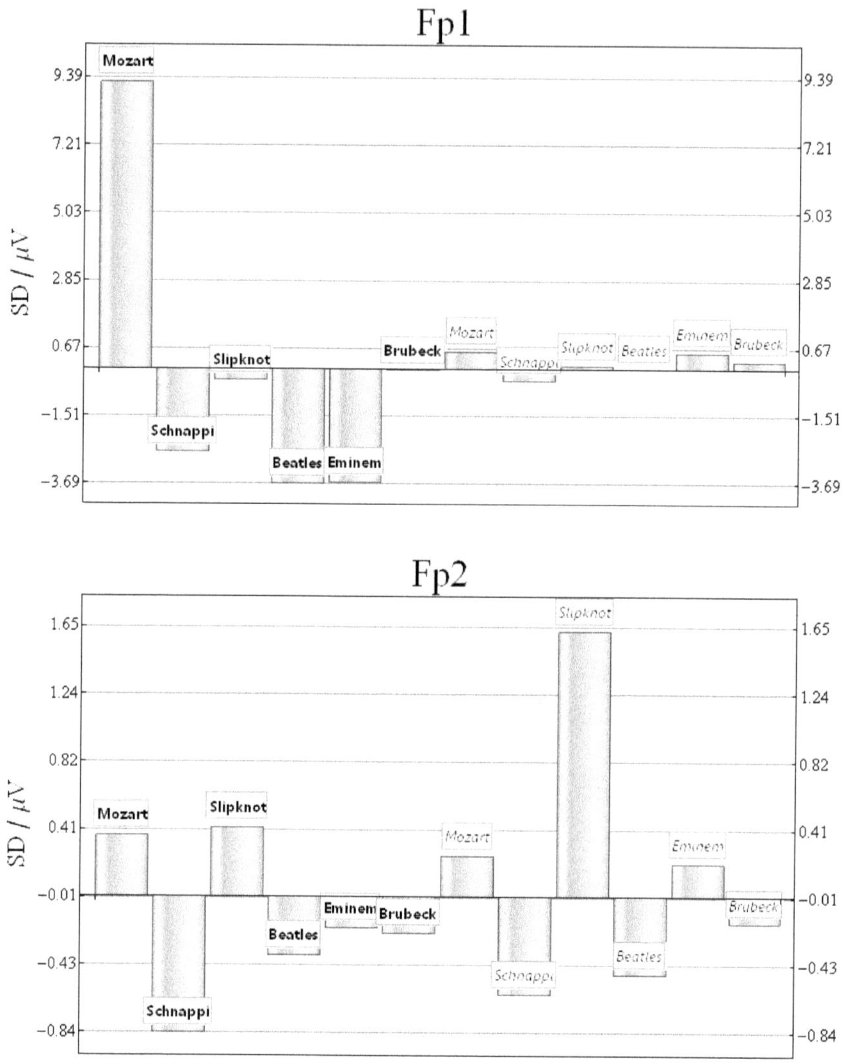

Fig. 3: Standard deviation of frontal partial Fp1 and Fp2 activity compared to mean activity averaged over five subjects with three trials each and individual mean activity substracted.

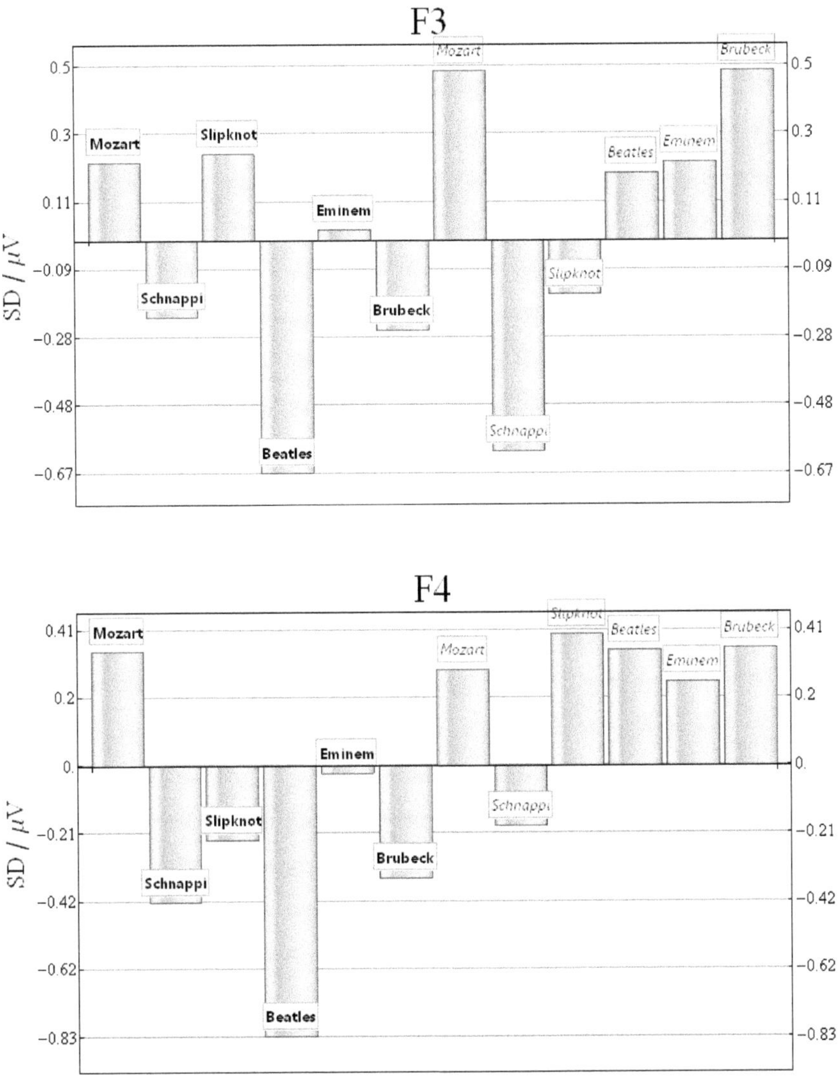

Fig. 4: Standard deviation of frontal F3 and F4 activity compared to mean activity averaged over five subjects with three trials each and individual mean activity substracted.

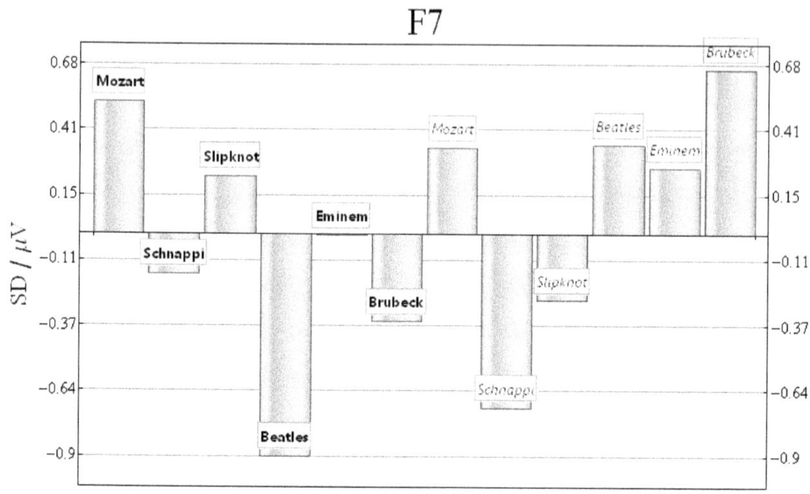

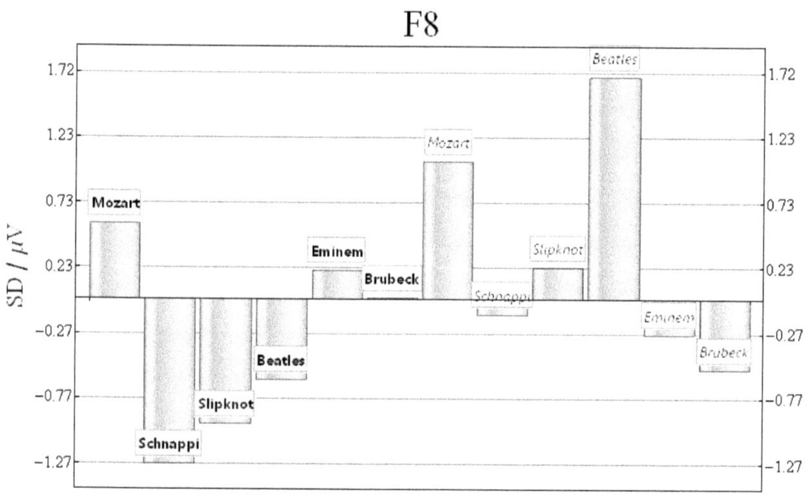

Fig. 5: Standard deviation of frontal F7 and F8 activity compared to mean activity averaged over five subjects with three trials each and individual mean activity substracted.

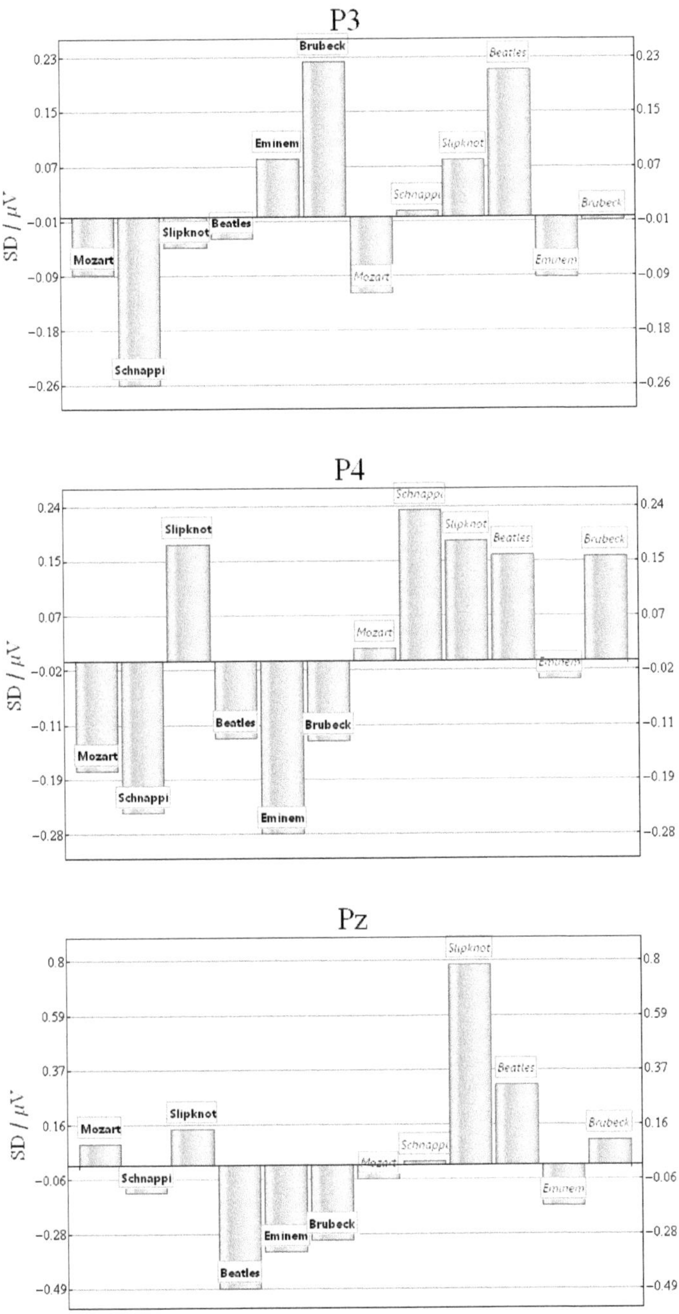

Fig. 6: Standard deviation of Parietal P3, P4 and Pz activity compared to mean activity averaged over five subjects with three trials each and individual mean activity substracted.

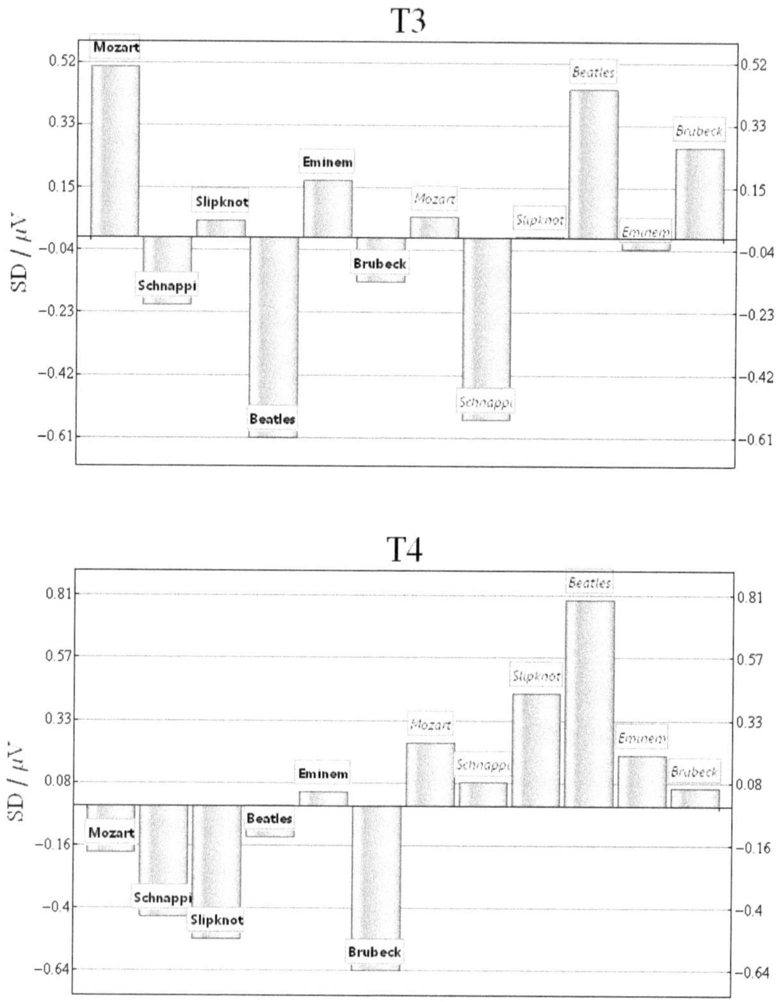

Fig. 7: Standard deviation of temporal T3 and T4 activity compared to mean activity averaged over five subjects with three trials each and individual mean activity substracted.

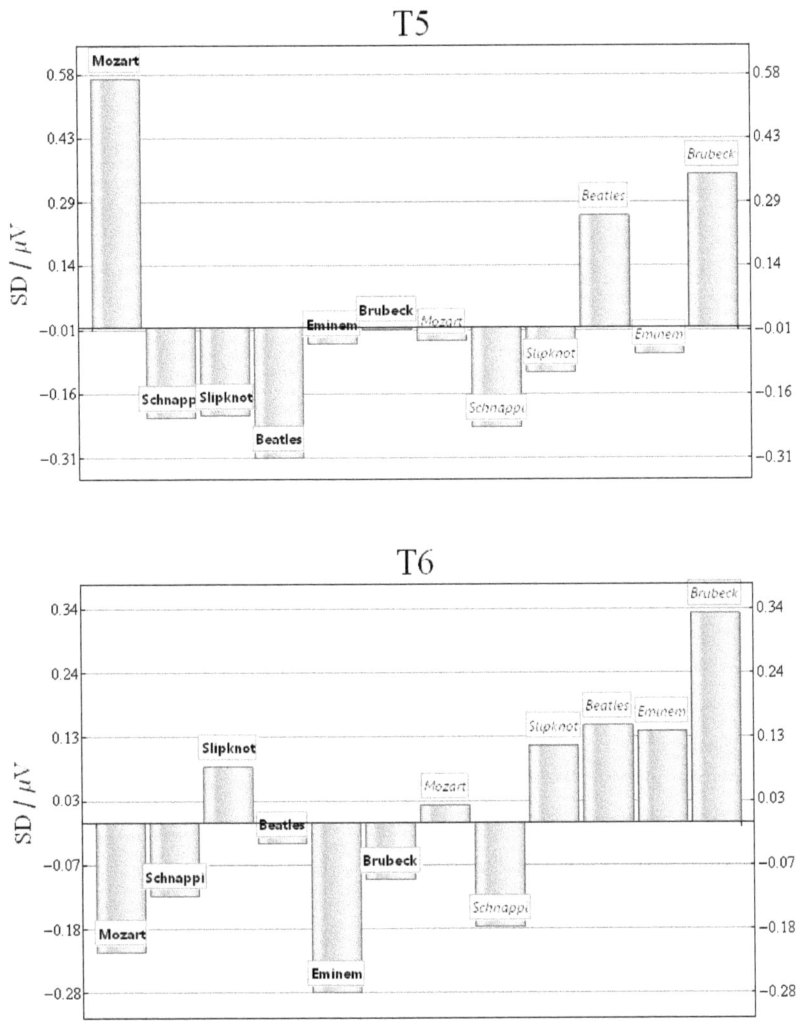

Fig. 8: Standard deviation of temporal T5 and T6 activity compared to mean activity averaged over five subjects with three trials each and individual mean activity substracted.

Acknowledgement

We thank Jukka Luovouri, Petri Toiviainen and all at the Center of Excellence for Systematic Musicology at the University of Jyväskylä, Department of Music, and all the participants of the International Summer School for Systematic Musicology for their help and support, especially Katrine Pram Nielsen, Christina Helm, and Berit Janssen.

References

Adorno, Th. W. (1949), *Philosophie der Neuen Musik*, Tübingen: Mohr.

Blood, A. J., R. J. Zatorre (2001), Intensely pleasurable responses to music correlate with activity in brain regions implicated in reward and emotion, *Proc. Natl. Acad. Sci. USA 98*, 11818-11823.

Eco, U. (2002), *Einführung in die Semiotik*, München: UTB Taschenbuch.

Kant, Immanuel (1990), *Kritik der praktischen Vernunft*, Hamburg: Meiner.

Karbusicky, Vladimir (1986), *Grundriß der Musikalischen Semantik*, Darmstadt: Verlag der Wissenschaftlichen Buchgesellschaft.

Halpern, A. R., R. J. Zatorre (1999), When that tune runs through your head: a PET investigation of auditory imagery of familiar melodies, *Cereb. Cortex*, 9, 697-704.

Janata, P. (2009), The Neural Architecture of Music-Evoked Autobiographical Memories, *Cerebral Cortex*, Advance Access 1-16.

Janata, P., J. Birk, J. D. Van Horn, M. Leman, B. Tillmann, J. J. Bharucha (2002), The Cortical Topography of Tonal Structures Underlying Western Music, *Science*, 298, 13, 2167-2170.

Leaver, A. M., J. Van Lare, B. Zielinski, A. R. Halpern, J. P. Rauschecker (2009), Brain Activation during Anticipation of Sound Sequences, *Journal of Neuroscience*, 29 (8), 2477-2485.

Pierce, Charles (1958), *Collected Papers, Vol V: Pragmatism and Pragmaticism*, Harvard: Harvard University Press.

Platon (1991), *Nomoi*, Frankfurt a. M.: Insel Verlag.

Plailly, J., B. Tillmann, J.-P. Royet (2007), The feeling of familiarity of music and odors: the same neural signature?, *Cereb. Cortex*, 17, 2650-2658.

Lévi-Strauss, C (1964), *Mythologica I. Le cru et le cuit*, Paris : Plon.

Lévi-Strauss, C. (1975), *Strukturale Antropologie*, Frankfurt a. M.: Suhrkamp.

Zatorre, R.J., A. R. Halpern, D. W. Perry, E. Meyer, A. C. Evens (1996), Hearing in the mind's ear: a PET investigation of musical imagery and perception, *Journal Cogn. Neuroscience*, 8, 29-46.

Albrecht Schneider
Arne von Ruschkowski
Rolf Bader

Klangliche Rauhigkeit, ihre Wahrnehmung und Messung

1. Summary: Perception and measurement of roughness

The present study on perception and measurement of roughness presents results from a number of experiments plus some theoretical considerations. We start from the concept of roughness in relation to sensory consonance and dissonance as developed by Hermann von Helmholtz (1863, 1870). In the following, we address the issue of Critical Bands (CB) and roughness sensation as well as the implementation of roughness measurement in computer models of the peripheral auditory system. To illustrate certain problems such models apparently can offer in regard to reliability and validity of measurement, some actual data from bell sounds are presented that indicate a very low degree of roughness (measured in asper, a unit explained in the article) whereas the roughness sensed by listeners appears to be more pronounced.

In the next section of the article, we report findings from an experiment in which subjects had to rate the degree of consonance and dissonance of 23 musical intervals (including some less common microtonal intervals) ranging from the unison (ratio of fundamental frequencies 1/1) to the octave (ratio of fundamental frequencies 2/1) in interval size. In order to offer stimuli that come close to real world sounds, all 23 intervals (Table 2) were generated with sawtooth waveforms on a Korg MS 20 synthesizer, stored as sound files on DAT, and played back from a CD and HiFi-equipment (loudspeakers, not headphones) to a group of 51 subjects who all had a musical background. The behavioral data yield a ranking list of the 23 intervals (Tab. 3) as well as two profiles that plot the means of consonance and dissonance ratings over interval number (the intervals are ordered according to size from no. 1 = unison to no. 23 = octave). The same 23 stimuli then were fed into a commercially available auditory model (ArtemiS 10; cf. Sottek 1993) that includes two options for roughness measurement, one of which is quite close to other such models known from psychoacoustics (e.g., Aures 1985, Daniel & Weber 1997). It should be noted that these models are based in the time domain insofar the amplitude modulation (AM) found in a number of cochlear filter channels is the decisive parameter from which roughness is computed. Taking into account Helmholtz' view that connects the sensation of roughness with sensory dissonance, one would expect the data obtained for roughness with the auditory model (Table 4) to correlate highly with the mean dissonance ratings resulting from the behavioral data. This, however, was not the case since the correlation is close to zero (r = 0.098).

In the next section, we present another experiment in which subjects were asked to listen to two versions of part of a Bach chorale („Wie schön leuchtet der Morgenstern") that were recorded, by Reinier Plomp (1998), in (a) equal temperament and (b) just intonation. Two groups of subjects had to rate the two versions on four variables, namely (1) consonance, (2) roughness/beats, (3) syntactic correctness, and (4) overall aesthetic impression. Descriptive statistics and inferential statistical analyses from the ratings are presented in Tables 5, 6, 7.

Again, the two versions were subjected to an analysis of roughness performed by an auditory model. Since the model yields uneven results, we tried computation of roughness in the frequency domain as an alternative, thus returning in certain respects to original ideas of Helmholtz as well as to more recent concepts that stem from similar considerations (Sethares 1993). Applying a straightforward and seemingly robust algorithm to the two versions of „Wie schön leuchtet der Morgenstern“, the roughness curves for 23 chords found in the first eight bars of the chorale correctly indicate considerably less roughness in the just intonation version as compared to equal temperament (Abb. 13).

In two final small experiment, we first analyzed part of a Bulgarian women song executed in a well-known style incorporating narrow intervals (typically, major seconds and minor thirds) sung in parallel with practically no vibrato, and intended as points of strong musical tension. From such musical styles it becomes obvious that the roughness found in simultaneous narrow intervals, though of course going along with AM as kind of a sensory side-effect, in fact is generated in the frequency domain since the AM found in the signal and the roughness perceived results from interaction of spectral components. Taking into account vocal and instrumental musical styles of the Balkans in which execution of narrow parallel intervals is central, one sees that roughness in these cases originates from the frequency domain, in particular from interaction of closely spaced spectral components found in two or more complex harmonic sounds. Of course, sensation of roughness may result from two signals that undergo AM (so that one acts as carrier, the other as modulator). Since the same degree of (sensed or measured) roughness may result from both spectral interaction and AM, and both can be considered equivalent in certain respects, one should seek to identify the actual source of roughness also in regard to analyzing cause and effect. Further, it is suggested that modeling of roughness in the time domain should be complemented by concepts of roughness based in the frequency domain. One reason that supports this view is that, as we found in a second ad hoc-experiment (in which 19 musically trained subjects took part), modulation depth as indicated by the temporal envelope profile alone apparently does not reliably indicate the amount of roughness sensed in a signal. Rather, complete amplitude modulation (that is, envelope varies in amplitude from 0 to peak value and back to 0) and amount of amplitude change as it occurs in the envelope of the signal used to define 1 asper (Abb. 18a) may be overestimated as an indicator of roughness. There are intervals composed of pairs of complex harmonic sounds such as the „Helmholtz minor second“ (the minor second B4 – C5 with a df of fundamental frequencies of ca. 33 Hz that Helmholtz gave as example for his concept of roughness and dissonance; Abb. 18b) where complete AM will not be observed but where musically trained listeners experienced a stronger sensation of roughness sensation than in the 1 asper reference signal that exhibits complete AM. Hence, it seems problematic to relate sensation of roughness exclusively to AM parameters.

2. Einführung: Rauhigkeit in der Hörtheorie von Helmholtz

Zu den aus zahlreichen psychoakustischen Experimenten wie auch aus der normalen Hörerfahrung bekannten Klangparametern gehört die Rauhigkeit von Klängen, die eingehend bereits von Hermann von Helmholtz (1863, 1870, 1896 u. ö.) beschrieben wurde. Er beobachtete und beschrieb u. a. Schwebungen, die sowohl zwischen zwei einfachen Tönen (Sinustönen) eng benachbarter Frequenz wie auch zwischen den Teiltönen komplexer Töne (d. h. Tönen mit harmonischen oder inharmonischen Spektren, die sich aus jeweils *k* Teiltönen zusammensetzen) einstellen. Dabei gab Helmholtz (1896, S. 273) zutreffend die Anzahl der Schwebungen (f_s in Hz) als absolute Differenz der Schwingungsfrequenzen, mithin als $f_s = | f2 - f1 |$ an. Die Maxima der Schwingungsamplitude bzw. der Tonintensität im Verhältnis zu den dazwischen liegenden Minima nannte Helmholtz *Schläge* (engl. Beats). Der Sachverhalt lässt sich rechnerisch und graphisch auf einfache Weise demonstrieren und als Ergebnis einer Amplitudenmodulation (nachfolgend: AM) erklären, die aus der Überlagerung zweier Schwingungen entsteht (vgl. Schneider 1997b, S. 95-99).

Nach eigenen Hörversuchen von Helmholtz ist unser Gehör in der Lage, langsamen Schwebungen (mit ca. 4-6 Hz) genau zu folgen. Solche Schwebungen werden zudem bis zu einer gewissen oberen Grenzfrequenz als angenehm, darüber aber wegen des intermittierenden Charakters der Töne, die nun in eine Sequenz kurzer Tonpulse aufgelöst erscheinen, zunehmend als störend empfunden. Die Rauhigkeit von Klängen ist nach Helmholtz (1913, S. 280ff.) am größten, wenn die Differenz zweier einfacher Schwingungen (d. h. von Sinustönen) etwa 30-40 Hz ausmacht. Dies entspricht nach seinen Beispielen etwa der kleinen Sekunde h' – c''. Die Schwingungsfrequenzen der beiden Töne liegen (bei einem Stimmton von a' = 435 Hz) bei 489.375 und 522 Hz, die Differenz mithin bei 32.625 bzw. aufgerundet 33 Hz. Geht man nicht von sinusförmigen Schwingungen, sondern von komplexen Tönen aus und bildet diese jeweils aus den ersten sechs Teilschwingungen bzw. Teiltönen nach (Frequenzverhältnisse jeweils 1:2:3:4:5:6, Amplitudenverhältnisse A = 1/n, also 1, 0.5, 0.33, 0.25, 0.2, 0.166), so resultiert ein stark schwebender Zusammenklang, dessen Amplitudenminima nur noch ungefähr 1/3 bis 1/2 des Scheitelwerts betragen. Die Zeitfunktion des durch additive Synthese hergestellten Zusammenklangs ist in Abb. 1 für 0.2 Sekunden gezeigt. Der Zusammenklang enthält also folgende Komponenten:

(1) y(t) = [(Sin[2Pi 489.375t]+0.5Sin[2Pi 978.75t]+0.33Sin[2Pi 1468.12t]+
0.25Sin[2Pi 1957.5t]+0.2Sin[2Pi 2446.87t]+0.166Sin[2936.25t])+
(Sin[2Pi 522t]+0.5Sin[2Pi 1044t]+0.33Sin[2Pi 1566t]+0.25Sin[2Pi 2088t]+
0.2Sin[2Pi 2610t]+0.166Sin[2Pi 3132t]), {t, 0, 0.2}]

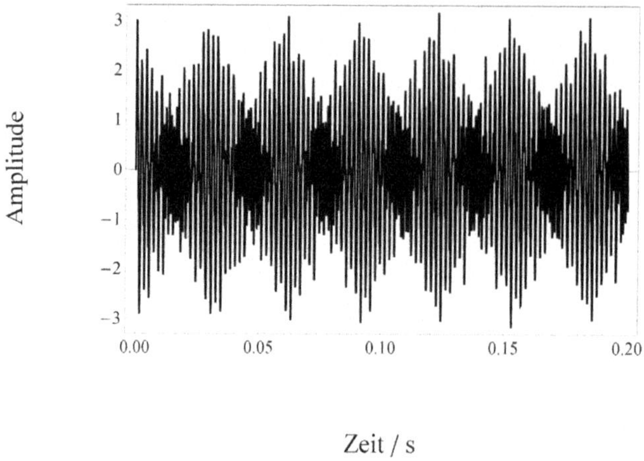

Abb. 1: Zeitfunktion der kl. Sekunde h' – c'', 0.2 Sekunden

Da die Anzahl der Schwebungen eines gegebenen musikalischen Intervalls von der Oktavlage abhängt, folgt bereits hieraus, dass die Empfindung der Rauhigkeit von Intervallen nicht über den gesamten Hörbereich konstant ist. Helmholtz bemerkte zudem, dass auch Intervalle wie die kleine und die große Terz, die (bei Intonation in reiner Stimmung) in der mittleren Lage konsonant erscheinen und fast keine Rauhigkeit aufweisen (mit Ausnahme des relativ geringfügigen Anteils, der durch Schwebungen zwischen Teiltönen auch zweier konsonierender Töne eines musikalischen Intervalls entsteht), in der tiefen Lage sehr rau klingen und deutliche Schwebungen geben. Helmholtz schloss aus diesen Befunden, dass die Rauhigkeit eines Zusammenklangs *in zusammengesetzter Weise von der Größe des Intervalls und von der Zahl der Schwebungen abhängt* (1913, S. 287).

Helmholtz machte für die Empfindung von Schwebungen die begrenzte Auflösungsfähigkeit des Innenohrs verantwortlich, wobei er sich das Corti'sche Organ als Kette parallel geschalteter mechanischer Resonatoren vorstellte. Nach seinen Überlegungen würden nahe beieinander liegende Schwingungsfrequenzen die gleichen Resonatoren anregen und so zu einer Überlagerung bzw. Modulation führen. Zwar ist die Resonator-Hypothese, die Helmholtz schon ganz zu Beginn seiner Forschungen zum musikalischen Gehör entwickelt hatte (Helmholtz 1857), längst widerlegt. Die Beobachtungen über die zunehmende Rauhigkeit an sich konsonanter Intervalle in tiefen Frequenzlagen waren jedoch korrekt. Sie fanden im 20. Jahrhundert eine Erklärung durch die Theorie der Frequenzgruppen (auch Kritische Bandbreiten benannt, vgl. Moore 1993, 1995, Zwicker & Fastl 1999. Kap. 6.2), die gleichfalls von einer begrenzten Frequenzauflösung des Innenohres bei (hydro)mechanischer Filterung eines komplexen Schallsignals ausgeht (Keidel 1992, Zenner 1994). Auf das Konzept der Frequenzgruppen bzw. Kritischen Bandbreiten kommen wir in Abschnitt 2 zurück.

Nach Helmholtz gibt es nun einen Zusammenhang von wahrgenommener Dissonanz und Rauhigkeit in der Hinsicht, dass bei zwei komplexen Tönen, die musikalische Intervalle (wie Oktave, Quinte, Quarte usw.) bilden, Schwebungen zwischen deren Teilschwingungen bzw. Teiltönen auftreten. Praktisch ausgenommen hiervon seien, so

Helmholtz (1913, S. 302ff.) nur die Oktave, die reine Duodezime und die Doppeloktave, die insoweit als perfekte Konsonanzen betrachtet werden dürfen. Es folgen dann die reine Quinte und die reine Quarte. Schon bei den anderen als konsonant geltenden Intervallen (den beiden Terzen und den beiden Sexten) kommen neben den in ihren Frequenzen koinzidierenden Partialtönen zunehmend solche vor, deren Frequenzen untereinander Schwebungen bilden. Diese sind je nach Intervall und Oktavlage z. T. wahrnehmbar, z. T. kaum. Helmholtz entwickelte dann aus der Betrachtung koinzidierender Teiltöne (bezogen auf den Klang einer Violine) einerseits, der Berechnung von Schwebungen bzw. von Rauhigkeit zwischen Teiltönen andererseits, seine eigene Konsonanztheorie, deren Ergebnisse Helmholtz in einer Art Rangreihe der Konsonanzen sowie in einer Graphik (1896, Figur 60A/B) zusammenfasste.

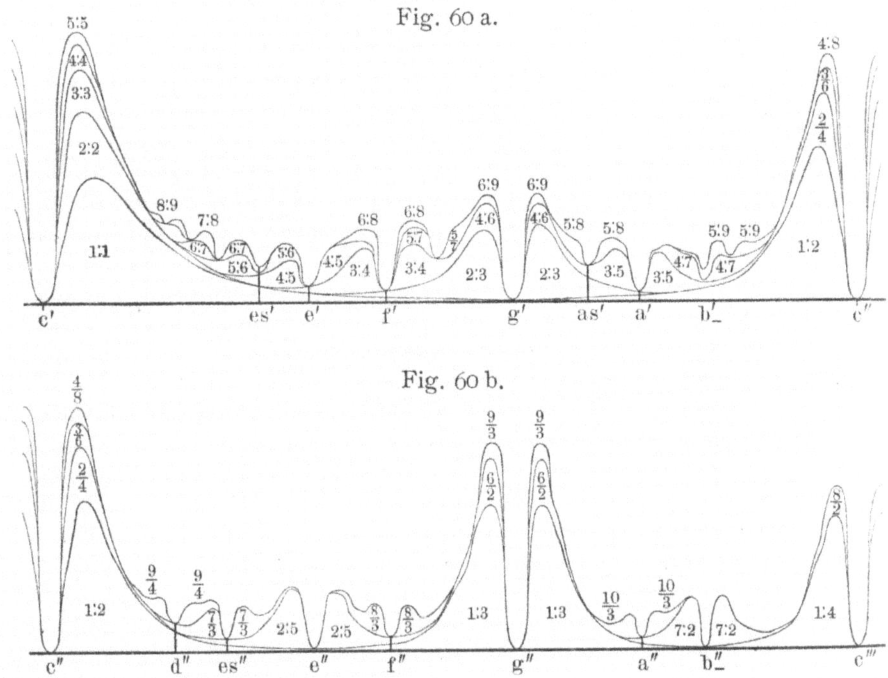

Abb. 2: Figur 60A/B aus Helmholtz 1913, 318.

Nach dieser Graphik, die, ausgehend vom Ton c', die musikalischen Intervalle bis zur Doppeloktave c''' darstellt, sind die Intervalle Oktav, Doppeloktav, Duodezime sowie Quinte diejenigen, die mit dem Grundton c' am höchsten konsonieren bzw. umgekehrt beim Zusammenklang mit diesem die geringste Rauhigkeit ergeben. Diese nimmt schon bei der großen Sexte a' (5/3) etwas zu und wächst geringfügig weiter bei der Quarte f' (4/3) sowie der Terz e' (5/4). Relativ gesehen, schneidet die Naturseptime b' (7/4) etwas besser ab als die kleine Terz es' (6/5) und die kleine Sexte as' (8/5). Die Punkte maximaler Rauhigkeit sind nach der Graphik innerhalb einer Oktave c' – c'' bei der kleinen Sekunde (16/15) und der großen Septime (15/8) zu erwarten. Der relative Konsonanzcharakter der Septime (7/4) wie auch des Tritonus (45/32 bzw. 7/5) wurde im Übrigen auch von Stumpf (1890, 1897) betont.

3. Rauhigkeit und Kritische Bandbreite: neuere Versuche und Befunde

Die Überlegungen von Helmholtz wurden im zwanzigsten Jahrhundert wiederholt aufgegriffen und in Versuchen überprüft, die im Wesentlichen Befunde ergeben haben, die mit der Graphik aus Helmholtz (Abb. 2) bzw. mit dessen Rangreihe der Intervalle gut übereinstimmen (s. etwa Plomp & Levelt 1965; Kameoka & Kuriyagawa 1969a/b). Kameoka und Kuriyagawa fanden dabei für ein aus zwei Sinustönen gebildetes Signal, das den Versuchspersonen (nachfolgend Vpn) mit einem Schalldruckpegel von 57 dB dargeboten wurde, den höchsten Dissonanzgrad bei einem Frequenzverhältnis von 484:440 = 1.1 bzw. 180 Cent, also beim Intervall ungefähr eines kleinen Ganztons (10/9 = 182 Cent). Bei komplexen Tönen mit mehreren Harmonischen gab es bei der Beurteilung musikalisch relevanter Intervalle gewisse Schwankungen, die von der spektralen Zusammensetzung der Stimuli abhängen (Anzahl n der Harmonischen mit n = 2 - 6; Intervalle aus Tönen, die entweder aus geradzahligen oder aus ungeradzahligen Teiltönen gebildet wurden). Insgesamt aber erbrachten diese Experimente das kaum überraschende Ergebnis, wonach *the consonance characteristics of chords of two harmonic complex tones showed peaks at simple frequency ratios. This was due to the coincidence of harmonics.* (Kameoka & Kuriyagawa 1969b, S. 1468). Die Koinzidenz der Obertöne war von Helmholtz selbst als Kriterium der Konsonanzempfindung erörtert und empirisch ausgewertet worden. Das gleiche Kriterium wurde in Konsonanztheorien des 19. und 20. Jahrhunderts immer wieder diskutiert (so u. a. von Th. Lipps, Felix Krueger, Heinrich Husmann). Der Sachverhalt ist dabei im Grunde einfach und doch – zumal im Hinblick auf eine notwendige Aktualisierung von Konsonanztheorien – keineswegs trivial. Bildet man einen vollkommen rein gestimmten Dur-Akkord 3:4:5 aus komplexen harmonischen Tönen, deren Teilschwingungen wiederum exakt harmonischen Frequenzverhältnissen entsprechen und deren Amplituden dem Verhältnis A = 1/n entsprechen (vgl. Schneider 1997a, S. 125f., Fig. 1), so resultiert eine Schwingungsform, die (a) streng periodisch und (b) durch steile Flanken zu Beginn sowie (c) eine differenzierte Binnengliederung innerhalb jeder Schwingungsperiode charakterisiert ist. Der so (z. B. mit den Programmen Mathematica oder MatLab) hergestellte Akkord erklingt (d) praktisch völlig frei von Schwebungen oder Rauhigkeit und führt (e) zu größtmöglicher ‚Verschmelzung' im Sinne der Konsonanztheorie von Carl Stumpf, weil es (f) zu einer (auch rechnerisch leicht überprüfbaren) Koinzidenz etlicher Teilschwingungen bzw. Teiltöne kommt. Gleichwohl bleiben, gerade weil diese Teiltöne nicht durch Modulation verwischt werden, (g) viele einzelne Komponenten in jedem der drei Akkordtöne als solche hörbar und identifizierbar, wenn man die Aufmerksamkeit auf diese richtet. Diesem Befund entspricht (h) die Analyse eines solchen Akkords mit Computermodellen des Innenohres, bei denen die Periodizität des streng harmonischen Signals zu einer neuronalen Repräsentation führt, die neben den Harmonischen selbst auch noch eine vollständige Reihe von Subharmonischen enthält (vgl. Schneider & Frieler 2009).

Die Überlegungen von Helmholtz (und mittelbar auch die von Stumpf) wurden vor nicht langer Zeit durch Berechnungen von Sethares (1993, 1998) bestätigt. Diese ergaben bei Zugrundelegung eines harmonischen Spektrums ($f_n = n \times f_1$, $n = 1, 2, 3, ..., k$), dessen Komponenten in ihrer Amplitude jeweils um einen bestimmten Faktor (0.88) abfallen, eine Dissonanzkurve ganz ähnlich der von Helmholtz (Abb. 2). Die absoluten Minima in jeder Oktave liegen erwartungsgemäß bei kleinen ganzzahligen Tonverhältnissen bzw. Intervallen, also bei 1/1 (1.0), 6/5 (1.2), 5/4 (1.25), 4/3 (1.33), 7/5 (1.4), 3/2 (1.5), 5/3 (1.67), 7/4 (1.75), 2/1 (2.0). Will man also Spektren und Skalen so aufeinander abgleichen, dass die Dissonanz minimal und die Konsonanz maximal wird, so müssen in beiden harmonische Verhältnisse realisiert werden. Die Dissonanzskala mit ihren lokalen Minima *provides a physical basis for justly intoned scales* (Sethares 1993, S. 1222). Solche hat William Sethares dann in eigenen Aufnahmen vorgeführt, bei denen praktisch in Echtzeit das Spektrum einzelner Töne bzw. von Akkorden und Zusammenklängen nach Maßgabe der Skala modifiziert, oder umgekehrt die Skala nach Maßgabe der spektralen Zusammensetzung von Tönen angepasst wird.[1] Dies führt im harmonischen Fall nicht nur zu einem Verschwinden jedweder Rauhigkeit, sondern umgekehrt auch zu einem Eindruck denkbar großer ‚Verschmelzung' im Sinne Stumpfs.

Die Diskussion um Schwebungen und Rauhigkeit wurde im zwanzigsten Jahrhundert durch die Beschreibung eines Sachverhalts neu belebt, der als ‚Kritische Brandbreiten' bzw. als ‚Frequenzgruppen' bekannt ist (vgl. Moore 1993, 1995, Zwicker & Fastl 1999, ch. 6, Hartmann 1998, ch. 10). Dabei geht es um die Tatsache, dass die Basilarmembran und die auf ihr aufruhenden Strukturen, die man zunächst statisch als eine Kette parallel geschalteter Bandpassfilter bestimmter Charakteristik, Bandbreite und Mittenfrequenz betrachten kann, nur ein begrenztes Auflösungsvermögen für Frequenzkomponenten aus komplexen Signalen besitzen (vgl. Plack & Oxenham 2005, S. 20ff.). Liegen also etwa zwei Schwingungsfrequenzen eines Tones dicht beieinander, so fallen sie nach dieser Lehre erfahrungsgemäß in ein auditorisches Filter und können in diesem nicht mehr zureichend diskriminiert oder gar einzeln identifiziert werden. Die Basilarmembran wird in der Regel in 24 oder 25 Frequenzgruppen bzw. Kritische Bandbreiten unterteilt, als deren Einheit das ‚Bark' (*z*; nach dem Elektroakustiker Heinrich Barkhausen, 1881-1956) definiert wurde. In erster Näherung entspricht eine Kritische Bandbreite ungefähr 1,3 mm bzw. (nach neueren Berechnungen) 0.9-1 mm der Basilarmembran. Die Bandbreite ist bei tiefen Frequenzen ($f < 150$ Hz) relativ groß; sie beträgt ab etwa 1 kHz ungefähr 260-280 Cent (die Werte hängen etwas vom Berechnungsverfahren ab). Das fragliche Intervall liegt also zwischen einer großen Sekunde (9/8) und einer kleinen Terz (6/5), wobei diese merklich unter- und jene deutlich überschritten wird. Die Verhältnisse für die ersten 14 Frequenzgruppen bis zu einer oberen Grenzfrequenz von 2.5 kHz sind in Tab. 1 zusammengefasst (modifiziert nach Zwicker & Fastl 1999). Ab etwa 2.5 kHz nimmt die Weite der Frequenzgruppen (ausgedrückt in Cents) zu höheren Frequenzen hin wieder zu.

[1] Vgl. die CD im Anhang von Sethares 1998 und dessen CD *Xentonality: matching timbre and tuning.* 1997

Tab. 1: Kritische Bandbreiten (modifiziert und ergänzt nach Zwicker & Fastl 1999)

Bark	untere/obere Grenzfrequenz		Mittenfrequenz	Bandbreite (Hz)	Intervall (Cent)
0	0	100			
			50	100	unklar
1	100	200			
			150	100	1200
2	200	300			
			250	100	702
3	300	400			
			350	100	498
4	400	510			
			450	110	421
5	510	630			
			570	120	366
6	630	770			
			700	140	347
7	770	920			
			840	150	308
8	920	1080			
			1000	160	278
9	1080	1270			
			1170	190	281
10	1270	1480			
			1370	210	265
11	1480	1720			
			1600	240	260
12	1720	2000			
			1850	280	261
13	2000	2320			
			2150	320	257
14	2320	2500			

usw.

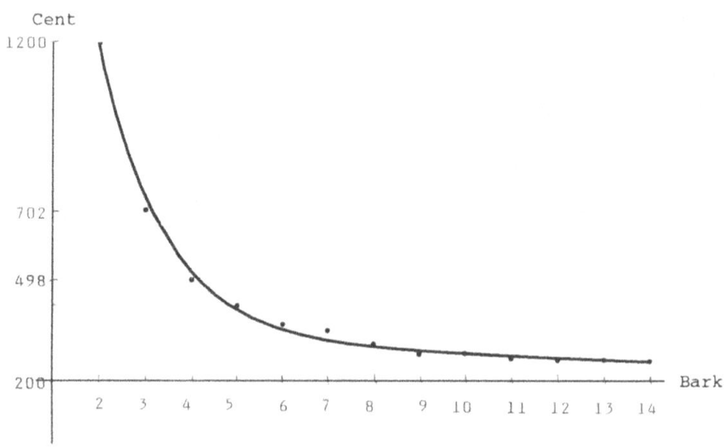

Abb. 3: Kritische Bandbreiten (2-14) in Cent

Da es hinsichtlich der Wahrnehmung um Intervallgrößen geht, wird die Bandbreite der Frequenzgruppen hier in Cents ausgedrückt; man sieht, dass die wahrnehmungsrelevante Größe der ersten Frequenzgruppe (0-100 Hz) nicht geklärt ist, während die zweite den Umfang einer Oktave, die dritte den einer Quinte und die vierte den einer Quarte haben soll. Zumal die Größenschätzung dieser Frequenzgruppen z(2-3) und z(3-4) dürfte mit Erfahrungen musikalischer Hörer nicht ohne weiteres zu vereinbaren sein und bedarf jedenfalls einer erneuten Prüfung. Die Bandbreite der Frequenzgruppen 2-14 kann durch eine fallende Exponentialfunktion sehr gut angenähert werden[2]; die Funktion lautet dann:

$$(2) \qquad CB(Cents) = 324.954 + \frac{1729.63}{Exp^{0.7x}} - 5.362\,x$$

Der Graph der Funktion ist in Abb. 3 für die Frequenzgruppen 2-14 mit ihren Centwerten dargestellt. Das Konzept der ‚Kritischen Bandbreiten' (Critical Bands, CB) bzw. der Frequenzgruppen ist in den letzten beiden Jahrzehnten gewissen Modifikationen unterworfen sowie auch kritisch betrachtet worden (vgl. Moore 1993, 1995, Schneider 1997a), da manche Leistungen des menschlichen Gehörs bzw. Leistungen von Musikern in Diskriminierungsversuchen deutlich besser sind als die von Nichtmusikern. Man hat daher ein revidiertes Modell aus Beobachtungsdaten und Berechnungen erstellt (sog. „Cambridge Critical Bands" oder Cams, Moore 1993, 1995), in dem die Frequenzgruppen enger als bei den älteren „Munich Critical Bands" (Zwicker & Fastl 1999; Abb. 3) bzw. der Bark-Skalierung gefasst sind; näherungsweise gilt etwa 5 Cams ≈ 3.5 Bark.

Geht man von den CB als Filterbank aus, so kann man annehmen, dass sich zwei Schwingungen nahezu gleicher Frequenz und Amplitude, die in einem Filterband liegen, dort als wechselseitige Störung und am Ausgang des Filters bei der mechanisch-elektrischen Transduktion des Signals in neuronalen Impulsmustern ausprägen, die im Prinzip die schon im Signal selbst vorliegende Amplitudenmodulation widerspiegeln. Mithin schlägt sich dann die örtliche Nähe zweier Erregungen sowie deren Stärke in einem spezifischen zeitlichen Erregungsmuster nieder, das auf einer höheren Station der Hörbahn, wahrscheinlich dem Colliculus inferior, ausgewertet wird (s. unten).

Das Modell einer peripheren Filterung in der Cochlea als Basis der auditorischen Signalverarbeitung ist in zahlreichen Hörmodellen algorithmisch implementiert und liefert verwertbare Ergebnisse. Die Ausgangsdaten der peripheren Filterung können beispielsweise in Form eines sogenannten Cochleagramms ausgegeben werden, dessen Ordinate meist 25 Frequenzgruppen umfasst und dessen Abszisse die Zeitachse bildet. Das Cochleagramm zeigt daher die zeitliche Entwicklung des durch gehörgerechte Filter (z.B. Gammatone-Filter oder ERB-Filter; vgl. Terhardt 1998, 248ff., Hut et al. 2006) analysierten Signals. Abb. 4 zeigt die Analyse eines Glockenklanges. Man sieht deutlich, dass die spektrale Energie beim Anschlag über etwa 18 Frequenzgruppen (skaliert in Bark) verteilt ist, innerhalb derer fünf Frequenzkomponenten einzeln identifizierbar

[2] Die Intervallgröße der Frequenzgruppen 2 – 14 nimmt von ca. einer Oktave bis zu einem Minimum von 257 Cent in z(13-14) ab; beginnend mit der Frequenzgruppe z(14-15) steigen die Intervallgrößen dann zu höheren Frequenzen hin wieder an: z(14-15) = 262.5 Cent, z(15-16) = 267 Cent, z(16-17) = 279 Cent, z(17-18) = 300 Cent usw.

sind. Diese stellen die stärksten Teiltöne des vom Gehörmodell analysierten Glockenklanges dar, der per Mikrophon in der Nähe der Glocke mit 48 kHz/16 Bit digital in einen Computer aufgezeichnet wurde. Es ist ersichtlich, dass die Teiltöne nach der Auflösungsfähigkeit des gewählten Modells untereinander keine Schwebungen bilden.

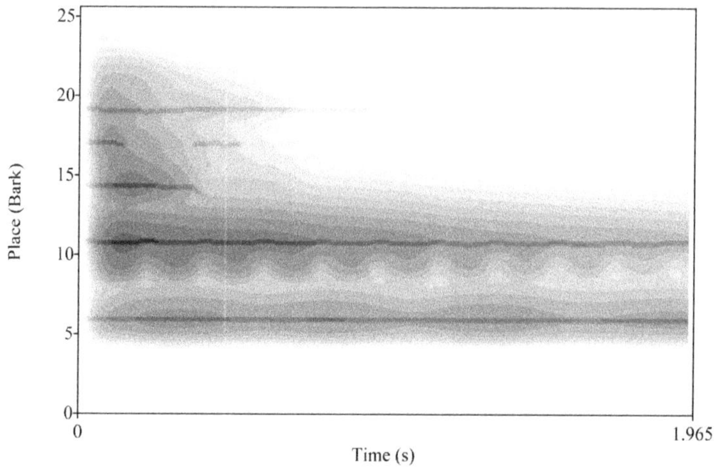

Abb. 4: Cochleagramm eines Glockenklangs

Allerdings deutet das Cochleagramm bei der zweiten Komponente (bei 11-12 Bark bzw. 1.2 kHz gelegen) selbst eine Frequenz- und auch Amplitudenmodulation an. Diese geht auch aus dem Intensitätsverlauf des Signals (Abb. 5) hervor. Eine akustisch präzise Analyse des Glockensignals zeigt, dass im Spektrum dieser Glocke zwei sehr eng beieinander liegende Teilschwingungen bzw. Teilfrequenzen (1190.5 und 1197 Hz) vorhanden sind, die auch im von der Glocke abgestrahlten Klang zu einer hörbaren Amplitudenmodulation (AM) bzw. zu Schwebungen führen.

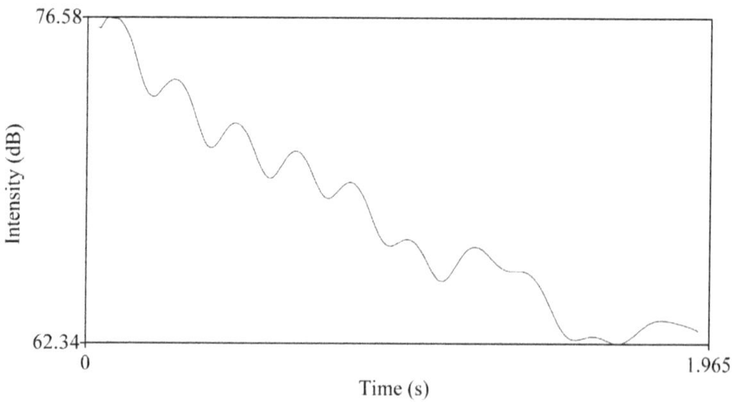

Abb. 5: Intensitätsverlauf des Signals mit Amplitudenmodulation

Schon das Cochleagramm, das insoweit nur das Ergebnis der ersten Filterung in der Cochlea darstellt, lässt Aufschlüsse über die Art des Signals sowie über Effekte, die diese im Gehör bzw. für das Hören verursachen, zu. Man muss bei der Betrachtung solcher Filtermodelle, wie sie hier und in zahlreichen Computermodellierungen (vgl. etwa Sottek 1993, Meddis & O'Mard 1997, Hut et al. 2006) als Nachbildung der Basilarmembran verwendet wurden, allerdings im Vergleich zum Innenohr der Säugetiere deren Fähigkeit in Betracht ziehen, die Charakteristik von Filtern durch efferente Steuerung praktisch in Echtzeit zu verändern und so nach Bedarf an die Spezifik von Signalen anzupassen. Durch diese bisweilen als „zweites Filter" bezeichneten Mechanismen (vgl. Keidel 1992, Zenner 1994) wird offenbar die Durchlassbreite und Flankensteilheit von auditorischen Filtern innerhalb weniger Millisekunden nachreguliert; durch das ‚zweite Filter' sowie den sog. kochleären Verstärker, der zu einer Anhebung des Signals um bis zu 40 dB führt, wird die tatsächlich sehr gute Leistungsfähigkeit des Gehörs auch bei der Unterscheidung nahe beieinander liegender Stimulusfrequenzen bewirkt, sofern das Signal innerhalb gewisser Pegelgrenzen liegt.

4. Computermodelle zur Berechnung der Rauhigkeit

Die für das Cochleagramm eingesetzte Filterbank führt die Analyse im Frequenzbereich durch, während die nachfolgende Verarbeitung auditorischer Signale vorwiegend im Zeitbereich angesiedelt ist und auch entsprechend modelliert wird (vgl. Meddis & O'Mard 1997). Die Analyse der Amplitudenmodulation kann gleichfalls im Zeitbereich ansetzen, nämlich an der mehr oder minder periodischen Schwankung der Hüllkurvenamplitude eines Signals (vgl. Abb. 1). Unser Gehör ist augenscheinlich in der Lage, diese Information jedenfalls innerhalb gewisser Grenzen auszuwerten (vgl. Schneider 1997b, 93-142). Schon Helmholtz machte geltend, dass er als Hörer die Schwankungen bzw. Schläge bis zu einer relativ hohen Modulationsfrequenz verfolgen und auszählen könne; er sagte, dass er die 66 Schwebungen zwischen h'' und c''' und sogar noch die 132 Schwebungen zwischen den einfachen Tönen h''' und c'''' hören könne, die Empfindung in den höheren Lagen allerdings zunehmend schwächer werde (Helmholtz 1913, S. 285). Bei der Verwendung einfacher Sinustöne treten in der Tat Schwebungen bzw. Beats mit vollständiger Interferenz des Signals ein, so dass die Amplitude periodisch zwischen dem Scheitelwert und dem Betrag 0 schwankt und das Ohr hierdurch eine Folge von Druckstößen empfängt (vgl. Abb. 5, 6). Dies macht die Auswertung einfacher als bei Signalen, bei denen die Modulation nur zu teilweiser Interferenz führt (Abb. 1).

Aus neurophysiologischen Versuchen an Tieren schon in den 1980er Jahre wurde gefolgert, dass in der Hörbahn auf der Höhe des Colliculus inferior spezialisierte und tonotop (in Karten räumlich angeordnete) Zellstrukturen sitzen, die zur Auswertung von Modulationsfrequenzen geeignet sind (Schreiner & Langner 1988, Langner 1992). Anknüpfend an diese Befunde wurde später ein Computermodell entwickelt, das die Analyse von amplitudenmodulierten Signalen mit bestimmten Eigenschaften leistet (Dau et al. 1997a/b). Bei diesem Modell folgt der Filterbank (Basilarmembran) und einem Halbwellen-Gleichrichter mit nachgeschaltetem Tiefpass pro Kanal (Haarzellen), die das periphere Gehör simulieren, eine nichtlineare Anpassung sowie eine weitere Filterbank zur Erfassung der Modulationsfrequenzen pro Kanal. Dieses Modell kann u. a. zur Untersuchung von Rauhigkeit in Klängen eingesetzt werden.

Es gibt eine Reihe weiterer Modelle zur Berechnung der Rauhigkeit aus Schallsignalen. Bekannt wurde ein Verfahren von Aures (1985a), bei dem gleichfalls zunächst in einer Filterbank die Zerlegung des Stimulussignals in 24 Frequenzgruppen erfolgt. Die relativen Schwankungen des Pegels der Erregung innerhalb jeder Frequenzgruppe und die entsprechende Modulation der zeitlichen Hüllkurve bilden die Grundlage der Berechnung von Teilrauhigkeiten pro Filterband, die (nach einer Gewichtung) schließlich zur Gesamtrauhigkeit aufsummiert werden. Sowohl das Modell von Sottek (1993) als auch das von Daniel und Weber (1997) sind bestrebt, der tatsächlich nicht konstanten Bandbreite der Frequenzgruppen (s. Tab. 1) und der dynamisch variablen Charakteristik von Filtern in der Cochlea Rechnung zu tragen. Dies geschieht durch Erhöhung der Anzahl von Filtern; bei Sottek beträgt der Abstand zwischen den Mittenfrequenzen benachbarter Bandpässe konstant 0,5 Bark, bei Daniel und Weber sind in ähnlicher Weise 47 überlappende Filter implementiert, jeder mit einer Bandbreite von 1 Bark. Die weiteren Verarbeitungsschritte zur Auswertung des Filter-Outputs bestehen bei Sottek (1993, S. 134ff., 1994) in der Berechnung der Schwankungen, denen die Hüllkurve unterliegt (mit Hilfe der Hilbert-Transformation), sowie in der Berechnung der Autokorrelationsfunktion (AKF) der nichtlinear verzerrten Hüllkurvenschwankungen zur Ermittlung der Teilrauhigkeiten in jedem Kanal. Die Teilrauhigkeiten werden schließlich zur Gesamtrauhigkeit aufsummiert. In diesem Modell gibt es weitere Verarbeitungsschritte, so etwa eine Tiefpassfilterung der Hüllkurvenschwankungen in den einzelnen Kanälen, die dem Umstand Rechnung trägt, dass unser Gehör solchen Schwankungen nicht beliebig schnell und nur bis zu einer bestimmten oberen Grenzfrequenz folgen kann.

Das Modell von Daniel und Weber (1997) folgt im Prinzip dem Ansatz von Aures (1985a), verbessert dieses Modell aber in verschiedenen Details. Bei dem Modell von Aures konnten bei Rauschsignalen zu hohe Rauhigkeitswerte auftreten (für Weißes Rauschen sollte die Rauhigkeit R nahe 0 liegen). Für diesen Fehler bieten Daniel und Weber eine Korrektur an.

Das Modell von Pressnitzer (1998) folgt im Wesentlichen den Überlegungen von Aures sowie Daniel und Weber, knüpft also an der durch Modulation zeitlich veränderlichen Hüllkurve eines Signals an.

5. Eigene Messungen und Experimente

Nachfolgend soll über einige Messungen und Hörversuche berichtet werden, die mit dem Phänomen der Rauhigkeit von Klängen und deren Wahrnehmung direkt oder mittelbar zu tun haben. Im Einzelnen geht es (a) um Messungen an Glockenklängen, (b) um den Vergleich von Messwerten, die ein Hörmodell geliefert hat, mit den Ergebnissen aus einem Hörversuch, an dem 51 Vpn teilnahmen. Schließlich geht es (c) noch einmal um Messungen der Rauhigkeit und der Schärfe bei einem Musikbeispiel, das in zwei unterschiedlichen Stimmungen realisiert wurde.

Vorab sollen die in ihrer Ausprägung abgestuft erscheinenden Klangmerkmale Rauhigkeit und Schärfe, die mithin einer Skalierung (auf ordinalem oder höherem Skalenniveau) zugänglich sind, in Bezug auf die Einheit, in der sie jeweils gemessen werden, definiert werden. Da der Eindruck klanglicher Rauhigkeit wesentlich (aber nicht ausschließlich) von Parametern der Amplitudenmodulation (Trägerfrequenz, Modulationsfrequenz, Modulationsgrad) abhängt, sei diese kurz erläutert (vgl. Meyer & Guicking 1974, 56ff.).

Im einfachsten Fall wird eine sinusförmige Trägerschwingung S_T mit der Frequenz f_T durch eine gleichfalls sinusförmige Modulierschwingung S_M mit der Frequenz f_M moduliert; die resultierende Zeitfunktion y(t) ist dann

(3) $y(t) = (A + a \, Sin[2 \, Pi \, f_M \, t]) \, Sin[2 \, Pi \, f_T \, t]$

mit dem Amplitudenmittelwert A und dem Amplitudenhub a. Dabei ergibt a/A den Modulationsgrad *m*. Dieser ist in der Regel $m < 1$, selten $m \geq 1$.
Durch trigonometrische Umformung erhält man aus Gl. 3

(4) $y(t) = A \, Sin[2 \, Pi \, f_T \, t] + a/2 \, Cos[2 \, Pi \, (f_T - f_M) \, t] – a/2 \, Cos[2 \, Pi \, (f_T + f_M) \, t]$

Das Fourier-Spektrum enthält demnach neben der Trägerschwingung S_T mit der Frequenz f_T und der Amplitude A durch die Modulation zwei Seitenfrequenzen ($f_T - f_M$) und ($f_T + f_M$), beide mit den Amplituden a/2. Die spektrale Betrachtung der Amplitudenmodulation (AM) ist deshalb von Belang, weil die Trägerfrequenz und die Seitenfrequenzen entweder in eine Frequenzgruppe bzw. CB oder aber in zwei benachbarte fallen können. Dieser Sachverhalt gewinnt bei komplexen Klängen bzw. Modulationen zusätzlich Bedeutung.

Die Rauhigkeit *R* mit der Einheit *asper* (von lat. asper = rauh) hat man wie folgt definiert: ein mit dem Modulationsgrad $m = 1$ bzw. zu 100% amplitudenmodulierter Sinuston mit der Trägerfrequenz f_T von 1 kHz und einer Modulationsfrequenz f_M von 70 Hz, der mit einem Schallpegel von 70 dB dargeboten wird, habe die Rauhigkeit 1 asper.

Der Zusammenhang von Rauhigkeit und Modulationsgrad wird durch ein Potenzgesetz (wie es in der Psychophysik von S.S. Stevens auch für die Empfindung der Tonhöhe und der Lautstärke postuliert wurde) jedenfalls für sinusförmige Signale recht gut beschrieben (Vogel 1975); es ist dann

(5) $R \sim m^k$, wobei in der Regel $1.2 \leq k \leq 2$.

Ein Sinuston von 1 kHz, der mit $f_M = 70$ Hz moduliert wird, besitzt die in Abb. 6 für einen Ausschnitt von 70 ms gezeigte Zeitfunktion sowie das zugehörige Spektrum mit den drei Komponenten bei 930, 1000 und 1070 Hz. Die Amplituden der beiden Seitenbänder liegen dabei –6 dB unter der Amplitude der Trägerfrequenz. Wird das Signal mit einem Pegel von 70 dB abgespielt, ruft es bei Vpn eine Rauhigkeitsempfindung von $R = 1$ asper hervor.

Die sensorisch empfundene Schärfe eines Klanges hängt erkennbar mit der spektralen Energieverteilung und auch mit dem Pegel zusammen, mit dem Klänge dargeboten werden (vgl. v. Bismarck 1974, Aures 1985b/c). Die Schärfe nimmt mit dem relativen Anteil höherer Frequenzen zu. Verschiebt man ein Schmalbandrauschen auf der Frequenzachse, nimmt die Schärfe mit steigender Mittenfrequenz zu. Nimmt man ein solches Rauschsignal mit der Mittenfrequenz 1 kHz und einer Bandbreite kleiner als der zugehörigen Frequenzgruppe (160 Hz; vgl. Nr. 8 in Tab. 1), so habe dieses Signal bei

Darbietung mit einem Pegel von 60 dB die Schärfe 1 acum (von lat. acumen = Spitze, Schärfe).

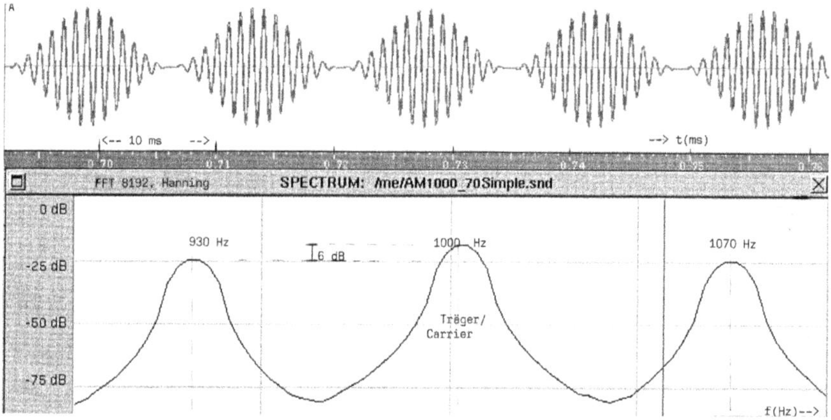

Abb. 6: Zeitfunktion und Spektrum eines amplitudenmodulierten Sinustons (f_T = 1 kHz, f_M = 70 Hz, $m = 1$)

6. Experiment 1: Messung der Rauhigkeit bei Glockenklängen mit einem Hörmodell

In unserem Institut untersuchen wir seit langem die Akustik von Idiophonen und die Wahrnehmung von Klängen mit inharmonischen Spektren, so wie sie etwa von europäischen Glocken und mehr noch von Stab- und Gongspielen in außereuropäischer Musik erzeugt und abgestrahlt werden (vgl. etwa Schneider 1997b, 1998, 2000, 2001a/b/c, Schneider & Müllensiefen 1999, Schneider & Leman 2002, Bader 2002a, 2002b, 2008). Im Zuge der Erforschung von Glockenspielen wurden im Jahr 2000 die Klänge der 42 Glocken (Schilling, Apolda, 1936) der Christianskirche in Hamburg-Altona mit 16 bit, 48 kHz Sampling auf DAT aufgezeichnet.[3] Die Klänge, vor allem der größeren Glocken, waren mit verschiedenen Anschlagstärken erzeugt worden, um die Abhängigkeit der spektralen Energieverteilung und die Dauer des Ausklingvorgangs von der Stärke der Erregung analysieren zu können.

Aus dem digitalen Klangmaterial wurden Klänge unterschiedlicher Anschlagstärke bzw. Amplitude der Glocken 2-6 mit dem Algorithmus bzw. Hörmodell von Daniel und Weber (1997) untersucht.[4] Dabei wurde jeweils die Rauhigkeit (in asper) als Funktion der Zeit berechnet und als Kurve aufgetragen. Durch Zuordnung der Rauhigkeitskurve zum Oszillogramm jedes Klanges ist leicht zu erkennen, wie das akustische Signal mit dem über das Hörmodell errechneten Rauhigkeitsverlauf zusammenhängt.

---

[3] Das Glockenspiel wurde von Eberhard Köther (Kantor und Organist an der Christianskirche) bedient, dem für seine Mitwirkung herzlich gedankt sei.

[4] Die Untersuchung wurde von Reinhard Weber (FB Physik, Abtl. Akustik, Carl-von-Ossietzky-Universität Oldenburg) durchgeführt, dem auch an dieser Stelle noch einmal für seine freundliche Unterstützung gedankt sei.

Die Rauhigkeit ist in vielen Glockenklängen objektiv durch die Modulation jeweils zweier eng benachbarter Eigenfrequenzen bedingt (vgl. oben, Abb. 4); diese Zwillingstöne sind Folge kleiner Abweichungen des Glockenkörpers von einer idealen rotationssymmetrischen Schale, bei der die Paare von Eigenmoden bzw. Eigenfrequenzen der Bedingung $\Omega_{mn} \sim \Omega_{nm}$ genügen (vgl. Schneider & Leman 2002). Je größer die durch den Gussvorgang bedingten Abweichungen von der perfekten Rotationssymmetrie sind, desto stärker treten Paare von Eigenfrequenzen auseinander (d.h., $\Omega_{mn} \neq \Omega_{nm}$). Die Zwillingstöne unterschiedlicher Frequenz verursachen dann deutlich hörbare Amplitudenmodulationen. Bei den Klängen, die vom Glockenspiel der Christianskirche aufgezeichnet wurden, war in den Oszillogrammen z.T. keine Amplitudenmodulation, z.T. deutliche ausgeprägte Modulation zu erkennen. Demnach durfte mit unterschiedlichen Rauhigkeitskurven für diese Klänge gerechnet werden.

Die Ergebnisse fielen in der Tat sehr unterschiedlich aus, allerdings mit Befunden, die nicht alle leicht zu erklären sind. Zwar gab es in mehreren Fällen Rauhigkeitskurven, deren modulierter Verlauf recht gut der physikalischen Amplitudenmodulation entspricht. Daneben wurde aber z.B. zu einem Klang der Glocke Nr. 4, dessen Oszillogramm eine sehr regelmäßige Amplitudenmodulation zeigt (s. Abb. 7a), eine Rauhigkeitskurve ermittelt, die dieser Modulation nicht korrespondiert, sondern nach ca. 3.8 Sekunden einen plötzlichen Anstieg der Rauhigkeit und einen ebenso plötzlichen Abfall derselben zeigt (Abb. 7b).

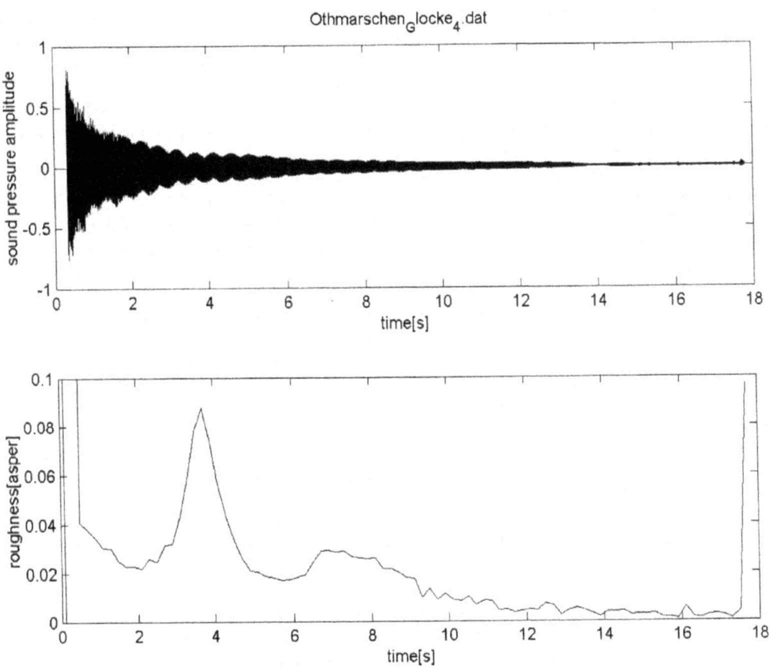

Abb. 7a/b: Glocke 4 des Glockenspiels der Christianskirche HH-Altona, Oszillogramm und zugehörige Rauhigkeitskurve

Was zu diesem Verlauf der Kurve geführt hat, war aus dem Signal direkt nicht zu ersehen.

Was zudem auffiel, war die Tatsache, dass die Rauhigkeitswerte sämtlicher gemessener Glockenklänge (darunter auch solche des berühmten Glockenspiels von Brügge; du Mery 1742-48) relativ niedrig ausfielen und nur selten den Wert von 0.1 asper überschritten, meist aber weit darunter blieben. Dies war deshalb erstaunlich, weil die hörbare Amplitudenmodulation etlicher Glockenklänge subjektiv deutlich „rauher" erscheint als etwa der Klang (Abb. 6), dem definitionsgemäß die Rauhigkeit 1 asper zugeschrieben wird. Das Ergebnis ist gleichwohl systematisch dadurch zu erklären, dass der für diese Rauhigkeitsempfindung erforderliche Modulationsgrad von $m \approx 1 - 1.2$ bei natürlichen Klängen wie etwa denen von Glocken selten erreicht wird. Wegen der Vielzahl von Spektralkomponenten in diesen komplexen teil- bzw. inharmonischen Klängen (vgl. Schneider & Müllensiefen 1999, Schneider 2001a, Schneider & Leman 2002) ist eine vollständige Amplitudenmodulation (AM), wie sie zwischen zwei Sinustönen leicht herstellbar ist (Abb. 6), kaum zu erwarten. Eine fast vollständige AM tritt allerdings beim großen Gong (*gong gede*) des javanischen Gamelan ein (vgl. Schneider 1997b, S. 259ff., S. 533-536), und zwar mit einer bewusst sehr niedrig gehaltenen Schwebungsfrequenz (*ombak*) von meist 2-5 Hz, die – ähnlich wie bei Glocken – durch die Aufspaltung einer Eigenmode in zwei Zwillingstöne verursacht ist. Bei den javanischen Gongs wird diese Schwebung nach dem Gussvorgang durch Bearbeitung des Gongs hergestellt (vgl. Jacobson & van Hasselt 1907, S. 34). Auf Grund dieser langsamen Schwebungen erscheint der Klang aber durchaus nicht „rauh", während die Rauhigkeit in Klängen von Glocken sehr ausgeprägt sein kann. Der Rauhigkeitseindruck solcher Glockenklänge hängt dabei nicht nur von der AM zwischen Paaren von Zwillingstönen, sondern auch von weiteren Spektralkomponenten ab, die (a) entweder dicht genug beieinander liegen, um sich i.S. der Kritischen Bandbreiten wechselseitig zu „stören" oder (b) untereinander inharmonische Frequenzrelationen bilden, ohne in die gleiche Frequenzgruppe zu fallen. Die Bedingung (b) trifft bei europäischen Mollglocken (vgl. Schneider & Müllensiefen 1999, Schneider & Leman 2002) für die meist als dritter Teilton auftretende Kleinterz und die oft in der nächsthöheren Oktave zu findende Großterz zu, die untereinander und im Zusammenwirken mit anderen Teiltönen stark dissonant wirken.

7. Experiment 2: Rauhigkeit und Konsonanzempfinden

Bei Helmholtz (s. o.) gibt es eine reziproke Beziehung von Rauhigkeit und Konsonanzempfinden, indem dieses abnimmt, wenn jene als akustische Eigenschaft und als psychoakustische Empfindungsgröße von Klängen zunimmt (et vice versa). Der Zusammenhang beider ist zwar bei Helmholtz und anderen Forschern (vgl. Plomp 1976, Kap. 3 und 4) beschrieben worden, verdient jedoch eine weitere empirische Überprüfung mit musikalisch relevantem Klangmaterial. Dazu wurden verschiedene Zusammenklänge (Zweiklänge aus komplexen Tönen) mit Hilfe eines Synthesizers (s. u.) hergestellt, die mehrheitlich die bekannten musikalischen Intervalle (kl. Sekunde, gr. Sekunde, kl. Terz, gr. Terz usw.) repräsentieren. Außerdem kamen einige Zusammenklänge zum Einsatz, bei denen der Abstand der Grundfrequenzen zweier komplexer Klänge kleiner als eine kleine Sekunde ist. Ferner wurden für manche Intervallkategorien je zwei alternative Realisierungen angeboten, so etwa bei der kl. Terz die pythagoräische Mollterz 32/27

(294 C.) und die kl. Terz 6/5 (316 C.), bei der großen Terz die reine große Terz 5/4 (386 C.) und die pythagoräische große Terz 81/64 (408 C.), beim Tritonus das Intervall 45/32 (590 C.) und der temperierte Tritonus als √2 (600 C.) sowie bei der kleinen Septime die ‚Naturseptime' 7/4 (969 C.) und die Sept 16/9 (996 C.) Die vollständige Liste der 23 Intervalle bzw. Zusammenklänge vom Einklang (Prime) bis zur Oktave ist der Tabelle 2 zu entnehmen.

Tab. 2: Intervallgröße der 23 Zweiklänge

Intervall Nr.	Abstand (Cent)	Intervallkategorie
1	0	Unisono
2	+20	
3	+50	temp. Viertelton
4	+75	
5	+100	temp. Halbton
6	+150	temp. Dreivierteltton
7	+204	großer Ganzton 9/8
8	+250	temp. Fünfvierteltton
9	+294	pythag. Kleinterz 32/27
10	+316	kl. reine Terz 6/5
11	+386	gr. reine Terz 5/4
12	+408	gr. pythag. Terz
13	+498	reine Quarte 4/3
14	+590	Tritonus (45/32)
15	+600	temp. Tritonus (√ 2)
16	+702	reine Quinte (3/2)
17	+813	kl. reine Sexte (8/5)
18	+884	gr. reine Sexte (5/3)
19	+969	kl. reine Septime (7/4)
20	+996	kl. Septime (16/9)
21	+1088	gr. reine Sept (15/8)
22	+1158	his-3 (dritte Oberterz von c)
23	+1200	Oktave (2/1)

8. Material und Methode

Um Klangreize zu gewinnen, die – anders als die in der Psychoakustik häufig verwendeten Sinustöne – der Lebenswirklichkeit entsprechen und insoweit als „ökologisch valide" betrachtet werden können, wurden Klänge eines monophonen Synthesizers KORG MS 20 verwendet. Dieser unter Pop- und Rockmusikern sowie bei Techno-Freaks noch immer beliebte Synthesizer besitzt zwei Sägezahngeneratoren. Diese wurden zunächst bei gleicher Fußtonlage (16') in ihrer Grundfrequenz so abgeglichen, dass keine Schwebungen mehr hörbar sowie messbar waren. Dazu wurden die Ausgangsspannungen der beiden Generatoren abwechselnd einem digitalen Stimmgerät (BOSS TU-100) zugeführt, das aus der Periodizität des Zeitsignals die Grundfrequenz des komplexen Tones ermittelt und optisch über ein Zeigerinstrument auf einer Cent-

Skala zur Anzeige bringt, wie weit die in Cent umgerechnete Grundfrequenz von der nach Maßgabe der gleichstufig temperierten Stimmung zu erwartenden Frequenz eines musikalischen Tones (z.B. a' = 440 Hz) abweicht. Der Messbereich beträgt ±50 Cent um die Frequenzposition jedes Tones einer chromatischen Skala; die Mess- bzw. die Ablesegenauigkeit dürfte bei ca. ±3 Cent liegen.

Die Größe jedes Simultanintervalls bzw. jedes Zusammenklanges, bei denen der tiefere Ton jeweils konstant auf eine Frequenz von 130.6255 Hz (Ton c) eingestellt blieb, wurde so genau wie mit dem beschriebenen Gerät möglich eingeregelt. Die entsprechenden Zusammenklänge wurden im Synthesizer über die Taste c ausgelöst und über ein Mischpult (ohne Benutzung von Filtern oder Regelverstärkern) auf DAT mit 48 kHz/16 Bit digital aufgezeichnet. Das Spektrum des aus der Überlagerung zweier Sägezahngeneratoren gleicher Fußtonlage (16') und Grundfrequenz c = 130.62 Hz geformten Unisono zeigt Abb. 8. Die spektrale Darstellung ist hier auf den Bereich 0-10 kHz begrenzt; die im Klang enthaltenen Teilschwingungen bzw. Teiltöne reichen weit über 10 kHz hinaus.

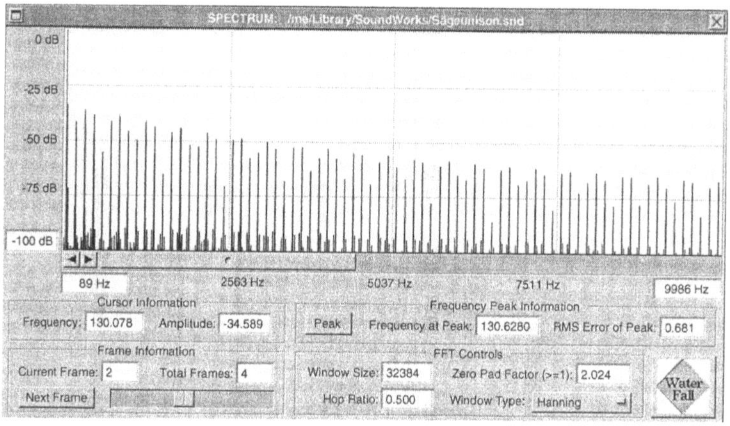

Abb. 8: Spektrum des Sägezahnklanges (Unisono, 2 Generatoren, Korg MS 20)

Die 23 nach dem hier beschriebenen Verfahren hergestellten Zusammenklänge wurden digital von 48 kHz/16 bit auf das CD-Format (44.1 kHz/16 bit) umgerechnet und in randomisierter Folge so auf eine CD kopiert, dass jeder der 23 Zusammenklänge zweimal hintereinander erklang. Die Dauer des Erklingens betrug bei jedem Zusammenklang ca. 4 Sekunden; zwischen den je ein Paar bildenden gleichen Klängen lag eine Pause von ca. 2 Sekunden, zwischen den Klangpaaren selbst jeweils eine Pause von ca. 5 Sekunden.

Die Darbietung erfolgte über eine gute HiFi-Anlage (Tandberg Integrated 3000 Amplifier; Electrovoice Sentry 500 Midfield Studiomonitore) mit einem Schalldruckpegel von ca. 70 dB (gemessen in 1 m Abstand vor dem Monitor).

Die n = 51 Vpn wurden gebeten, die 23 Zusammenklänge auf einer siebenstufigen bipolaren Skala Konsonanz ← → Dissonanz einzustufen. Die Skala war aus sieben gleichabständigen Alternativen gebildet, enthielt aber keinerlei Zahlangaben oder andere eine Metrik bildende Werte:

Konsonanz ( ) ( ) ( ) ( ) ( ) ( ) ( ) Dissonanz

Die 23 Skalen für die 23 Zusammenklänge wurden auf einem DIN A4 –Blatt untereinander stehend ausgedruckt. Jede Vpn erhielt also nur ein Blatt für die Einstufungen der Stimuli auf den Skalen.

9. Auswertung

Die Skalen wurden nach dem Versuch mit jeweils zwei Zahlen ausgestattet, nämlich einer, die den relativen Grad der Ausprägung von Konsonanz von 1 – 7 angibt, und einer zweiten, die den relativen Grad der Ausprägung von Dissonanz 1 – 7 angibt; die so aufgeteilten Skalen

Konsonanz	( )	( )	( )	( )	( )	( )	( )	Dissonanz
	1	2	3	4	5	6	7	
	7	6	5	4	3	2	1	

sind aus der Natur der Sache gegenläufig und verhalten sich komplementär zueinander. Wenn also eine Vpn z.B. das zweite Kästchen von links für einen der 23 Zusammenklänge angekreuzt hat, bedeutet dies einen Wert von „6" für die empfundene Konsonanz und von „2" für die bei diesem Zusammenklang empfundene Dissonanz.

Die mit entsprechenden Skalen versehenen Rohdaten wurden in das Statistikprogramm SPSS eingegeben. Es resultieren also zwei verschiedene Dateien, eine für Konsonanz- und eine für Dissonanzdaten. Die statistischen Kennwerte aM, SD und Range beider Dateien sind der Tab. 3 und der Tab. 4 zu entnehmen. Da es sich bei den Daten genau genommen nur um eine ordinale Messung handelt, ist das aM ebenso wie die Berechnung des SD aus statistischen Gründen im Grunde nicht zulässig oder zumindest mit Vorbehalten zu versehen. Diesem Umstand wurde durch Berechnung des auch für ordinalskalierte Daten gut geeigneten Kennwertes Median Rechnung getragen. Die 23 Medianwerte sind für die entsprechenden Stimuli (23 Zusammenklänge aus Tab. 2) nachfolgend in Tab. 3 aufgelistet.

Man sieht leicht, dass die Mittelwerte der Konsonanz- bzw. der Dissonanzurteile sich jeweils komplementär verhalten und die Standardabweichung für jede Intervallkategorie identisch ist, da ja auch nur ein Urteil auf jeder zweipoligen Skala Konsonanz ←→ Dissonanz abgegeben wurde. Aus den 2 x 23 Mittelwerten lassen sich zwei Kurven herstellen, die das Konsonanz- bzw. Dissonanzprofil der Intervalle im Sinne des von Helmholtz berechneten Profils (Abb. 2) darstellen und mit diesem sowie ähnlichen Profilen anderer Forscher (z.B. Sethares 1993) direkt vergleichbar sind.

Tab. 3: Intervalle und Mittelwerte für Konsonanz und Dissonanz

			Konsonanz			Dissonanz		
Nr.	Cent	Intervall	aM	SD	Med	aM	SD	Med
1	0	Unisono	6.66	0.93	7	1.33	0.93	1
2	+ 20		4.80	1.65	5	3.2	1.65	3
3	+ 50	temp. Viertelton	2.55	1.25	2	5.45	1.25	6
4	+ 75		1.80	1.22	1	6.2	1.22	7
5	+100	temp. Halbton	1.94	1.03	2	6.06	1.03	6
6	+150	temp. Dreivierteltton	1.76	0.86	2	6.23	0.86	6
7	+204	großer Ganzton 9/8	2.96	1.11	3	5.04	1.11	5
8	+250	temp. Fünfviertelton	2.43	1.24	2	5.57	1.24	6
9	+294	pythag. Kleinterz 32/27	3.51	1.25	4	4.49	1.25	4
10	+316	kl. reine Terz 6/5	3.43	1.76	3	4.57	1.76	5
11	+386	gr. reine Terz 5/4	4.2	1.8	4	3.8	1.8	4
12	+408	gr. pythag. Terz (81/64)	5.23	1.11	6	2.76	1.11	2
13	+498	reine Quarte 4/3	5.23	1.48	6	2.76	1.48	2
14	+590	Tritonus (45/32)	2.71	1.27	3	5.29	1.27	5
15	+600	temp. Tritonus (√ 2)	2.51	1.36	2	5.49	1.36	6
16	+702	reine Quinte (3/2)	6.1	1.17	6	1.9	1.17	2
17	+813	kl. reine Sexte (8/5)	4.59	1.33	5	3.41	1.33	3
18	+884	gr. reine Sexte (5/3)	5.96	0.94	6	2.04	0.94	2
19	+969	kl. reine Septime (7/4)	3.94	1.41	4	4.06	1.41	4
20	+996	kl. Septime (16/9)	3.51	1.25	3	4.49	1.25	5
21	+1088	gr. reine Sept (15/8)	2.49	1.19	2	5.51	1.19	6
22	+1158	his-3 (dritte Oberterz)	3.76	1.69	4	4.23	1.69	4
23	+1200	Oktave (2/1)	6.22	0.95	6	1.78	0.95	2

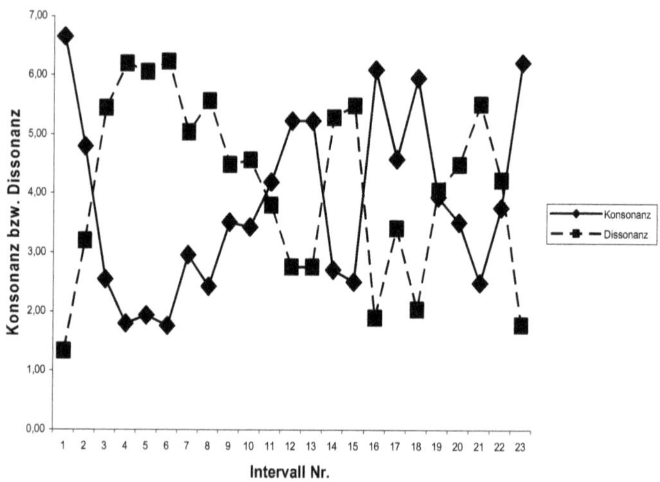

Abb. 9: Konsonanz- und Dissonanzprofil aus den Mittelwerten der Tab. 3

Die Mittelwerte aus der Tab. 3 ergeben folgende Rangreihe der Intervalle gemäß ihres Konsonanzgrades: Einklang, Oktave, reine Quinte, große reine Sexte, reine Quarte, pythagoräische große Terz, leicht schwebender Einklang, kleine reine Sext, reine große Terz, kleine reine Septime, his-3 als Intervall zwischen großer Sept und Oktave usw. An der Rangreihe der Konsonanzen ist vielleicht erstaunlich, dass die (als Ton- bzw. Frequenzverhältnis sowie als Wellenform kompliziertere) pythagoräische große Terz 81/64 deutlich besser abschneidet als die reine große Terz 5/4.

Die Intervalle mit dem höchsten Dissonanzgrad liegen ausweislich der Mittelwerte einmal im Bereich eines temperierten Vierteltons bis zu einem temperierten Fünfviertelton (Frequenzdistanz der Grundtöne 50 – 250 Cent), was mit dem Konzept der Kritischen Bandbreite gut übereinstimmt. Zweitens finden sich hohe Dissonanzwerte bei der großen reinen Sept (was bei diesem Simultanintervall bzw. Zweiklang nicht verwundert) sowie bei beiden Tritoni, von denen der temperierte noch etwas dissonanter bewertet wurde.

Helmholtz (1863 u. ö.) und ähnlich Stumpf (1890, 1897, 1898) hatten die kleine reine Sept (7/4) und auch den Tritonus (45/32 ≈ 7/5) als nicht sonderlich dissonante Intervalle bzw. sogar als ‚Halbkonsonanzen' empfunden. Während die vorliegenden Mittelwerte für Konsonanzen und Dissonanzen diese Einschätzung der kleinen reinen Sept bestätigen, fallen die Ergebnisse unseres Hörversuchs beim Tritonus deutlich ungünstiger aus.

10. Messung der 23 Zusammenklänge mit einem Hörmodell

Die 23 Zusammenklänge (Tab. 2) wurden zwecks Messung der Rauhigkeit und der Schärfe in ein als Software implementiertes Hörmodell (ArtemiS 10 [Head Acoustics]; nach Sottek 1993, 1994; Auflösung der Filterbank: 0.5 Bark = 47 Filter) gegeben, wobei für jeden Zusammenklang eine Rauhigkeitskurve als Funktion der Zeit resultiert, die gespeichert sowie als Plot ausgegeben wurde. Aus den 23 Rauhigkeitskurven wurden die zeitlich gemittelten Durchschnittswerte (in asper) berechnet; diese finden sich in Tab. 4:

Tab. 4: Mittelwerte der Rauhigkeit von 23 Zusammenklängen (Hörmodell)

Nr.	Cent	Intervall	Mittelwert (asper)
1	0	Unisono	0.828
2	+20		1.03
3	+50	temp. Viertelton	0.899
4	+75		0.922
5	+100	temp. Halbton	0.748
6	+150	temp. Dreivierteiton	0.555
7	+204	großer Ganzton 9/8	0.618
8	+250	temp. Fünfviertelton	0.553
9	+294	pythag. Kleinterz 32/27	0.473
10	+316	kl. reine Terz 6/5	0.629
11	+386	gr. reine Terz 5/4	0.565
12	+408	gr. pythag. Terz	0.437
13	+498	reine Quarte 4/3	0.436
14	+590	Tritonus (45/32)	0.339
15	+600	temp. Tritonus (√ 2)	0.329
16	+702	reine Quinte (3/2)	0.5
17	+813	kl. reine Sexte (8/5)	0.341
18	+884	gr. reine Sexte (5/3)	0.278
19	+969	kl. reine Septime (7/4)	0.259
20	+996	kl. Septime (16/9)	0.124
21	+1088	gr. reine Sept (15/8)	0.091
22	+1158	his-3 (dritte Oberterz)	0.121
23	+1200	Oktave (2/1)	0.309

Die Ergebnisse dieser Messung sollen nun einmal für sich interpretiert und dann mit den Daten zur empfundenen Dissonanz der Zusammenklänge unseres Hörversuchs in Beziehung gesetzt werden. Die in Tab. 4 gelisteten Mittelwerte dürften zunächst Erstaunen hervorrufen, da die Rauhigkeit eines Unisono-Intervalls sehr klein sein sollte, hier aber ein relativ hoher Wert erscheint. Dieser ist nur so zu erklären, dass von den zahlreichen höheren Partialen, die auf Grund der verwendeten Wellenform (2 x Sägezahn) im Spektrum erscheinen (s. Abb. 8), jeweils mehrere in eine Frequenzgruppe fallen und – da die Amplituden bzw. Pegel dieser Partiale auch in höheren CB relativ stark sind – dort jedenfalls zu einer so starken Modulation führen, dass sie bei dem Algorithmus dieses Hörmodells als Teilrauhigkeiten in etlichen Filterbändern zu Buche schlagen. Das Modell hat zwar Gewichtungsfunktionen, die bestimmte Effekte wie z.B. den Einfluss der Frequenzlage einer (sinusförmigen) Trägerfrequenz auf die Rauhigkeitsempfindung nachbilden. Im Ergebnis aber summiert das Hörmodell die Teilrauhigkeiten aus den einzelnen Frequenzgruppen bzw. CB zu einer Gesamtrauhigkeit *R* auf. Aus Experimenten mit einem anderen Hörmodell wurde gefolgert, dass sich die empfundene Gesamtrauhigkeit komplexer Klänge bzw. Zusammenklänge nicht einfach additiv aus den Teilrauhigkeiten vorhersagen lässt (vgl. Pressnitzer 1998).

Die mit dem Hörmodell berechneten Rauhigkeiten sind durch den Algorithmus des Modells bedingt und numerisch wohl korrekt. Sie spiegeln indessen schwerlich die

Hörerfahrung durchschnittlicher Musikhörer wider, die eine reine Quinte 3/2 in der hier gewählten Frequenzlage (mit Grundton c ~ 130.6 Hz) wahrscheinlich weniger rauh empfinden als eine pythagoräische Kleinterz, ein mit 294 Cent relativ enges Intervall, das wegen der deutlich hörbaren Amplitudenmodulation rauh erscheint. Das Modell hingegen errechnet für die reine Quinte, die ohne hörbare Modulation erklingt, eine etwas höhere Rauhigkeit als für die pythagoräische Kleinterz.

Die gemittelten Rauhigkeitswerte des Hörmodells für die 23 Zusammenklänge (Tab. 4) korrelieren ferner praktisch nicht mit den gleichfalls gemittelten Dissonanzwerten (Tab. 3); der Produkt-Moment-Korrelationskoeffizient beider Datenreihen ergibt $r_{xy} = 0.098$ (n. s.).

Demnach korrespondieren vorliegend die Beobachtungsdaten zur empfundenen Dissonanz der 23 Zusammenklänge offenbar nicht mit den algorithmisch in einem Hörmodell ermittelten Rauhigkeitswerten. Hierzu ist anzumerken, dass – worauf Stumpf (1898) gegen Helmholtz insistiert hat – musikalische Zusammenklänge herstellbar sind, die einerseits sehr dissonant wirken, ohne dass hörbare Rauhigkeit eine Rolle spielt[5]; dieser Sachverhalt ist in der Regel über die weite Lage der zu einem Zusammenklang verbundenen Töne herzustellen, so dass sich diese untereinander nicht im Sinne der Frequenzgruppen bzw. CB stören.

Andererseits sind auch Klänge möglich, die rauh wirken, ohne merklich dissonant im Sinne der Intervallgröße zu sein. Für Stumpf war daher die Rauhigkeit im Verhältnis zur gesamten wahrgenommenen Dissonanz nur ein *akzessorisches Merkmal*, während Helmholtz Rauhigkeit als eine bei dissonanten Zusammenklängen häufig auftretende, bei konsonanten Akkorden hingegen fehlende Empfindung beobachtet und beschrieben hat.

11. Experiment 3: Stimmung, Konsonanz und Rauhigkeit

Aus den bei Sethares (1993) im Anschluss an Helmholtz getroffenen Feststellungen ergibt sich, dass eine klanglich möglichst konsonante Realisierung musikalischer Intervalle bzw. Akkorde der tonalen Musik ein passendes Stimmungssystem erfordert, und zwar das der Reinen Stimmung, bei der Schwebungen und Rauhigkeit minimiert werden.[6] Bei einer akustischen Realisierung der gleichen Intervalle bzw. Akkorde in gleichstufig temperierter Stimmung nehmen Schwebungen sowie – je nach Intervall- und Akkordstruktur – auch die hörbare Rauhigkeit zu.

Die Differenz beider Realisierungen (Reine Stimmung ./. gleichstufige Temperierung) vermögen musikalisch geschulte HörerInnen in der Regel gerade deshalb zu erfassen, weil in einem Fall die Schwebungen gering sind oder ganz fehlen, so etwa bei gelungenen Einspielungen und Aufführungen von Madrigalen oder bei professionellen Barbershop-Ensembles.[7] Im anderen Fall aber erscheinen merkliche Schwebungen z. B. in homophonen Akkordsätzen von Streichern und ausgeprägte Rauhigkeit bei vollgriffigen Akkorden in Orgelwerken, wo zudem die Stimmungsdifferenzen zwischen der ‚horizontalen' Tastenstimmung (gleichstufig temperiert) und der ‚vertikalen' Stimmung

[5] Zu Stumpfs Kritik an Helmholtz (die durchaus möglich, aber etwas überzogen war) und zu den von Stumpf vorgeschlagenen Zusammenklängen s. Schneider 1997a mit Hörbeispielen.

[6] Vgl. Sethares 1993, 1998 und auch Messergebnisse bei Schneider 2008, 43-45.

[7] Zu verweisen wäre etwa auf die Aufnahmen der Madrigale Gesualdos durch das *London Consort of Musicke* einerseits, Aufnahmen des Ensembles *Take Six* andererseits.

der Mixturen und Farbregister (rein) zur Rauhigkeit beitragen (vgl. Schneider et al. 2004).

Reinier Plomp, einer der führenden Psychoakustiker des 20. Jahrhunderts (s. Plomp 1976), hat vor einigen Jahren noch einmal Fundamente der Hörtheorie erörtert und durch Klangbeispiele auf einer CD erläutert (Plomp 1998).[8] Eine der Aufnahmen, die das Konsonanzphänomen und die Hörbarkeit von Schwebungen bei gleichstufiger Temperierung demonstriert, bietet die ersten sechs Takte des Chorals ,Wie schön leuchtet der Morgenstern' in der Bearbeitung von J.S. Bach (BWV 763). Diesen vierstimmigen Choral hat Reinier Plomp mit synthetischen Sounds realisiert, die ein harmonisches Spektrum mit einer typischen Hüllkurve (6 dB Abfall der Teiltonamplituden pro Oktave) besitzen; das Spektrum aller Klänge wurde bei 8 kHz (per Tiefpass) abgeschnitten. Die sechs Takte erklingen in zwei Versionen, nämlich (1) in reiner Stimmung und (2) in gleichstufiger Temperierung. Reinier Plomp schreibt hierzu: *Tegenover de wat drogere eerste uitvoering klinkt de laatste door het optreden van zwevingen levendiger, hetgeen door de meeste luisteraars positief zal worden gewaardeerd.* Die bei der Version in gleichstufiger Temperierung auftretenden und je nach Zusammenklang unterschiedlichen Schwebungen lassen also vermutlich diese Darbietung für die meisten HörerInnen lebendiger erscheinen, wodurch diese Version dann auch angenehmer wirkt als die durch stationäre Akkorde geprägte Version in reiner Stimmung.

Einer der Autoren dieser Abhandlung (A.S.) hat die Aufnahmen Plomps mehrfach bei Hörversuchen präsentiert, bei denen die beiden Versionen auf vier unipolaren Skalen zu beurteilen waren, nämlich (1) Konsonanzwirkung, (2) Schwebungen, (3) Richtigkeit des Satzes und (4) ästhetischer Gesamteindruck.

Von den vier Skalen, die auch bei anderen Hörversuchen in gleicher oder ähnlicher Kombination eingesetzt wurden (Schneider & Müllensiefen 1999, Schneider 2001), knüpfen zwei direkt an psychoakustisch begründete Wahrnehmungsgrößen an, eine weitere bezieht sich eher auf die syntaktische Ebene der Musik, während die vierte die Gesamtwirkung abfragt, in die – dies kann hypothetisch vermutet werden (s. u.) – die anderen drei anteilig mit eingehen. Nachfolgend seien die Ergebnisse zweier Hörversuche aus dem WS 2002/03 und aus dem WS 2007/08 mitgeteilt, an denen $n_1 = 23$ und $n_2 = 30$ Studierende der Musikwissenschaft und anderer Fächer auf freiwilliger Basis teilnahmen. Die Ergebnisse der deskriptiven Statistik finden sich in der Tab. 5:

---

[8] Prof. Plomp hat einem der Autoren (A.S.) die nicht leicht zu findende Veröffentlichung (CD plus 2 Booklets) geschenkt, wofür ihm ganz herzlich gedankt sei.

Tab. 5: Deskriptive Statistik, *Wie schön leuchtet der Morgenstern*, $n_1 = 23$ und $n_2 = 30$ Vpn

	Version A (reine Stimmung)				Version B (gleichst. Temper.)			
Daten n_1	Range	aM	SD	Median	Range	aM	SD	Median
Konsonanzwirkung	1-7	3.78	1.68	3	2-7	5.13	1.46	6
Schwebungen	1-7	4.43	1.62	5	1-6	3.3	1.67	3
Richtigk. des Satzes	1-7	3.78	1.91	4	1-7	5.39	1.64	6
Ästhetischer Gesamt.	1-7	2.83	1.75	2	1-6	3.91	1.38	4

Daten n_2	Range	aM	SD	Median	Range	aM	SD	Median
Konsonanzwirkung	1-7	3.17	1.31		2-7	4.6	1.38	
Schwebungen	2-7	5.53	1.07		2-6	3.77	1.35	
Richtigk. des Satzes	2-7	4.03	1.47		3-7	5.07	1.14	
Ästhetischer Gesamt.	1-5	2.4	1.16		1-7	3.57	1.48	

12. Inferenzstatistik Hörversuch *Wie schön leuchtet der Morgenstern*, Stichprobe n_1

Zur Durchführung von Mittelwertsvergleichen wurde zuvor die Varianzhomogenität durch F-Tests und Bartlett-Tests geprüft. Das Kriterium der Varianzhomogenität einander korrespondierender Dateien wurde erfüllt. Die Mittelwertskontraste aller vier Skalen bzw. Urteilsdimensionen erwiesen sich bei t-Tests als signifikant. Bei den Daten des ersten Versuchs ergaben die t-Tests (gepaarte Daten) folgende Ergebnisse:

Konsonanz	t = 2.252 (2.074, p = 0.05), signifikant
Schwebungen	t = 2.208 (2.074, p = 0.05), signifikant
Richtigkeit des Satzes	t = 3.23 (2.819, p = 0.01), sehr signifikant
Ästhet. Gesamteindruck	t = 2.737 (2.074, p = 0.05; 2.819, p = 0.01), signifikant

Eine einfaktorielle ANOVA, bei der die beiden Tunings als Stufen des Faktors bzw. als Treatments genommen werden, ergibt folgenden Befund:

Konsonanz	F = 8.47 (7.264, F = 0.01), sehr signifikant
Schwebungen	F = 5.456 (4.065, F = 0.05), signifikant
Richtigkeit des Satzes	F = 9.394 (7.264, F = 0.01), sehr signifikant
Ästhet. Gesamteindruck	F = 5.478 (4.065, F = 0.05), signifikant

Auf Grund des Befunds wird die H1 angenommen, wonach sich die unterschiedlichen Treatments (vorliegend die Tunings) in ihrer perzeptorischen und ästhetischen Wirkung deutlich unterscheiden. Eine einfaktorielle MANOVA, bei der die Tunings als zwei Stufen des Faktors und die vier Skalen (Konsonanz, Schwebungen etc.) als abhängige Variablen genommen wurden, führt zum gleichen Ergebnis: Wilks-Lambda = 0.717; F = 5.245 (3.981, F = 0.01), sehr signifikant.

Schließlich wurde – praktisch als Gegenprobe zu den Mittelwertskontrasten – die Korrelation der Variablen untereinander überprüft; sind die Kontraste signifikant, fallen die Korrelationen i. d. R. nicht signifikant aus.

13. Inferenzstatistik Hörversuch *Wie schön leuchtet der Morgenstern*, Stichprobe n_2

Die meisten statistischen Prozeduren des ersten Versuchs wurden auch hier durchlaufen. Die Mittelwertskontraste für die vier Variablen sind signifikant bzw. sehr signifikant. Demnach unterschieden sich die beiden Versionen auch für die Vpn dieser Stichprobe deutlich.

Anstatt einer Berechnung der multiplen Korrelation zwischen den Variablen wurde hier für beide Versionen (reine Stimmung, gleichstufige Temperierung) ein Regressionsmodell berechnet, bei dem die Konsonanzwirkung, die Schwebungen und die Richtigkeit des Satzes als unabhängige Variablen, der Ästhetische Gesamteindruck als abhängige Variable definiert wurden. Die Modellzusammenfassung ergab für die Version A (rein) ein R-Quadrat von 0.358 und ein korrigiertes R-Quadrat von 0.284, die ANOVA mit F = 4.831 eine Signifikanz von p = 0.008. Bei den Urteilen der Version in Reiner Stimmung bestehen nach der Berechnung der Koeffizienten folgende Zusammenhänge der Variablen:

Tab. 6: Koeffizienten (Regression, Version A, reine Stimmung)

Modell	Standardisierte Koeffizienten		
	Beta	T	Signifikanz
(Konstante)		1,366	0,184
A Konsonanzwirkung	0,039	0,228	0,822
A Schwebungen	-0,358	-2,117	0,044
A Richtigkeit Satz	0,444	2,756	0,011

Demnach spielt bei dieser Version die Konsonanzwirkung für den ästhetischen Gesamteindruck keine große Rolle, wohl aber die ‚Richtigkeit des Satzes', die in positiver Richtung den höchsten Betrag liefert, während die Variable ‚Schwebungen' in negativer Hinsicht stark zu Buche schlägt.

Bei der Version B (temperiert) lauten die entsprechenden Werte: R-Quadrat 0.199, korrigiertes R-Quadrat 0.107; bei der ANOVA mit F = 2.156 beträgt die Signifikanz p = 0.117. Die Werte der Koeffizienten für die Version B sind aus Tab. 7 zu ersehen:

Tab. 7: Koeffizienten (Regression, Version B, gleichstufig temperiert)

Modell	Standardisierte Koeffizienten		
	Beta	T	Signifikanz
(Konstante)		1,588	0,124
B Konsonanzwirkung	0,318	1,438	0,162
B Schwebungen	-0,175	-0,934	0,359
B Richtigkeit Satz	0,063	0,281	0,781

Bei der temperierten Fassung fallen die Koeffizienten weniger klar aus: die Variable ‚Schwebung' erzielt nur einen mäßig negativen Wert, die Konsonanzwirkung einen mittleren positiven; die ‚Richtigkeit des Satzes' ist hier nahezu ohne Bedeutung.

Die Ergebnisse der beiden Hörversuche bestätigen nicht allein die von Reinier Plomp schon erwartete Bevorzugung der Version in gleichstufiger Temperierung, die in den Variablen Konsonanzwirkung, Richtigkeit des Satzes und beim ästhetischen Gesamteindruck deutlich besser abschneidet als die Version in reiner Stimmung. Die Befunde zeigen offensichtlich erstens ein recht unsicheres Urteilsverhalten der ProbandInnen bei beiden Versionen und bei fast allen Variablen, was sich in den Streuungsmaßen (Range, SD) der Urteile niedergeschlagen hat. Zweitens dürften die meisten ProbandInnen (Studierende im ersten Studienjahr) auch noch nicht über eine hinreichend analytische Höreinstellung verfügt haben, da bei der Version in reiner Stimmung, bei der objektiv die sensorische Konsonanzwirkung durch Koinzidenz von Teiltönen bzw. fehlende oder stark reduzierte Interferenz von Schwingungsfrequenzen gefördert wird, die Konsonanz als deutlich geringer beurteilt wurde als bei der Version in gleichstufiger Temperierung, bei der zahlreiche Interferenzen auftreten.

Noch erstaunlicher ist der hohe Mittelwert für ‚Schwebungen' bei der Version in reiner Stimmung, die objektiv eben gerade keine hörbaren Schwebungen bzw. allenfalls marginale aufweist. Umgekehrt erhielt die temperierte Version, die reichlich Schwebungen enthält, eine durchschnittlich deutlich niedrigere Beurteilung auf dieser Variablen.

Bei der Variablen ‚Richtigkeit des Satzes' schneidet die Version in reiner Stimmung wahrscheinlich deshalb schlechter ab, weil hier mehr als zwölf Töne pro Oktave auftreten und Studierende in den Anfangssemestern, die noch nie in ihrem Leben eine solche Realisierung von Musik gehört haben, die unterschiedliche Intonation von Tönen und Akkorden in Kreuz- und B-Tonarten nicht apperzipieren, sondern für „falsch" halten.

Da ausweislich des Regressionsmodells sowie weiterer statistischer Befunde von einer Beteiligung der drei übrigen Variablen am ästhetischen Gesamturteil auszugehen ist, erklären sich die divergierenden Gesamturteile für beide Versionen eben zu einem Teil aus dem Einfluss der Variablen 1-3, zum anderen in der Tat aus dem weniger analytisch als vielmehr „ganzheitlich" geleiteten Gesamteindruck.

Das Urteilsverhalten kann insgesamt zu einem erheblichen Teil auf den Umstand zurückgeführt werden, dass viele Studierende auch der musikwissenschaftlichen Fächer heute ganz überwiegend mit Pop- und Rockmusik aufgewachsen sind, in der z.B. in praktisch allen Studioproduktionen fortwährend Geräte bzw. Algorithmen zur Zeitachsenmanipulation durch Phasing, Flanging, Chorus zum Einsatz kommen. Alle so produzierten Aufnahmen zeigen klanglich spektrale bzw. zeitliche Modulationen, die zu einer Vielzahl unterschiedlich schneller und unterschiedlich starker Schwebungen führen. Das in Studiokreisen als „Phasengewaber" bekannte Klangbild ist mithin der Normalzustand von Pop- und Rockmusik, die seit den ‚Sixties' vielfach von solchen Effekten lebt.

Hingegen bilden stationäre Klänge mit einer nahezu reinen Teiltonstruktur, wie sie bei von mehreren Saxophonen gespielten Akkorden auftreten, eher die Ausnahme.

Ganz offensichtlich erscheinen jüngeren HörerInnen, die mit dem „Phasengewaber" der Popmusik aufgewachsen und mit anderen Musikgattungen (z.B. Madrigal, Streichquartett, Saxophonquartett, aber auch etwa mit vokaler Mehrstimmigkeit des Balkans) nie in Berührung gekommen sind, Akkorde und Zusammenklänge um so konsonanter, je mehr sie durch die Stimmung einerseits, die Effekte (z.B. den in Popkreisen überaus beliebten Chorus) andererseits in beständiger Modulation und damit in Schwebungen erklingen.

Umgekehrt erscheinen diesen HörerInnen stationäre Klänge und „stehende" Akkorde mit starken harmonischen Teiltönen als wenig angenehm, weil einmal die gewohnte Modulation fehlt und zweitens diese Klänge durch Koinzidenz von Teilschwingungen bzw. Teiltönen sensorisch an Schärfe gewinnen können. Bei fehlender Interferenz von Teiltönen nimmt die Helligkeit von harmonischen Klängen (messbar als normalisierter spektraler Zentroid) und damit auch ihre Schärfe in der Regel zu.

14. Messung von ‚Wie schön leuchtet der Morgenstern' mit dem Hörmodell (ArtemiS 10)

Die beiden Versionen der Choralbearbeitung (Plomp 1998) wurden hinsichtlich ihrer Rauhigkeit und ihrer Schärfe mit dem Hörmodell (ArtemiS 10, nach Sottek 1993, 1994) untersucht. Nach den Überlegungen, die im Verlauf dieser Studie mit Bezug auf den Einfluss der Ton- und Stimmungssysteme auf beide Empfindungsparameter angestellt wurden, sollte bei der Version in reiner Stimmung die Rauhigkeit geringer ausfallen als bei der Version in gleichstufiger Temperierung, während die Schärfe bei jener etwas höher als bei dieser sein sollte. Die zu den beiden Versionen (a) rein und (b) temperiert gehörenden Rauhigkeitskurven zeigt die Abb. 10:

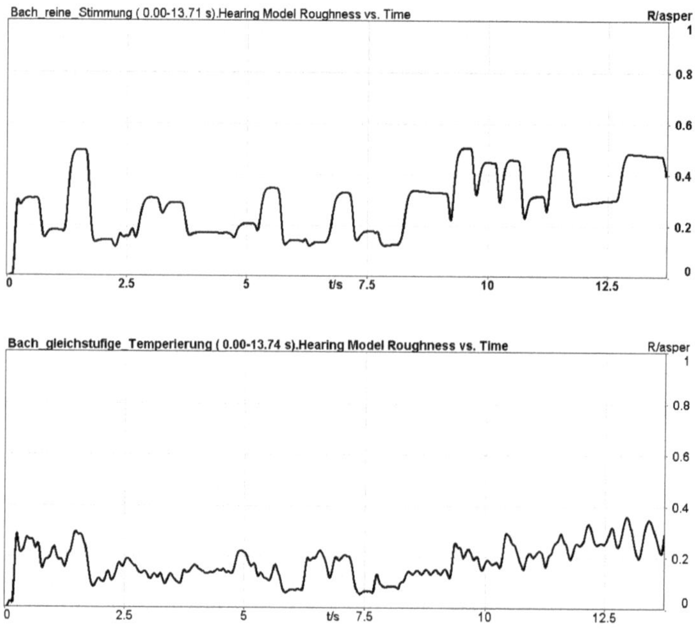

Abb. 10: Messung von ‚Wie schön leuchtet der Morgenstern' mit ArtemiS 10 (Hearing Model Roughness vs. Time)

Die Version in reiner Stimmung zeigt deutlich weniger Modulation als die temperierte. Allerdings betragen die zeitlich gemittelten Durchschnittswerte der Rauhigkeit für die Version in reiner Stimmung 0.267 asper und für die in gleichstufiger Temperierung 0.173 asper. Damit liegt, was auch die beiden Kurven (Abb. 10) bereits erkennen lassen, der vom Hörmodell gelieferte Mittelwert für die reine Version über dem der temperierten Version. Dieses unerwartete Ergebnis kann analog zu den Ausführungen, die oben (S. 122) zu den mit diesem Hörmodell an 23 Zweiklängen erzielten Messergebnissen gemacht wurden, am ehesten noch so erklärt werden, dass höhere starke Partiale, die jeweils in eine CB bzw. Frequenzgruppe fallen, als Verursacher von Amplitudenmodulationen und damit von Rauhigkeit erfasst werden. Die Rauhigkeit dieser höheren Teilschwingungen bzw. Teiltöne geht hier jedoch relativ zum objektiven Klangverlauf und auch im Vergleich zum Höreindruck offenbar zu stark in die Gesamtrauhigkeit *R* ein.

Bei der Schärfe, die gleichfalls für beide Versionen als Funktion der Zeit berechnet und als Plot mit den entsprechenden Kurven ausgegeben wurde[9], sind die Unterschiede zwischen beiden Versionen relativ gering. Die Mittelwerte lauten 4,23 acum für die reine und 4,12 acum für die temperierte Version. Dieser geringe Unterschied bestätigt immerhin im Grundsatz die relative Zunahme der Schärfe bei einer Realisierung in reiner Stimmung. Dieser Unterschied zeigt sich ganz ähnlich in den Verläufen des spektralen Zentroids bzw. der aus dem normalisierten spektralen Zentroid berechneten Helligkeit beider Versionen[10]; letztere liegt durchschnittlich etwa bei 80. Man muss vor allem bei den Werten für die Schärfe und die Helligkeit beider Versionen bedenken, dass bei ‚Wie schön leuchtet der Morgenstern' das Spektrum aller Klänge auf eine obere Grenzfrequenz von 8 kHz begrenzt und bei dieser abgeschnitten wurde (vgl. Plomp 1998).

15. Experiment 4: ‚Wie schön leuchtet der Morgenstern', Rauhigkeitsmessung nach Helmholtz

Die einfache Definition, die Hermann von Helmholtz (1863, 1870 u.ö.) für Rauhigkeit als Ergebnis der absoluten Differenz zweier Schwingungen gegeben hat (S. 105f.), wurde mit Hilfe eines unmittelbar an Helmholtz anknüpfenden Algorithmus für musikalische Klänge experimentell untersucht. Hierbei wurde entsprechend Helmholtz von der Voraussetzung ausgegangen, dass die Rauhigkeit R zweier Sinustöne bzw. zweier Spektralkomponenten f_1 und f_2 für $|f_1 - f_2| = 0$ Hz verschwindet (R=0) und für $|f_1 - f_2| =$ 33 Hz maximal wird (R=1). Für $|f_1 - f_2| > 33$ Hz nimmt die Rauhigkeit mit wachsender Differenz der beiden Frequenzen zunächst allmählich und dann asymptotisch gegen Null ab.

---

9 Parameter der Messung: Sharpness vs. Time; Calculation method: FFT/Head, Sharpness method: Aures; resolution: 0.5 Bark; Window function: Hanning, spectrum size 4096 pts, overlap 50 pts; sound field: free.

10 Berechnet mit dem Programm sndan (Beauchamp 1995) per Phase Vocoder mit der Analysefrequenz 26 Hz, FFT: 4096 pts, 848 harmonics, Hamming-Gewichtung.

Ein solcher Ansatz wurde bereits von Sethares (1993) anhand der experimentell gefundenen Rauhigkeitskurven von Plomp & Levelt (1965, S. 556, Fig. 10) entwickelt. Seine Rauhigkeitskurve berechnet sich als Subtraktion zweier e-Funktionen wie

(6) $d(x) = e^{-ax} - e^{-bx}$.

Hierbei ist $x = d^*/ (s_1\, f1 + s_2)\, (f_2\text{-}f_1)$ mit $f_2 > f_1$. Maximale Rauhigkeit ergibt sich daher für $x = 1/ (a\text{-}b)\, \ln (a/b)$. Mittels der Konstanten $s_1 = .021$ und $s_2 = 19$ wird $d^* = .22$ ($d^* = .24$ bei Sethares) berechnet. Hieraus folgt eine Abhängigkeit von f_2- f_1 für maximale Rauhigkeit von f_1 wie sie in Abb. 11 zu sehen ist.

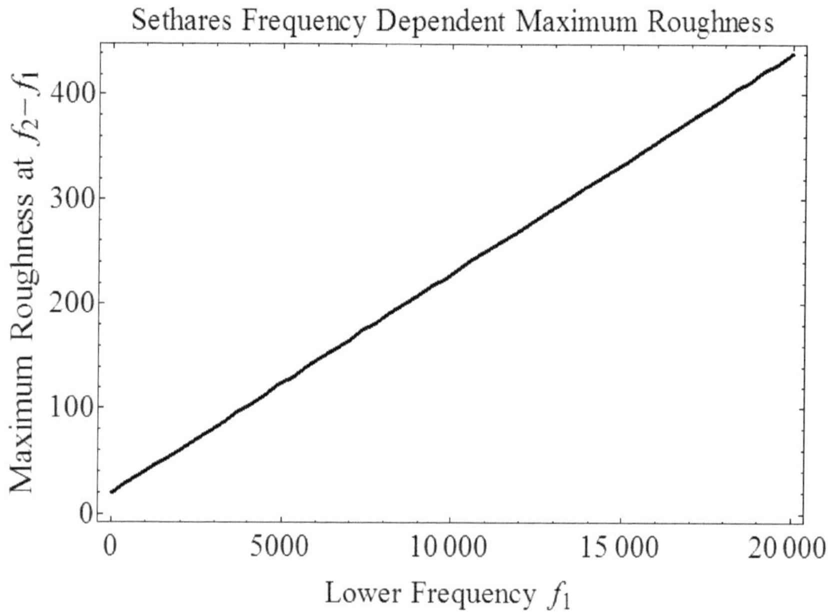

Abb. 11: Abhängigkeit der maximalen Rauhigkeit für f_2-f_1 von der Grundfrequenz f_1 bei f_2>f_1.

Helmholtz schlug eine maximale Rauhigkeit bei f_2-f_1 = 33 Hz vor. Im Algorithmus von Sethares variiert dieser Wert und ist etwa für f_1 = 20 Hz bei f_2-f_1 = 19.42 Hz, für f_1 = 100 Hz bei f_2-f_1 = 21.1 Hz, für f_1 = 1000 Hz bei f_2-f_1 = 40 Hz oder für f_1 = 20 kHz bei f_2-f_1 = 439 Hz. Allerdings ist diese Angleichung an die kritischen Bandbreiten nur gering und so ist dieser Ansatz von dem Helmholtz'schen nicht grundlegend verschieden.

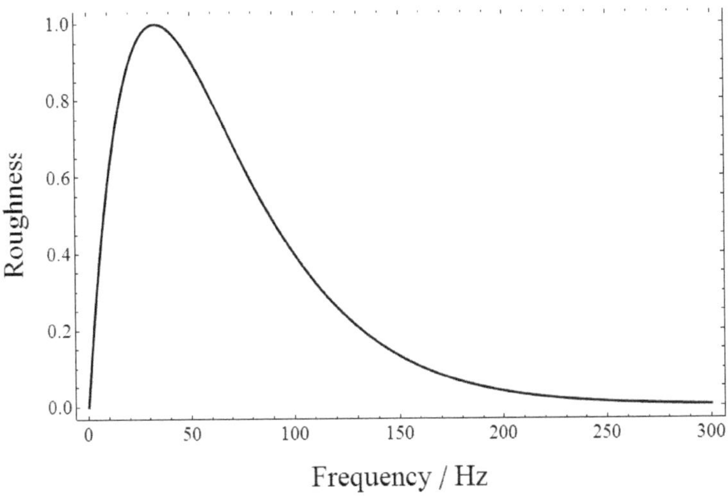

Abb. 12: Rauhigkeit vs. Frequenzdifferenz |df| für ein Frequenzpaar mit df = $f_1 - f_2$.

Die nun in der vorliegenden Arbeit zur Berechnung der Rauhigkeit von Akkorden aus komplexen harmonischen Tönen verwendete Formel[11] lautet

(7)

$$\mathrm{R(n)} = \mathrm{A_1 A_2} \frac{|\,\mathrm{df_n}\,|}{\mathrm{f_r}\; e^{-1}} e^{-|\mathrm{df_n}|/\mathrm{f_r}}$$

mit den Amplituden A_1 und A_2 der Frequenzen f_1 und f_2 mit $df_n = f_1 - f_2$ und $f_r = 33$ Hz. Um die Gesamtrauhigkeit eines komplexen Klangs zu errechnen, wird dieser in seine Spektralkomponenten zerlegt und für alle *n* möglichen Kombinationen die Einzelrauhigkeit R(n) berechnet. Die Gesamtrauhigkeit ist dann die Summe der Einzelrauhigkeiten nach

(8)

$$\mathrm{R} = \sum_{i=1}^{n} \mathrm{R(n)}$$
.

Die untersuchten Klänge waren 23 einzelne Akkorde (Takt 1-6) des Bach'schen Orgelchorals „Wie schön leuchtet der Morgenstern" (BWV 763) in einer Realisierung von Reinier Plomp (1998), welche als Audiofile (von der CD) vorlag. Die beiden Versionen der Takte 1-6 des Chorals wurden von Plomp (a) in reiner und (b) in gleichstufig temperierter Stimmung eingespielt. Aus den oben (S. 126) genannten Gründen war zu

[11] Die Formel wurde von Rolf Bader entwickelt, der auch das Programm zur Berechnung der Rauhigkeit nach Helmholtz in Mathematica geschrieben hat.

erwarten, dass die objektiv messbare Rauhigkeit in der Version in reiner Stimmung kleiner sei als in der temperierten.

Zusätzlich wurden die spektralen Zentroide der Klänge berechnet, um einen möglichen Zusammenhang zwischen Rauhigkeit und Helligkeit darzustellen. Es ist zu erwarten, dass hellere Klänge relativ gesehen auch rauher sind, da sie mehr Obertöne enthalten, welche sich bei hinreichend kleinen Frequenzabständen wechselseitig stören können. Dies muss aber nicht der Fall sein. Treten viele Oberschwingungen bzw. Obertöne auf, welche perfekt aufeinander liegen, so ist der Klang zwar hell, aber nicht rauh.

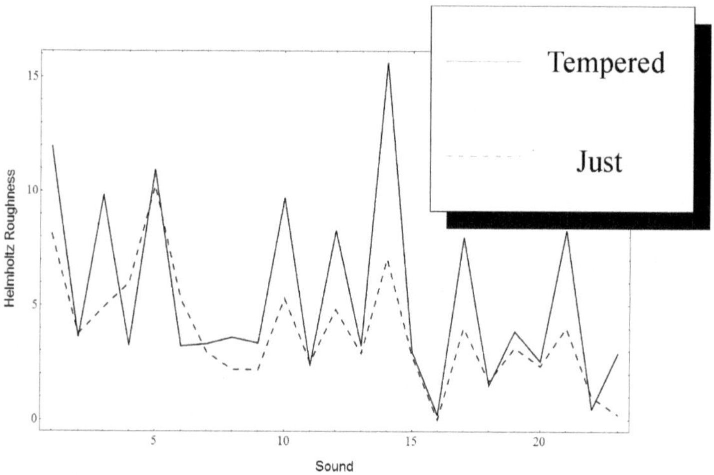

Abb. 13: Rauhigkeiten der 23 Akkorde für reine und temperierte Stimmung, berechnet nach Helmholtz. Die Akkorde in reiner Stimmung sind deutlich weniger rauh als die in temperierter Stimmung.

Abb. 13 zeigt die Rauhigkeiten der 23 Klänge in der reinen und in der gleichstufig temperierten Stimmung. Die reine Stimmung wird als deutlich weniger rauh berechnet als die temperierte. Dies bestätigen auch die Mittelwerte über alle 23 Klänge. In der Folge der 23 rein gestimmten Akkorde reicht lediglich der fünfte in etwa an die Rauhigkeit des entsprechenden Akkords aus der temperierten Version heran.

Tab. 8: Mittelwerte für Rauhigkeit und Zentroid: Reine und temperierte Stimmung

	temperierte Stimmung	reine Stimmung
Rauhigkeit	5.35	3.80
Spektraler Zentroid	2338.6 Hz	2243.1 Hz

In der Tabelle 8 zeigt sich deutlich, dass die reine Stimmung als durchschnittlich weniger rauh berechnet wurde, während die spektralen Zentroide sich nur wenig voneinander unterscheiden. Dies hat vorliegend vor allem seinen Grund darin, dass die Aufnahme von R. Plomp lediglich den Frequenzbereich von 0 – 8 kHz umfasst. Die

Beziehungen zwischen den Zentroiden und den Rauhigkeiten sind in Abb. 14 gezeigt. Der Verlauf der Zentroidkurve ist in beiden Versionen relativ ähnlich; es gibt jedoch einige Ausreißer. So wird der spektrale Zentroid für den Klang 8 in der temperierten Version mit 3463.9 Hz (temp) und in der reinen mit 3015.3 Hz (rein) berechnet; die entsprechenden Rauhigkeitswerte sind R = 3.59 (temp) und R = 2.19 (rein). Die Maximalwerte der Rauhigkeit in diesem Akkord bzw. Klang Nr. 8 betragen aber 10.64 (temp) und 10.2 (rein).

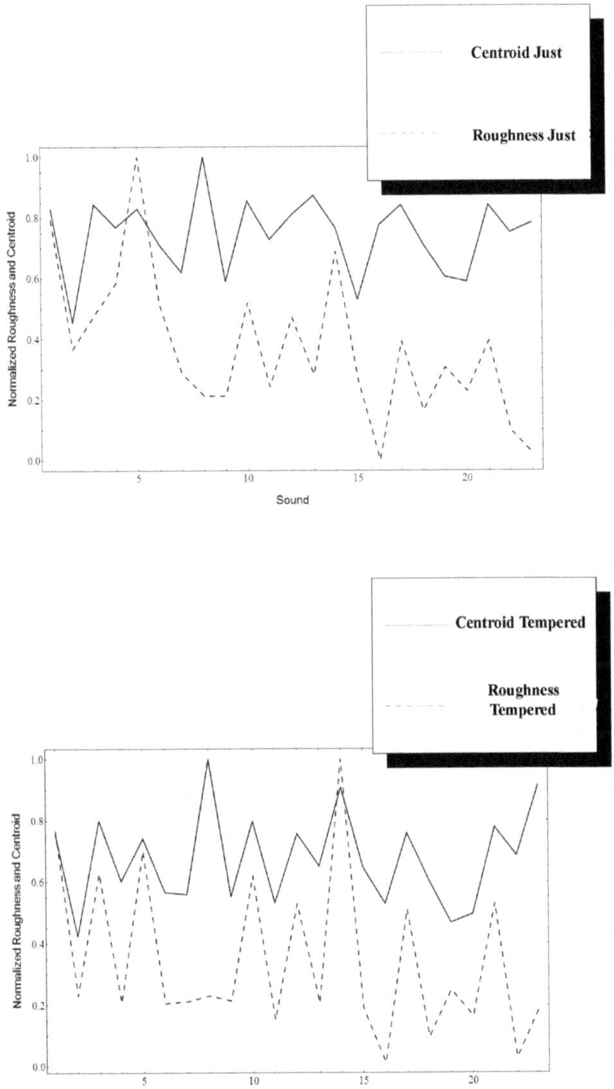

Abb. 14: Rauhigkeit und spektraler Zentroid für die Versionen in reiner und temperierter Stimmung.

Der unmittelbar an Helmholtz anknüpfende Algorithmus für die Berechnung der Rauhigkeit gibt nach den gezeigten Befunden sehr gut die durchaus feinen Unterschiede der Rauhigkeit im musikalisch relevanten Bereich wieder. Das Problem der Oktavlage, auf das Helmholtz selbst hingewiesen hat (S. 106) und das bei Klängen mit tiefen Frequenzen wegen der CB von Bedeutung ist, spielt offenbar bei den hier analysierten Sounds keine Rolle. Erst bei tiefen Tönen mit Grundfrequenzen unter 100 Hz kommt das Problem auf, dass z.B. die Oktave 33 Hz – 66 Hz nach dem verwendeten Rauhigkeitsalgorithmus (Gl. 6, 7) wegen df = 33 Hz maximale Rauhigkeit zeigen würde, obwohl die Oktave als musikalisches Intervall im Prinzip die geringste Rauhigkeit aufweisen müsste. Freilich zeigt die Breite der CB bzw. Frequenzgruppen (Tab. 1; Abb. 3), dass in sehr tiefer Lage eine mit Sinustönen oder anderen nicht direkt musikalisch relevanten Schallarten gemessene CB die Größenordnung einer Oktave erreicht. Es ist nun zu prüfen, inwieweit die Frequenzspannen von Kritischen Bandbreiten – bleibt man im Grundsatz bei diesem Konzept – von der Teiltonstruktur und von der spektralen Energieverteilung komplexer Töne abhängt, die jeweils miteinander Intervalle bilden. Der hier vorgeschlagene Algorithmus kann dann ggf. entsprechend modifiziert und insbesondere auf Zusammenklänge in tiefer Lage angepasst werden. Bei Rauhigkeitsmessungen in mittleren Lagen, um die es auch Helmholtz vorrangig ging, liefert der Algorithmus jedenfalls Ergebnisse, die mit der objektiv gegebenen Klangstruktur der untersuchten Stimuli gut übereinstimmen.

16. Experiment 5: Klanganalyse und Rauhigkeitsmessung an bulgarischer Schwebungsdiaphonie

Während die meisten historischen Musikstile in Mitteleuropa auf der Vorherrschaft konsonant klingender Akkorde beruhen, sind aus der Volksmusik des Balkans und auch aus anderen Kulturen etliche Musikformen bekannt, in denen Schwebungen und Rauhigkeit eine wichtige Rolle spielen (vgl. Brandl 1992, Schneider 1997b, 2001c). Dies sei hier an einem Beispiel aus Bulgarien verdeutlicht. Die Lieder der Frauen bestehen in mehreren Regionen aus einer Diaphonie, die parallele Kleinterzen, aber auch große Sekunden sowie „neutrale“ Terzen (Abstand der beiden Grundfrequenzen ≈ 345 Cent) verwendet. Außerdem kommt es zu Schwebungen durch ein „Beinahe-Unisono“, bei dem die Grundfrequenzen beider Vokalstimmen sich nur wenig unterscheiden.

Im nachfolgenden Beispiel liegen bei einem solchen Schwebungsvorgang die beiden Grundfrequenzen f_n und f_m bei etwa 378-380 Hz sowie bei 389-397 Hz. Beide Grundfrequenzen sind, über den etwa 3.3 Sekunden dauernden Schwebungsvorgang betrachtet, nicht ganz stabil, wobei die Grundfrequenz der tieferen Stimme allerdings viel weniger fluktuiert als die der höheren. Der Abstand der beiden Grundfrequenzen f_n und f_m beträgt vorliegend zwischen etwa 10 Hz und fast 18 Hz; bezogen auf die tatsächlich gemessenen Frequenzen[12] bedeutet dies einen Unterschied von ca. 46 Cent bis zu etwa 82 Cent, mithin die Größenordnung eines Vierteltons bis zu einem chromatischen Halbton. Durch den geringen und zudem zeitlich etwas veränderlichen Frequenzabstand von f_n und f_m tritt eine Amplitudenmodulation auf, die in Abb. 15 durch den Text ‚Beats („Unisono“)‘ kenntlich gemacht ist. Abb. 15 zeigt das Spektrogramm

[12] Bestimmt mit der ‚Frequency at Peak‘-Funktion in Spectro 3.01; Fenster: 16384 pts, Hanning-Gewichtung.

bzw. Sonagramm (100-1000 Hz) der ersten Phrase aus dem Lied *Vetar Vee*, das 1968 in der Region Shope in Bulgarien aufgenommen wurde.[13] Zu sehen ist der Verlauf der beiden Stimmen an Hand ihrer Grundfrequenzen sowie jeweils auch der erste Oberton, der gegenüber jenen erheblich stärker in der Amplitude ausfällt (kenntlich gemacht durch die Graustufe: je dunkler eine Klangkomponente gezeichnet wird, desto stärker ist ihre Amplitude bzw. ihr Pegel).

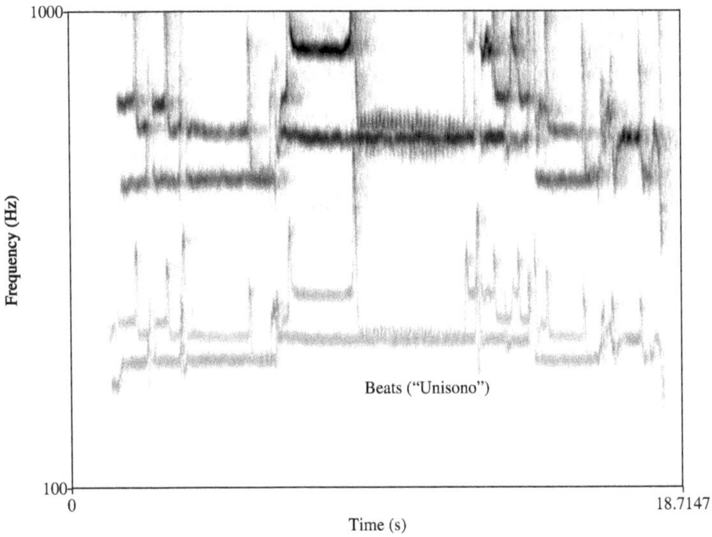

Abb. 15: Diaphonie, 2 Frauenstimmen, *Vetar Vee* (Raim & König 1988), Track 1

Die Amplitudenmodulation wirkt deshalb so eindringlich, weil die fraglichen Zusammenklänge auf Vokalen mit nasaler Stimmgebung (wie etwa /è/ oder /ä/) gesungen werden. Hierdurch treten die entsprechenden Vokalformanten sehr stark in Erscheinung und besitzen höhere Teiltöne im Spektrum deutlich mehr Energie als tiefere. Der Sachverhalt ist aus Abb. 16 zu entnehmen; das Sonagramm (200 – 2400 Hz) zeigt über den bei ≈ 380-395 Hz liegenden Grundfrequenzen des Zusammenklangs die durch intensive Modulation gekennzeichneten Oberschwingungen bzw. Teiltöne.[14]

[13] Nähere Angaben bei Raim & König 1988 (CD-booklet).

[14] Die Klanganalysen wurden mit den Programmen Praat Version 4.5.14 (Spektrogramm, Cochleagramm und Formantberechnung nach Burg) sowie Spectro 3.01 (Spektrum und genaue Frequenzbestimmung durch parabolische Interpolation) durchgeführt.

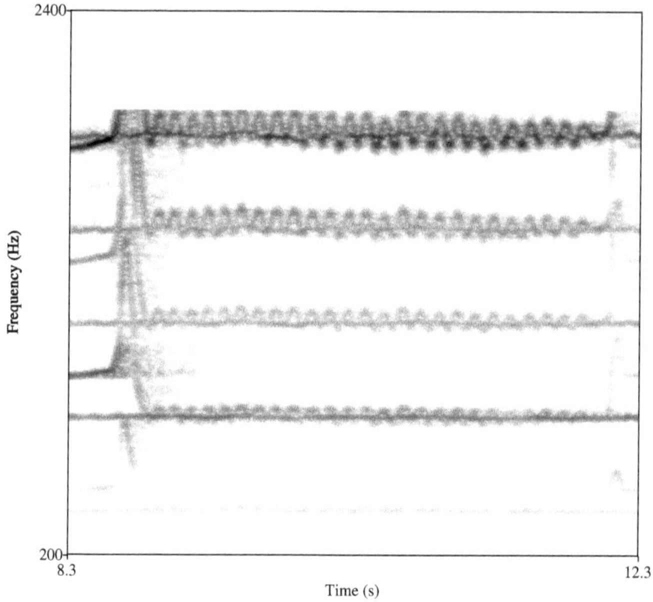

Abb. 16: Zusammenklang zweier Frauenstimmen, AM, höhere Teiltöne sehr stark

Die per Frame des Zeitsignals berechneten Formantwerte ergeben zwei Formantbereiche, von denen der erste gut abgegrenzt bei ca. 750-800 Hz liegt, der zweite weit weniger klar bei etwa 1100-1300 Hz. Eine dritte Konzentration spektraler Energie findet sich dann bei 1.9-2.1 kHz.

Während die Rauhigkeit dieses Zusammenklangs und auch anderer im Verlauf dieses diaphonischen Gesangs nicht zu überhören ist, geht die mit dem Programm Artemis 10 berechnete Rauhigkeitskurve (Abb. 17) nicht über einen Maximalwert von 0.225 asper hinaus. Dies ist angesichts der Signalstruktur unseres Beispiels ein unerwarteter Befund auch deshalb, weil vorliegend sowohl die beiden Grundfrequenzen f_n und f_m wie auch jeweils Paare ihrer Teiltöne wegen des geringen Frequenzabstand in je eine Frequenzgruppe/Kritische Bandbreite bzw. in je ein diesen entsprechendes Filter des Ohrmodells fallen müssten. Da bei merklich ausgeprägter Modulation (s. Abb. 16) die Teilrauhigkeit pro Filterband einen bestimmten Betrag erreicht und die Gesamtrauhigkeit durch Summation der Teilrauhigkeiten ermittelt wird, wären eher höhere Rauhigkeitswerte als die mit dem Modell gemessenen zu erwarten. Zumindest erscheint die sensorische Rauhigkeit bei dem hier erörterten Beispiel aus der Schwebungsdiaphonie subjektiv viel ausgeprägter als im Referenzsignal zur Definition von 1 asper (Abb. 6).

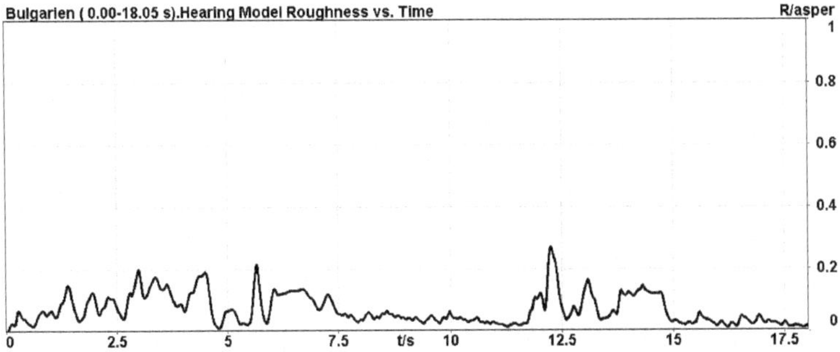

Abb. 17: Rauhigkeitskurve zum Beispiel der Schwebungsdiaphonie (Abb. 15).

17. Diskussion

Die in diesem Beitrag anhand theoretischer Überlegungen sowie eigener empirischer Untersuchungen erörterte Rauhigkeit gehört zu den psychoakustischen Grundeigenschaften von Schallen und damit auch von musikalischen Klängen. Die hier vorgelegten Befunde aus Messungen und aus Hörversuchen sind bereits oben erörtert worden. Ergänzend hierzu sei angemerkt, dass verschiedene Experimente, darunter Verhaltensbeobachtungen an Vpn und Messungen auf der Basis von Computermodellen, gerade bei der Rauhigkeit eine nicht unerhebliche Abhängigkeit der Ergebnisse von den Versuchsbedingungen und von den Methoden der Messung (vgl. Plomp & Levelt 1965, Terhardt 1974, Vogel 1975, Aures 1985a, Daniel & Weber 1997, Pressnitzer 1998, de Baene et al. 2004, Vassilakis 2005) zeigen. Ein wesentlicher Grund hierfür scheint in der unterschiedlichen Ausgangsbasis der Konzepte zu liegen, die teils in der Frequenzebene („frequency domain"), teils in der Zeitebene („time domain") angesiedelt sind. Dies bedeutet im konkreten Fall, dass entweder (a) vom Frequenzabstand und der Frequenzlage zweier oder mehrerer mit Bezug auf die Kritischen Bandbreiten hinreichend eng benachbarten Spektralkomponenten oder aber (b) von der sinusförmigen Amplitudenmodulation eines i. d. R. periodischen Trägersignals (Sinus- oder Cosinusschwingung, seltener komplexe Töne) ausgegangen wird. Beide Signalarten erzeugen sensorisch Rauhigkeit, die sogar in ihrer Wirkung als sehr ähnlich empfunden werden kann.

Auch lassen sich beide Betrachtungsweisen – die spektrale der ‚Beats' und die in der Zeitebene angesiedelte der Amplitudenmodulation – formal sowie inhaltlich ineinander überführen (vgl. Plomp 1976, ch. 4, Terhardt 1974 und 1998, Kap. 10.2, Hartmann 1998, 462f.). Demnach kann man mit einem Paar von Sinustönen gleicher Amplitude bei geeignetem Frequenzverhältnis die gleiche zeitliche Hüllkurve wie mit einem AM-Sinuston erzeugen; auch die sensorische Rauhigkeit ist in beiden Stimuli dann gleich groß, wenn der Modulationsgrad des AM-Sinustons mit $m \sim 0.67$ gewählt wird.

Die Korrespondenz beider Ebenen sei durch das nachfolgende Schema verdeutlicht:

a) Frequenzebene

Basiskomponenten	Signalentstehung	Hüllkurve	Spektrum	Empfindung
Sinus/Cosinus				
} →	Superposition	→ AM-Signal	→ Paare von	→ Rauhigkeit
Sinus/Cosinus	(additive Synthese)		Komponenten	

b) Zeitebene

Sinus/Cosinus (f_c) →	AM	→ AM-Signal	→ Trägerfrequenz	→ Rauhigkeit
Sinus/Cosinus (f_{mod})	(Träger*Modulator)		Seitenfrequenzen	

Diese Korrespondenz, die sich bei elementaren Signalen bzw. Stimuli leicht herstellen lässt, macht eine Analyse komplexer Klänge, wie sie insbesondere in musikalischen Kontexten entstehen, keineswegs überflüssig. Rauhigkeit findet sich als Klangphänomen und als sensorisches Merkmal auch in „westlicher" Kunstmusik (vgl. Voigt 1985); in etlichen Stilen der europäischen Volksmusik tritt sie markant in Erscheinung (s.o.), ebenso in bestimmten Gattungen des javanischen und des balinesischen Gamelan (z.B. gong kebyar). Die Ausgangsbasis zur Erzeugung von Schwebungen und von Rauhigkeit ist dabei unterschiedlich: Einerseits gibt es Instrumente, bei denen Schwebungen bereits durch die Eigenmoden des schwingenden Systems und durch das Auftreten von ‚Zwillingstönen' (degenerierte Paare von Eigenmoden) verursacht werden. Dies ist vor allem bei Gongs und Glocken sowie ähnlichen Schwingern mit nicht ganz perfekt eingehaltener Radialsymmetrie der Fall. Auch bilden sich Schwebungen und Rauhigkeiten häufig in Orgeln durch Stimmungsdifferenzen zwischen Pfeifenreihen (Grundregister ./. Mixturen; vgl. Schneider et al. 2004). In allen vorgenannten Fällen findet man in der Zeitfunktion des abgestrahlten Klanges und schon in der Analyse einzelner Schwingungsmoden vielfältige Formen von teils regelmäßiger, teils unregelmäßiger AM (s. Beispiele in Schneider 1997b, Bader 2002a).

Andererseits gibt es Musikstile, bei denen Schwebungen und Rauhigkeit durch Überlagerung diskreter Stimmen bewirkt werden, indem der Abstand zweier oder mehrerer Grundfrequenzen sehr gering gehalten und so gezielt ‚Beats' erzeugt werden (vgl. Abb. 15, 16). Dies ist insbesondere bei den zahlreichen Formen der ‚Schwebungsdiaphonie' (vgl. Brandl 1992) der Fall. Da hier z.B. von zwei Sängern oder Sängerinnen parallel und synchron Töne mit relativ stabiler Grundfrequenz produziert werden, findet die Erzeugung der Zusammenklänge im Frequenzbereich statt. Die ‚Beats' und damit die sensorische Rauhigkeit kommen mithin durch Superposition zustande, die mit einer AM der Hüllkurve einhergeht. Die AM ist hier aber nicht die *Ursache*, sondern die *Folge* der Klangerzeugung.

Es war diese Situation, die Helmholtz (1863 u. ö.) zu seinen Überlegungen mit Bezug auf die Rauhigkeit und die durch sie hervorgerufene sensorische Dissonanzempfindung geleitet hat. Helmholtz' Konzept ging also von der Frequenzebene aus und schließt naturgemäß dann auch eine Betrachtung der zeitlichen Charakteristika solcher Klänge, die als schwebend oder rauh empfunden werden, ein. Obwohl einige Forscher die Überlegungen von Helmholtz fortgeführt haben (etwa Plomp & Levelt 1965, Kameoka & Kuriyagawa 1969a/b, Plomp 1976, Sethares 1993, 1998), hat sich zwischenzeitlich die

Untersuchung der Rauhigkeit und vor allem die algorithmische Modellierung der Rauhigkeitswahrnehmung mit Hilfe von Computern stark in den Zeitbereich verlagert (vgl. Sottek 1993, 1994; Daniel & Weber 1997, Pressnitzer 1998). Wie die in diesem Beitrag mitgeteilten Messungen mit solchen Modellen erkennen lassen, sind die Ergebnisse im Einzelfall nicht leicht zu interpretieren. Es scheint jedoch, dass eine allein im Zeitbereich angesiedelte Betrachtung die tatsächlichen Verhältnisse eher unterschätzt. Nimmt man als Beispiel den Klang aus Beispiel 1, so lässt sich eine Hüllkurve über die Zeitfunktion legen, die in guter Näherung mit

(9) $y(t) = 0.6\text{Cos}(2\text{Pi}\ 33t)+2.2$

zu bestimmen und in der nachfolgenden Abbildung 18b gezeigt ist.

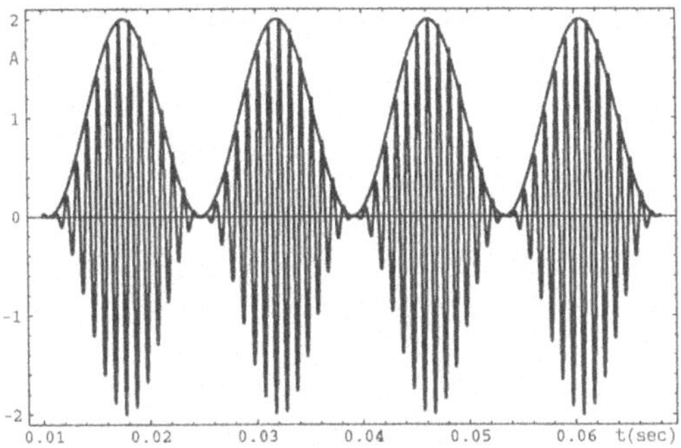

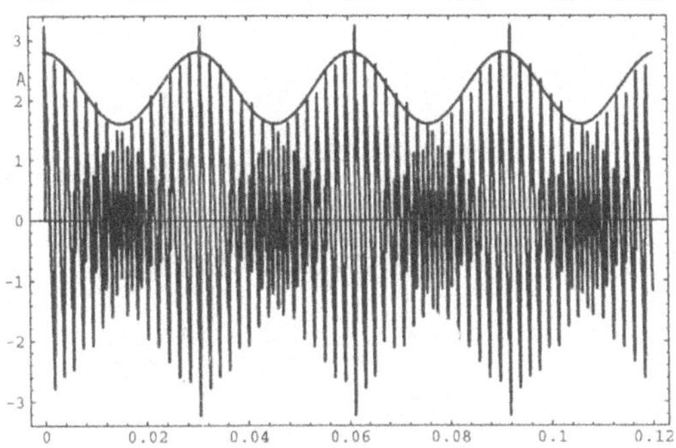

Abb. 18a: 1 asper-Referenzsignal, Zeitfunktion mit Hüllkurve
Abb. 18b: „Helmholtz-Sekunde" h' – c": Zeitfunktion mit Hüllkurve

Der durch die Amplitude der Hüllkurve umschriebene Modulationshub beträgt hier 1.2; bei noch genauerer Anpassung der Hüllkurve erreicht deren Amplitude dann etwa 1.3-1.5, was aber immer noch unter dem Amplitudenwert der Hüllkurve liegt, die sich über die Zeitfunktion aus Gl. 3 bilden lässt und zu dem Referenzsignal von 1 asper führt (s. Abb. 6 und 18a). Hier erreicht die Amplitude und so auch der durch die Hüllkurve bestimmte Abstand zwischen Schwingungsmaximum und 0 den Betrag 2. Mithin übertrifft die an der Hüllkurvenschwankung festgemachte Amplitudendifferenz ($A_{max} - A_{min}$)

beim 1 asper-Referenzsignal den Betrag der „Helmholtz-Sekunde“, wenn auch lediglich um den Betrag 0.5-0.8. Zudem fallen definitionsgemäß beim 1 asper-Referenzsignal alle drei Komponenten in eine Frequenzgruppe, nämlich in die von z = 8-9 Bark mit der Bandbreite von 920-1080 Hz und der Mittenfrequenz 1000 Hz. Zum Vergleich: bei der „Helmholtz-Sekunde“ (Gl. 1 und Abb. 1, 18b) fallen die Komponenten 978.75 und 1044 Hz in die Frequenzgruppe von 8-9 Bark, die beiden Komponenten 2446.87 Hz und 2610 Hz in die Frequenzgruppe z = 14; dabei haben die beiden tieferen Komponenten in unserem synthetisch erzeugten, aber nicht unrealistischen Klang lediglich Amplituden von 0.5, die beiden höheren sogar jeweils nur 0.2, wodurch die resultierende Reibung bzw. AM entsprechend gemildert wird. Gleichwohl erscheint die „Helmholtz-Sekunde“ subjektiv nicht weniger rauh als das 1 asper-Referenzsignal, sondern wird eher als rauher empfunden.

Dies bestätigen jedenfalls Daten aus Hörversuchen, bei denen 3 Klänge jeweils auf einer siebenstufigen Skala (1 = geringe, 4 = mittlere, 7 = starke Rauhigkeit) zu beurteilen waren.

Klang 1 war das 1 asper-Referenzsignal (m = 1, f_c = 1000 Hz, f_{mod} = 70 Hz), bei Klang 2 wurden die Trägerfrequenz und die Modulationsfrequenz halbiert (also f_c = 500 Hz, f_{mod} = 35 Hz), der Modulationsgrad m = 1 beibehalten. Dieser Klang liegt mit seiner Trägerfrequenz und mit seiner Modulationsfrequenz in etwa in dem Bereich der „Helmholtz-Sekunde“, deren Frequenzlage gleichfalls bei 500 Hz angesiedelt ist und deren Modulationsfrequenz ca. 33 Hz beträgt (s. oben). Klang 3 war die synthetisch hergestellte ‚Helmholtz-Sekunde‘ (Abb. 1). Die drei Klänge wurden jeweils mit einem Pegel von ca. 75 dB(A) über hochwertige HiFi-Komponenten (Tandberg Integrated Amp 3012, ElectroVoice Sentry 500 Studiomonitore) wiedergegeben. Die n = 19 Vpn hörten jeden Klang (Dauer: 3 Sekunden) drei Mal.
Die Rohdaten des Hörversuchs ergaben folgende Kennwerte:

Tabelle 9: Subjektiv empfundene Rauhigkeit von drei Klängen (n = 19)

Kennwert	Klang 1	Klang 2	Klang 3
Arithm. Mittel (aM)	5	3.81	5.368
Standardabw. (SD)	0.667	0.602	0.895
Median (Med)	5	4	6

Ausweislich der Basisstatistik erscheint Klang 3 („Helmholtz-Sekunde“) rauher als der 1 asper-Referenzton. Der entsprechende t-Test (Klang 1 ./. Klang 3) ergibt t = 2.689 (p [0.05] = 2.101; p [0.01] = 2.878), d. h. die Mittelwertsdifferenz ist deutlich signifikant. Dies zeigt ebenfalls ein multipler Mittelwertsvergleich über alle drei Dateien durch Scheffé-Test; der Kontrast D der Mittelwerte von Klang 1 und und Klang 3 ist hoch; Prüfwert D = 1.559 (Grenzwert 0.916, p = 0.001). Zudem sind die Urteile für die beiden Stimuli konsistent, was sich durch Korrelation der Urteilsreihen zeigen lässt. Die Korrelation der Urteile zu Klang 1 und 3 ist hochsignifikant (r_{xy} = 0.745, Prüfwert t = 4.602, p [0.001] = 3.965).

Der Klang 2 fällt sowohl gegenüber dem 1 asper-Referenzsignal wie gegenüber der „Helmholtz-Sekunde“ in der Beurteilung der subjektiv empfundenen Rauhigkeit deutlich ab.

Die „Helmholtz-Sekunde" ist mithin im Vergleich mit zwei AM-Signalen ungeachtet der geringer ausgeprägten Amplitudenschwankung der Hüllkurve der Klang, der am rauhesten erscheint. Die drei Klänge sowie ein vierter, bei dem der Ton h' der ‚Helmholtz-Sekunde' aus sechs phasenstarr gekoppelten Sinuskomponenten, der Ton c'' hingegen aus sechs phasenstarr gekoppelten Cosinuskomponenten gebildet sind, wurden mit den beiden in ArtemiS 10 zur Verfügung stehenden Algorithmen (Standard und Hörmodell) berechnet. Die vier Klänge waren zuvor mit Hilfe des Programms CoolEdit auf einen Pegel von –3dB (bezogen auf 0 dBfs) normalisiert worden. Dabei ergaben sich die in Tab. 10 zusammengefassten Rauhigkeitswerte, die als Mittelwerte über der Zeit bestimmt wurden.

Tabelle 10: Rauhigkeit von vier Klängen, Berechnung (in asper) nach ArtemiS 10

Sound		Standard	Hörmodell
(1a)	Helmholtz-Sekunde (Sinus/Sinus)	3.55	0.39
(1b)	Helmholtz-Sekunde (Sinus/Cosinus)	3.45	0.39
(2)	1 asper-Referenzsignal (1000 Hz*35 Hz)	2.88	2.36
(3)	AM-Signal (500 Hz*35 Hz)	1.08	1.16

Die mit dem Standardalgorithmus aus ArtemiS 10 erzielten Ergebnisse stimmen zumindest mit der Rangfolge der subjektiven Rauhigkeitsurteile überein, d.h. die ‚Helmholtz-Sekunde' erreicht Platz 1, das 1 asper-Referenzsignal Platz 2 und das AM-Signal mit halbierten Werten für Träger- und Modulationsfrequenz Platz 3. Die Abwandlung der ‚Helmholtz-Sekunde' durch Kombination von Sinus- und Cosinuskomponenten, die hier zur Erkundung von etwaigen Phaseneffekten vorgenommen wurde, hat offenbar nur einen geringen Einfluss; sie führt im Ergebnis zu einer etwas anderen Wellenform, aber nur zu einem marginalen Unterschied des durch $A_{max} - A_{min}$ ausgedrückten Modulationshubs. Die Schwankungen des Schalldruckpegels betragen bei der nur aus Sinuskomponenten gebildeten Helmholtz-Sekunde 2.1 dB, bei der aus Sinus- und Cosinuskomponenten synthetisierten 2.13 dB. Es hat den Anschein, dass Veränderungen der Phasenlage von Spektralkomponenten (vgl. auch Pressnitzer & McAdams 1999) dann zu perzeptorischen Effekten führen, wenn hierdurch die Hüllkurve des jeweils resultierenden Signals hinsichtlich der Differenz $A_{max} - A_{min}$ merklich beeinflusst wird.

Die vier Klänge wurden sodann mit den Algorithmen von Bader (nach Helmholtz) und Sethares analysiert. Bei diesen wird die gemessene Rauhigkeit **nicht** in asper normiert, d.h. die Messwerte können nur für einen Vergleich der Signale untereinander (ohne Bezug auf einen vorab definierten, evtl. nicht selbst gemessenen Standard) herangezogen und interpretiert werden. Die Messwerte gehen aus Tab. 11 hervor:

Tabelle 11: Rauhigkeit von vier Klängen, Berechnung nach Bader/Helmholtz und Sethares

Sound		Bader	Sethares
(1a)	Helmholtz-Sekunde (Sinus/Sinus)	1.508	0.3317
(1b)	Helmholtz-Sekunde (Sinus/Cosinus)	1.502	0.3087
(2)	1 asper-Referenzsignal (1000 Hz*35 Hz)	0.734	0.1486
(3)	AM-Signal (500 Hz*35 Hz)	1.172	0.1977

Ausweislich der Werte ergibt sich die Reihenfolge der Signale nach ihrer Rauhigkeit bei Bader/Helmholtz und bei Sethares übereinstimmend als 1a, 1b, 3, 2. Auch hier rangiert also die ‚Helmholtz-Sekunde' klar vor dem 1 asper-Referenzsignal. Zudem rückt sogar das AM-Signal mit halbierter Modulierfrequenz vor das 1 asper-Referenzsignal, was mit der Tatsache zusammenhängt, dass die Modulierfrequenz f_{mod} = 35 Hz sehr nahe an die von Helmholtz als Wert maximaler Rauhigkeit definierte AM mit df = 33 Hz heranrückt.

Auffällig ist, dass sowohl beim Hörtest mit Vpn wie bei der Messung mit dem Standard-Algorithmus von ArtemiS 10 und den Algorithmen von Bader/Helmholtz und Sethares die ‚Helmholtz-Sekunde' als das rauheste Signal identifiziert wird, während bei der Messung mit dem Hörmodell-Algorithmus von ArtemiS 10 (und nur mit diesem) das 1 asper-Referenzsignal den mit weitem Abstand höchsten Rauhigkeitswert erzielt. Bei diesem Algorithmus (und ähnlichen Modellen von Aures, Daniel & Weber etc.) ist die entscheidende Größe die durch AM hervorgerufene Hüllkurvenschwankung $A_{max} - A_{min}$. Auf die Ursache der AM i.S. des oben (S. 140) gebotenen Schemas kommt es hier nur insoweit an, wie die Frequenzlage von Spektralkomponenten pro CB am Filterausgang zu einem AM-Signal führt, das als Teilrauhigkeit in diesem Filterband ausgewertet werden kann.

Die Vpn hingegen haben die ‚Helmholtz-Sekunde' als besonders rauh eingeschätzt, obwohl bei diesem Signal bzw. Stimulus die Hüllkurvenschwankung $A_{max} - A_{min}$ geringer ist als bei beiden AM-Signalen. Mithin dürfte die größere Rauhigkeit der ‚Helmholtz-Sekunde' mit der spektralen Zusammensetzung bzw. der Reibung von Spektralkomponenten im komplexen Zusammenklang zu tun haben. Für die resultierende Empfindung sind mehrere Erklärungsansätze denkbar: erstens ist nicht auszuschließen, dass auch die beiden Grundfrequenzen der Sekunde h'-c", die hier die weitaus stärksten Komponenten des Intervalls bilden, in eine Frequenzgruppe fallen (ungeachtet der in Tab. 1 genannten Werte, die lediglich eine Orientierung, aber keine präzisen Grenzen liefern). Zweitens kann eine neurophysiologische Erklärung versucht werden, die von einer ersten Filterung auditiver Signale auf der Ebene der Basilarmembran und der sich anschließenden mechanisch-elektrischen Transduktion der Signale auf der Ebene der Haarzellen im Innenohr ausgeht. Die neuronalen Impulsmuster (gemessen in Form sog. Interspike Intervalle und deren Häufigkeiten pro Zeiteinheit) ergeben eine spezifische Repräsentation einer kleinen Sekunde (etwa a' – b') im Hörnerven von Säugern, wobei in diesem Code auch die Rauhigkeit von Signalen enthalten ist (Tramo et al. 2001). Es wurde schon früher auf Grund von Befunden aus Tierversuchen vermutet, dass Stationen der Hörbahn wie etwa der Colliculus inferior auf Grund ihrer speziellen Anatomie und funktionalen Organisation zur Analyse von AM (und z. T. auch von FM) in der Lage seien (vgl. Langner 1992, Escabi & Read 2005). Drittens sei auf eher physikalisch begründete Modelle der aktiven Verarbeitung von Schallreizen im Innenohr verwiesen, bei denen die Haarzellen als Oszillatoren betrachtet und Rauhigkeit sowie andere Phänomene bei Zusammenklängen aus der Interaktion dieser Oszillatoren hergeleitet werden (Jülicher et al. 2001).

18. Zusammenfassung

Die in dieser Studie vorliegenden Befunde aus mehreren Experimenten verdeutlichen die Problematik einer dominant im Zeitbereich angesiedelten Betrachtung der sensorischen Rauhigkeit. Diese hat offenbar eine temporale und eine spektrale Dimension, die beide in die subjektiv empfundene Rauhigkeit eines Klanges bzw. allgemein eines Signals eingehen. Es scheint, dass die heute übliche Vorgehensweise, Rauhigkeit insbesondere als Funktion mehr oder minder periodischer Hüllkurvenschwankungen und mithin als sensorisches Korrelat von AM-Signalen zu betrachten, die jeweils spezifische spektrale Zusammensetzung von Klängen zu wenig berücksichtigt bzw. zu schwach gewichtet.

Der direkt an Helmholtz anknüpfende und im Frequenzbereich angesiedelte Berechnungsansatz der Rauhigkeit nach den Gl. 6 und 7 ist gegenüber den mehrstufigen, vornehmlich im Zeitbereich operierenden Hörmodellen wesentlich einfacher, jedoch mit Blick auf die Ergebnisse durchaus robust. Dieser Ansatz bedarf freilich noch einiger Präzisierungen im Hinblick auf die Frequenz- bzw. Oktavlage gerade auch musikalisch relevanter Zusammenklänge; hierzu ist es erforderlich, zuvor die Kritischen Bandbreiten und damit die Frequenzgruppen für Zusammenklänge aus Paaren komplexer Töne (natürliche oder synthetisch hergestellte) in Hörversuchen zu ermitteln.

19. Literatur und Klangquellen

Aures, W. (1985a), Ein Berechnungsverfahren der Rauhigkeit, *Acustica* 58, 268-281.

Aures, W. (1985b), Der sensorische Wohlklang als Funktion psychoakustischer Empfindungsgrößen, *Acustica* 58, 282-290.

Aures, W. (1985c), Berechnungsverfahren für den sensorischen Wohlklang beliebiger Schallsignale, *Acustica* 59, 130-141.

Bader, Rolf (2002a), *Fraktale Dimensionen, Informationsstrukturen und Mikrorhythmik der Einschwingungvorgänge von Musikinstrumenten*, Phil. Diss. Univ. Hamburg.

Bader, Rolf (2002b), Fractal correlation dimensions und discrete pseudo-phase plots of percussion instruments in relation to cultural world view, *Ingenieras* V, no. 17, 1-11.

Bader, Rolf (2008), Efficient auditory coding of a xylophone bar, *Hamburger Jahrbuch für Musikwissenschaft* 24, 197-212.

Beauchamp, James (1995), New methods for computer analysis and synthesis of musical sounds, *32nd Czech Conference on Acoustics*, Prague, Sept. 1995, 7-16.

Bismarck, G. v. (1974), Sharpness as an attribute of the timbre of steady-state sounds, *Acustica* 30, 159-172.

Brandl, Rudolf (1992), Die "Schwebungsdiaphonie" im Epiros und verwandte Stile im Lichte der Psychoakustik, In: R. Schumacher (Hrsg.), *Von der Vielfalt musikalischer Kultur. Fs. Für Josef Kuckertz,* Anif/Salzburg, 43-79.

Daniel, Peter, R. Weber (1997), Psychoacoustical Roughness: implementation of an optimized model, *Acustica* 83, 113-123.

Dau, Torsten, Birger Kollmeier, Armin Kohlrausch, (1997a), Modelling auditory processing of amplitude modulation. I. Detection and masking with narrow-band carriers, *Journal of the Acoust. Soc. of America* 102, 2892-2905.

Dau, Torsten, Birger Kollmeier, Armin Kohlrausch (1997b), Modelling auditory processing of amplitude modulation. II. Spectral and temporal integration, *Journal of the Acoust. Soc. of America* 102, 2906-2919.

De Baene, Wouter, André Vandierendonck, Marc Leman, Andreas Widmann, Mari Tervaniemi (2004), Roughness perception in sounds: behavioral and ERP evidence. *Biological Psychology* 67, 319-330.

Escabi, Monty, Heather Read (2005), Neural Mechanisms for spectral analysis in the auditory mid-brain, thalamus, and cortex, In: M.S. Malmiereca, Dexter Irvine (eds.), *Auditory spectral processing* (=Intern. Rev. of Neurobiology, Vol. 70), Amsterdam etc.: Elsevier, 207-252.

Hartmann, William (1998), *Signals, Sound, Sensation*, New York: Springer.

Helmholtz, Hermann von (1857), Über die physiologischen Ursachen der musikalischen Harmonien (Vortrag Bonn 1857), In: H. v. Helmholtz, *Vorträge und Reden*, 4. Aufl. Braunschweig, 1896, Bd 1, 119-155.

Helmholtz, Hermann von (1863), *Die Lehre von den Tonempfindungen als physiologische Grundlage für die Theorie der Musik*, Braunschweig: Vieweg (3. Aufl. 1870, 5. Auf. 1896, 6. Aufl. 1913).

Hut, Rolf, Marinus Boone, Andreis Gisolf (2006), Cochlear Modeling as Time-Frequency Analysis tool, *Acustica* 92, 629-636.

Jacobson, Edward, J.H. van Hasselt (1907), *De Gong-fabricatie te Semarang*, Leiden: Brill.

Jülicher, Frank, Daniel Andor, Thomas Duke (2001), Physical basis of two-tone interference in hearing, *Proc. Nat. Acad. Sciences of the U.S.A.* Vol. 98, no. 16, 9080-9085.

Kameoka, Akio, Mamoru Kuriyagawa (1969a), Consonance Theory Part I: Consonance of dyads, *Journal of the Acoust. Soc. of America* 45, 1451-1459.

Kameoka, Akio, Mamoru Kuriyagawa (1969a), Consonance Theory Part II: Consonance of complex tones and its calculation method, *Journal of the Acoust. Soc. of America* 45, 1460-1469.

Keidel, Wolf-Dieter (1992), Das Phänomen des Hörens. Ein interdisziplinärer Diskurs. *Naturwisssenschaften* 79, 300-310, 347-357.

Langner, Gerald (1992), Periodicity coding in the auditory system, *Hearing Research* 60, 115-142.

Meddis, Ray, Lowel O'Mard (1997), A unitary model of pitch perception, *Journal of the Acoust. Soc. of America* 102, 1811-1820.

Meyer, Erwin, Dieter Guicking (1974), *Schwingungslehre*, Braunschweig: Vieweg.

Moore, Brian (1993), Frequency Analysis and pitch perception, In: W.A. Yost, A.N. Popper, R.R. Fay (eds.), *Human Psychophysics*, New York, Berlin: Springer, ch. 3.

Moore, Brian (1995), Frequency Analysis and Masking, In: B. Moore (ed.), *Hearing*, San Diego, New York etc.: Academic Pr., ch. 5.

Plack, Christopher, Andrew Oxenham (2005), The Psychophysics of pitch. In: C.J. Plack, A.J. Oxenham, R.R. Fay, A.N. Popper (eds.), *Pitch. Neural coding and Perception*. New York: Springer, ch. 2.

Plomp, Reinier (1976), *Aspects of tone sensation. A psychophysical study*, London etc.: Academic Pr.

Plomp, Reiner (1998), *Hoe wij horen. Over de toon die de muziek maakt*, Breukelen: Plomp (CD + 2 booklets).

Plomp. Reinier, W.J. Levelt (1965), Tonal consonance and critical bandwidth, *Journal of the Acoust. Soc. of America* 38, 548-560.

Pressnitzer, Daniel (1998), *Perception de rugosité psychoacoustique: d'un attribut élémentaire de l'audition à l'écoute musicale*, Thèse de doctorat de l'Université Paris 6.

Pressnitzer, Daniel, Stephen McAdams 1999. Two phase effects in roughness perception. *Journal of the Acoust. Soc. Of America* 105, 2773-2782.

Raim, Ethel, Martin König (1988), *A Harvest, a shephard, a bride. Village music of Bulgaria*, New York: Elektra Records (CD + booklet).

Schneider (1997a), „Verschmelzung", tonal fusion, and consonance: Carl Stumpf revisited. In: Marc Leman (ed.), *Music, Gestalt, and Computing. Studies in Cognitive and Systematic Musicology*, Berlin, New York: Springer, 117-143.

Schneider, Albrecht (1997b), *Tonhöhe – Skala – Klang. Akustische, tonometrische und psychoakustische Studien auf vergleichender Grundlage*, Bonn: Orpheus-Verlag.

Schneider, Albrecht (1998), Autoregressive spectral analysis and wavelet gammatone filtering of idiophone sounds: implications for pitch perception, In: I. Zannos (ed.), *Music and Signs. Semiotic and cognitive studies in music*, Bratislava: ASCO, 99-116.

Schneider, Albrecht (2000), Inharmonic Sounds. Implications as to ‚Pitch', ‚Timbre', and ‚Consonance', *Journal of New Music Research* 29, 275-301.

Schneider, Albrecht (2001a), Complex inharmonic sounds, perceptual ambiguity, and musical imagery In: R.I. Godøy, H. Jørgensen (eds.), *Musical Imagery*, Lisse etc.: Swets & Zeitlinger, 95-116.

Schneider, Albrecht (2001b), Virtual pitch and musical instrument acoustics: the case of idiophones, In: B. Enders, J. Stange-Elbe (eds.), *Musik im virtuellen Raum*, Osnabrück: Universitätsverlag Rasch, 397-417.

Schneider, Albrecht (2001c), Sound, pitch, and scale: from ‚tone measurements' to sonological analysis in ethnomusicology, *Ethnomusicology* 45, 489-519.

Schneider, Albrecht (2008), Foundations of Systematic Musicology: a study in history and theory, *Hamburger Jahrbuch der Musikwissenschaft* 24, 11-61.

Schneider, Albrecht, Daniel Müllensiefen (1999), Systematische und Vergleichende Musikwissenschaft in Hamburg, *Hamburger Jahrbuch für Musikwissenschaft* 16, 43-63.

Schneider, Albrecht, Marc Leman (2002), Sonological and psychoacoustic Characteristics of Carillon bells, In: *The Quality of Bells* (=Proc. of the 16th Meeting of the FWO Research Soc. on Foundations of Music Research / Brugge Eurocarillon 2002), Ghent: IPEM.

Schneider, Albrecht, Andreas Beurmann, Richard von Busch, Lüder Schmidt, Eberhard Lauer (2004), Über Beziehungen von Orgelstimmungen zu klanglichen Merkmalen, *Acta Organologica* 28, 323-346.

Schneider, Albrecht, Klaus Frieler (2009), Perception of harmonic and inharmonic sounds: results from ear models, In: R. Kronland-Martinet, S. Ystad, K. Jensen (eds.), *CMMR: Genesis of meaning in sound and music*, Berlin, New York: Springer, 11-44 (=LNCS 5493).

Schreiner, Christoph, Gerald Langner (1988), Coding of temporal patterns in the central auditory nervous system, In: G.M. Edelman, W.E. Gall, W.M. Cowan (eds.), *Auditory function: neurobiological bases of hearing*. New York, Chichester: Wiley, 337-361.

Sethares, William (1993), Local consonance and the relationship between timbre and scale, *Journal of the Acoust. Soc. of America* 94, 1218-1228.

Sethares, William (1998), *Tuning, timbre, spectrum, scale*, London: Springer.

Sottek, Roland (1993), *Modelle zur Signalverarbeitung im menschlichen Gehör*, Diss. Ing. RWTH Aachen 1993.

Sottek, Roland (1994), Gehörgerechte Rauhigkeitsberechnung, *Tagungsbericht DAGA '94*, Dresden, 1197-1200.

Stumpf, Carl (1890), *Tonpsychologie*, Bd 2, Leipzig: J. Barth.

Stumpf, Carl (1897), Neueres über Tonverschmelzung, *Zeitschrift für Psychol.* 15, 280-303.

Stumpf, Carl (1898), *Konsonanz und Dissonanz*, Leipzig.

Terhardt, Ernst (1974), On the perception of periodic sounds (roughness), *Acustica* 30, 202-213.

Terhardt, Ernst (1998), *Akustische Kommunikation*, Berlin, New York: Springer.

Tramo, Mark, Peter Cariani, Bertrand Delgutte, Louis Braida (2001), Neurobiological Foundations for the theory of harmony in western tonal music. In: R. Zatorre, I. Peretz (eds.), *The Biological Foundations of Music* (=Annals of the New York Acad. Of Sciences Vol. 930), New York: Acad. of Sciences, 92-116.

Vassilakis, Pantelis (2005), Auditory roughness as a means of musical expression, In: R. Kendall, R. Savage (eds.), *Perspectives in Systematic Musicology*, Los Angeles: Dept. of Ethnomusicology, UCLA, 119-144.

Vogel, A. (1975), Über den Zusammenhang zwischen Rauhigkeit und Modulationsgrad, *Acustica* 32, 300-306.

Voigt, Wolfgang (1985), *Dissonanz und Klangfarbe. Instrumentationsgeschichtliche und experimentelle Untersuchungen*, Bonn: Verlag für Syst. Musikwiss.

Zenner, Hans-Peter (1994), *Hören. Physiologie, Biochemie*, Stuttgart: Thieme.

Zwicker, Eberhard, Hugo Fastl (1999), *Psychoacoustics. Facts and models*. 2nd ed. Berlin, New York: Springer.

Albrecht Schneider

Forensic Music Psychology[1]

Abstract: In the present article, aspects of forensic music psychology are discussed in regard of criminal law and civil law, respectively. Some areas are indicated where music psychology might be usefully employed to examine and to clarify certain aspects relevant for legal assessments. Music psychology (as part of systematic musicology) seems suited to support the court because of its methodology that allows to investigate issues on an empirical level, and with experimental designs that can be adapted to the needs specific to certain issues. To illustrate some of the problems the musicologist or psychologist working in the field of forensic music psychology is confronted with, a recent lawsuit concerning Authors' rights, and in particular aspects of "unconscious borrowing" as well as long term memory for music is reported and discussed in some detail.

1. Introduction: Forensic music psychology: the term and its meaning

The term 'Forensic Music Psychology' has probably never been used before. It can be regarded as an extension to the term 'forensic psychology' that denotes, most of all, *the direct provision of psychological information to the courts, that is, to psychology in the courts* (Blackburn 1996, Kapardis 2003, 2). Though in practice forensic psychology most often is involved in criminal law (for example, when assessing the credibility of victims or of eyewitnesses), there is no obvious reason that would exclude psychological expertise from other legal matter. In fact, as will be discussed in regard to one recent lawsuit in the field of music, psychologists may be called in for giving an expert witnesses statement on certain aspects relevant to author's right. If one accepts that the term 'forensic psychology' must not be restricted to criminal law, but can be used in a broader sense to refer to psychology in a legal context, one may, consequently, use the term forensic music psychology to indicate the application of music psychology in legal contexts. These may relate to both criminal and civil law, and will have a focus on music psychology as applied in lawsuits. However, one should take into account that debates on legal matters are not confined to the courtroom. For example, recent debates on the impact so-called 'killer games' as well as certain music videos (see below) might have on children and adolescents amply demonstrate that the areas in which legal matters are discussed include politics, media, and public opinion. In order to clarify issues and to substantiate arguments, music psychology should take part in such debates.

[1] This article was written for a symposium on Forensic Music Psychology as part of the 7th Triennal Conference of the European Society for the Cognitive Sciences of Music (ESCOM), Jyväskylä, Finland, August 2009.

2.1 Music videos, Rappers, and Criminal Law

On March 11th, 2009, a 17-year-old male person stormed into the school he had attended until 2008 at Winnenden (close to Stuttgart, Germany). He was armed with a 9 mm pistol with which he killed 9 pupils and 3 teachers, wounding several others. Within the next two hours, he shot another 3 adults to death, wounded two police officers, and finally killed himself. On his computer, so-called „killer games“ or "shooters" (e.g., „counter-strike“) were secured. This case of a young person killing fellow pupils and teachers in his former school had been preceded by similar events in the United States, in Germany, and also in Finland. The incident provoked a public discussion as to the negative influence those "shooters" exert on young male persons. In the course of investigations following the incident, it was found that the German rapper, Kaas had produced a video called "Amok Zahltag" ("Amok pay day") in which he narrates a story that can be taken as an instruction for a mad gunman of how to run amok in a public school. In fact, there are obvious parallels between the rapper's account and the actual behaviour of the murderer. The video apparently had been available on 'youtube' for a while before it was withdrawn on March 13th. It seems possible that the offender may have watched this (or a similar) music video in which hatred, violence, and, finally, running amok with a gun are the central topic. The question that was raised after the incident of course was if a music video of the type Rapper Kaas had produced can "trigger" the action of the young murderer as it occurred. The public prosecutor's office of Tübingen in fact started preliminary proceedings against Rapper Kaas and his "Amok Zahltag", which were based on § 131 StGB (glorification of violence; German: Gewaltverherrlichung)[2]. According to the news, the charge was dismissed by beginning of August, 2009, on grounds that glorification of violence could not be proved beyond doubt (sic). Consequently, the content of "Amok Zahltag", though certainly questionable from a moral as well as from an aesthetic point of view, would have to be tolerated as an "expression of art" (that is protected as such under article 5 of the German constitution). Certainly, "Amok Zahltag" and similar products provoke public discussion that seems necessary in assessing possible effects of "killer" videos on adolescents. It is quite clear that, in order to decide a complex and controversial issue such as that sketched here with appropriate care, the judges of the criminal court would call for an expert who would have to be familiar with music videos of the kind Kaas and other rappers (e.g., Swiss, Gzuz 187) had produced in case these rappers would indeed be charged.

2.2 How does suicide relate to heavy metal genres?

The issue whether a CD or music video and also live performances communicating feelings of hatred and violence can stimulate persons to commit crimes that include bodily harm in Germany has been raised in regard of musical activities of the NeoNazi scene (e.g., Dornbusch & Raabe 2002). Also, the violent actions of young people encountered in the *banlieues* of French cities have been related to messages of rappers (e.g., Tödt 2007). Already since the 1980s and 1990s, both heavy metal and rap genres

---

[2] Prosecution based on § 26 StGB (Instigation; German: Anstiftung) would require that the rapper acts *intentionally* so as to bring someone to commit a crime (again, intentionally). To prove intention with respect to an offence of a suspect or accused always is a delicate juridical matter.

(such as Death or Black metal and 'gangsta' rap) have come under scrutiny. A number of empirical studies found that adolescents who had a strong preference for heavy metal and rap genres apparently tended to prefer a more risky life style (including increased usage of drugs and alcohol, higher rates of delinquency, reckless or even violent behaviour, etc.) while their performance in school or high school seemingly was below average (Took & Weiss 1994 and literature reviewed in Scheel & Westefeld 1999, Isaacs 2008). Also, because of an increase of adolescent suicide in the U.S. a hypothetical relation between preferences for rock genres such as Death Metal or Black Metal and suicidality had been put forward. According to an investigation covering 50 states of the U.S., a relation was established statistically such that "the greater the strength of the metal subculture [in a certain area or region], the higher the youth suicide rate" (Stack et al. 1994). An in-depth study based on a rather small sample (n=121) of adolescents (students of a highschool in the Midwest) confirmed the positive correlation of preference for heavy metal and above average suicidal risk (Scheel & Westefeld 1999). These findings, though, cannot simply be taken as proving a cause and effect relationship in the sense that regular exposure to rap or heavy metal idioms will produce delinquent or troubled teenagers. Rather, adolescents who have already a history of family and school problems resulting in maladjustment apparently are attracted to such genres that "fit" and probably also reinforce their mood. There seems to be a risk that problematic messages contained in such genres in regard of violence (e.g., against women), suicide, etc. may amplify feelings and attitudes of adolescents who exhibit unstable and deviant behaviour.

Apart from the effects rap and heavy metal idioms may have on adolescents, or at least on certain groups of teenagers, a survey of research showed that, during the 1990s, increasing exposure of children and adolescents to media resulted in increased violent and aggressive behaviour, increase in substance and alcohol use as well as accelerated onset of sexual activity (cf. Villani 2001).

Taking such studies as well as a broad range of media reports on rock music and riotous or violent as well as suicidal behaviour, drug abuse and other symptoms of personal and social maladaption into account, one author has blamed rock and pop in general to further the development and actual use of violence among, in particular young people as well as decreasing sensitivity for violence and criminal behaviour (Miehling 2006). Whether or not such claims are correct (and if so, to which extent) needs careful empirical investigation and assessment in which music psychology should play an active part.

3.1 Some Fundamentals of Authors' Rights and Copyright Law

The past two decades saw an ever increasing number of so-called 'cover versions' in pop and rock music as well as a trend towards 'hit-recycling' (Pendzich 2004). Because preexisting material is used extensively, infringement of copyright is found frequently in music. While some infringements appear as outright plagiarism, the relation of 'source' and 'copy' in other cases is more difficult to determine. In legal cases where the 'similarity' of songs or instrumental pieces is at issue, a musicologist or other music specialist will be asked by the plaintiff, by the defendant, or by the court to serve as expert witness. Investigating whether or not 'similarity' exists between two or more songs is one task he or she must carry out. Another evolves from the question: What are the reasons that explain such 'similarity' in a given case? (see section 4.2). It is this question that extends well into the area of music psychology.

In this section devoted to civil law, I will briefly outline some fundamentals of Authors' rights and Copyright law, respectively. In particular, the notion of a 'work' (or 'original work') as distinguished from an 'adaptation' as well as a 'derivative work' will be specified. In addition, the concept of 'free use' will be discussed. In the subsequent chapter, I will turn to some copyright cases that involved music psychology in one way or another. I will refer to German law mainly because it offers a true *ius auctoris* (droit d'auteur, Authors' Rights; German: *Urheberrecht*) featuring the rights of creative individuals (authors, composers, painters, etc.). Authors' rights as part of civil law is found frequently in the legal systems of continental Europe whereas Copyright Law as part of the common law tradition has been established in Great Britain and the United States, respectively.

Authors' rights (as codified in the German *Urheberrechtsgesetz*; abbreviated: UrhG[3]), as its name implies, first of all protects the moral rights (droit moral) of an author, or of several authors working as a team. The UrhG includes regulations pertaining to contractual and economic aspects since the author shall benefit from his or her work. According to the German constitution (Art. 14 GG), and the jurisdiction developed in regard to Authors' rights, intellectual property is of the same rank as is material property (BVerfGE. 31, 229). Copyright Law differed historically from Authors' rights in that it had a focus on the 'copyright holder' rather than on the author or creator. During the past decades, the gap between Authors' rights and Copyright Law has been narrowed because the U.S. Copyright Act[4] had been adapted, in 1976, to international conventions that intend to protect authors and their works (e.g., the Berne Convention for the protection of literary and artistic works).

Both Authors' Right and Copyright Law protect the works authors create. German law as codified in the UrhG and similar laws of Austria and Switzerland relates to original works of literature, music, art, etc. Original works are such that can be considered "personal intellectual creations" (*…persönliche geistige Schöpfungen*, § 2 Abs. 2 UrhG) of an author. Likewise, adaptations and (re)arrangements of original works will be protected in case these are "personal intellectual creations" (§ 3 UrhG) of an author. There is one decisive restriction, though, since such adaptations can be published and marketed only with the prior consent of the original author (§ 23 UrhG) or of the holder of the respective rights (e.g., the heirs at law of an author).

Notwithstanding § 3 UrhG demands "personal intellectual creation" for an adaptation or (re)arrangement of a preexisting work to be protected, the German High court (BGH) did not request a significant level to which (re)arrangement has to be executed before protection is granted. A simple arrangement of a folk song ("Schwarzbraun ist die Haselnuss") for a brass band and male choir was considered sufficient for obtaining Authors' rights as an adaptation (BGH GRUR 1968, 321, 324 "Haselnuß"). In a more recent ruling (BGH GRUR 1991, 531, 532 "Brown Girl I"; BGH GRUR 1991, 533, 534 "Brown Girl II"), the BGH had set the standard even lower as a very faint rearrangement

---

[3] The UrhG, in its present shape, is in effect since 1965. There have been many amendments since, the last dating from December 17th, 2008.

[4] Copyright Act (Pub. L. 94-553, 19th of October, 1976, in effect since 1st of January, 1978).

of a traditional song known from the Caribbean was accepted as an adaptation that conforms to the requirements defined in § 3 UrhG. That (re)arrangement known as "Brown Girl" had been published (on LP) in 1975. Consequently, a very similar or almost identical version of "Brown Girl in the ring" that had been produced and issued by the German pop group, Boney M., in 1978, was found to infringe the rights of the arrangement that had been published before. The level of "intellectual creation" (or at least of manual effort) that could be attested to either the arrangement of the plaintiff, or to the imitation of that arrangement produced by the defendant, viewed in objective terms is close to zero. Therefore, it is more than problematic to treat such "arrangements" of a traditional tune as on a par with an original work (see Schneider 1992). Meanwhile, the German UrhG has been changed insofar § 3 no longer protects superficial arrangements of folk or traditional songs.

Since the version published by Boney M. was practically identical with another arrangement of "Brown Girl" published earlier, it would qualify legally as a so-called 'cover-version'. This term is not found in the UrhG, but is used frequently among musicians, producers, music publishers, and record companies to denote that an artist has produced his or her own version of a preexisting song that has been recorded by another artist, and has been published before. For example, both the Beatles and the Rolling Stones in their early careers recorded cover versions of songs of their idols (Chuck Berry, Bo Diddley, Carl Perkins, etc.). The BGH ruled that a cover-version is a new recording of an existing version that preserves the characteristics of that version. Hence, a cover-version is not an adaptation (in the sense of § 3 UrhG) but only a [mechanical, or otherwise technical] reproduction or copy (in the sense of § 16 UrhG: *Vervielfältigungsrecht*, the right to manufacture copies of a work) of a preexisting work (BGH GRUR 1998, 376, 378 "Coverversion"). A cover-version can be recorded and distributed with prior consent of the producer and/or record company (who own the so-called ancillary copyright; German: *Leistungsschutzrecht*, subsisting in the original recording). If permission is denied, any producer of phonograms can enforce a compulsory licence (§ 61 UrhG: *Zwangslizenz*) that entitles him to produce and publish a cover-version of a preexisting recording.

No prior consent with respect to publishing and economic utilization is needed for an individual work that has been created by making so-called "free use" (*freie Benutzung*) of another original work taken as an inspiration (§ 24 Abs. 1 UrhG). For example, Max Reger's Four Symphonic Poems op. 128 that the composer had conceived after paintings by Arnold Boecklin[5] are independent musical works inspired by the respective paintings that are works of art. Another example of "free use" is Arnold Schönberg's *Verklärte Nacht*, op. 4 (1902) that, as an instrumental piece, refers to a poem of same title of Richard Dehmel. Neither Reger nor Schönberg would have needed the consent of Böcklin or Dehmel, respectively, to publish their individual creations that resulted in original works.

According to § 24 Abs. 2 UrhG, "free use" excludes works of music insofar as it is not allowed to take a tune (*Melodie*) from a preexisting work as a basis for another such work. This clause that seems peculiar to the German UrhG is called *Melodienschutz* (melody or tune protection). The specific regulation as defined in § 24 Abs. 2 UrhG may appear somewhat strange in the light of a long tradition of "musical borrowing" that

5 Max Reger. Vier Tondichtungen op. 128 (1913) für großes Orchester (nach Arnold Böcklin).

ranges from the early Renaissance (cf. Meconi 2004) to modern times (e.g., Charles Ives; see Fenner 2005). As is well known, there are many works in European music history where composers have taken a tune or a theme from another composer's work to create new and original works of their own. One might point to, e.g., Max Reger's *Variationen und Fuge über ein Thema von J.S. Bach*, op. 81 (1904), or to Ralph Vaughan Williams' *Fantasia on a theme by Thomas Tallis* (1909/10). Adaptation of themes to works of music such as mentioned here at first glance could be viewed as copyright infringement in regard to the very strict and narrow formulation of § 24 Abs. 2 UrhG (for a critical assessment of the *Melodienschutz*, see Nordemann, Vinck, Hertin 1994, § 24, annotations no. 12-17). The historical and legislative background to the German *Melodienschutz* (cf. Fuchs 1983, 241-245; Schlingloff 1990, 92-102; Dieth 2000, 162-185) indicates that there was one essential aspect from a legal point of view that had been acknowledged, and there were also economic considerations that came into play. The legal argument, apparently first developed in France, was that if a work of art, literature, or music is protected as a whole, so must all its parts. Consequently, if for example a violin or piano concerto is based on a distinctive theme (that, typically, occurs many times in its original form or with small variations in such a concerto), this theme deserves to be protected as an individual artistic creation.

Economic reasons brought forward in favour of a strict *Melodienschutz* in Germany pointed to a possibility that, unless tight protection was installed, everybody could exploit melodies well-known from operettas, etc. by including them into a *Potpourri*, a most popular genre around 1900. Also, one has to remember that, as the history of "light music" amply demonstrates, composers could (and still can) make a fortune by inventing catchy melodies. Composers of operettas, *Salonmusik,* as well as of songs and of *Schlager* had an interest (in line with their publishers) to see their valuables – melodies that first achieve popularity, and then become 'evergreens' – protected against whatever plagiarism or unauthorized use there might be. The *Melodienschutz* came into effect in 1901/02 (§ 13 Abs. 2 LUG[6]), and was maintained in the UrhG of 1965 though by this time there had been proposals to abandon this special regulation. Objections against a strict *Melodienschutz* said that it would hamper musical composition, in particular of 'light music', and would prevent musical parody altogether. Further, it was argued that a special *Melodienschutz* was not necessary since the UrhG not only protects works in total but also their constitutive parts, which implies that melodies will be protected anyway (cf. Nordemann, Vinck, Hertin 1994, § 2, nos. 27, 46).

Leaving aside economical considerations for the moment, one reason suited to justify a specific *Melodienschutz* lies in the Gestalt properties of each distinct melody that will be recognized and remembered not just as a "part" of a musical piece but as a characteristic or even unique *Gebilde* (formation) integrating elements into one coherent pitch-time-structure.

This of course was the point of view elaborated by Christian von Ehrenfels (1890) and other Gestalt psychologists (e.g., Koffka 1936, 431-438). In brief, a melody can be regarded as a temporal formation comprising a sequence of tones organized into intervals whereby each tone of the melody represents a certain pitch relative to a scale. Further, in regard to time, tones can be defined by their onset time t_k (relative to the be-

[6] LUG = Gesetz betreffend das Urheberrecht an Werken der Literatur und der Tonkunst (1901); cf. Fuchs 1983, 239-248.

ginning of the melody, t_0) and their duration (in ms), respectively. In addition, a melody typically comprises a pattern of metrical accents (and possibly additional accents imposed by interval step, by the lyrics sung to a certain melody, or by other factors). A melody thus appears as a distinct contour in a pitch-time-space according to interval and time relations. Taking into account its accent structure, one may model also the dynamics of a melody (resulting in a three-dimensional representation). Since a contour can undergo a number of geometrical transformations (e.g., symmetry operations), it is possible to "translate" a given melody into a number of variants or derivatives. It has been demonstrated in many experiments that subjects recognize melodies under transposition (cf. Krumhansl 1997). Because of such properties, melody has been the prototype of a temporal Gestalt (as well as a paradigm for Gestalt psychology; see Wellek 1963). To be sure, protection of distinct rhythms or beat patterns is also possible due to Gestalt criteria (see Reinfeld 2001).

Summing up essentials of German Urheberrecht, it protects

a. original works created by individuals;
b. translations of literary works, adaptations and similar products if such are equivalent to original works; both (a) and (b) require that the works to be protected result from "individual intellectual creation". For an adaptation it is necessary that it is sufficiently independent of the original work that serves as a 'master' or as a model. Thereby, features of the original may be preserved somehow yet the adaptation must exhibit characteristic features of its own so that it becomes an original work.

In addition to original works and adaptations, a broad range of works seems possible that stem from "free use" of preexisting works. "Free use" may include all kinds of material not protected by the UrhG (e.g., the history of mankind as a source for literary and dramatic works, nature as an inspiration for works of art, etc.), however, in music tunes and themes fall under the restriction of the *Melodienschutz*, to the effect that one can not "borrow" a tune (if still protected as part of an original work, a condition that does not apply to many folk and vernacular songs).

Taking a look at basic concepts of U.S. Copyright law, § 102 of the U.S. Copyright Act (= U.S.C. 17 § 102) states that legal protection "subsists … in original works of authorship", such as literary or musical works. Copyright subsists also in motion pictures and other audiovisual works as well as in sound recordings. The Copyright Act in § 103 refers to compilations and derivative works as being included into copyright as specified in section 102. Derivative works "employing preexisting material in which copyright subsists" in principle will be protected. However, protection "does not extend to any part of the work in which such material has been used unlawfully". Since preexisting material is equally available to various authors of derivative works, it follows that "copyright in a compilation or derivate work extends only to the material contributed by the author of such work" (§ 103 b CA). In this respect, American Copyright law seems to differ from European codifications. The idea behind § 3 UrhG is that an adaptation, in order to be protected by Authors' rights, must reach such a level of creation that a new and coherent work emerges (though it may still be "derivative" to some extent in regard to the original). The European concept of 'adaptation' appears to be more holistic, whereas the American concept of a 'derivative work' seems to be more additive in nature (as the term 'compilation' used in § 103 a CA implies):

adaptation → creation of a new work based to some extent on concepts and material of a preexisting work; adaptation is regarded equivalent to an original work if the level and extent of creation are sufficient to justify such

derivative work → preexisting material plus material contributed by the author of the derivative work combine into a partly known, partly new texture that in total makes up the derivative work.

It should be added that the concept of 'Fair use' as defined in U.S.C. 17 § 107 is quite different from the concept of 'free use' explained above. 'Fair use' in American copyright[7] refers to the use of copyrighted material "for purposes such as criticism, comment, news reporting, teaching ..., scholarship, or research". This regulation thereby corresponds to regulations contained in chapter 6 of the German UrhG (Limits of authors' rights; §§ 44a – 63a) where, in § 51, the right of citation is stated. The portion of a protected work that can be used in a citation without permission from the author of the protected work is dependent on the purpose of the citation. Typically, one finds quotations from copyrighted works in scholarly works, in book reviews, etc. Such quotations will always be small relative to the length of the work from which the citation is taken. The same holds true in regard to works of music that have been published as sheet music, and from which excerpts might be taken to be included in a musicological research paper or book. According to § 51 no. 3 UrhG it is also possible to use a small number of delimited portions of a work of music that has been published within another 'independent work of music' created later. In effect, citation in music thus is allowed notwithstanding the *Melodienschutz* (see above).

3.2 Plagiarism, "unconscious borrowing", and "double creation" in music

Musicologists familiar with the repertoire of a certain era of western art music (e.g., music of the Renaissance, the Baroque, or the 'classical' era) usually know a considerable number of correspondences between pairs of individual works. Such correspondences often relate to themes and motives yet may also comprise harmonic schemes or other features relevant to musical structure. While in certain cases it is known that a given correspondence depends on borrowing that, to be sure, often was done to pay homage to a teacher, or to one of the "musical heroes" of the past, in other the cause for more or less obvious parallels that rest in certain elements and features is difficult to assess. One explanation could be to assume that a younger composer somehow knew about the work of an old master, and adapted some theme or chord progression he found of interest to his own composition. If parallels occur between two works composed at about the same time one might think of "double creation". This concept implies that the

[7] See information contained in http://en.wikipedia.org/wiki/Fair_use and links to additional resources offered there.

creation of two or more works, or of considerable parts thereof, has been achieved independently. Still another possibility would be that two composers who start to elaborate on a concept or idea acknowledged by both as challenging, create works that are quite similar, and in parts even identical. For example, one might point to certain works composed for the keyboard by the Dutch master Jan Pieterszoon Sweelinck, and the English master John Bull who had emigrated from England to live in Antwerp[8], It is probable that the two composers met in person, and so could have exchanged ideas or even actual works.

The notion of plagiarism in a strict sense implies intentional use of a work that has been created by another author, including pretension of authorship by the user who acts unlawfully, and in effect commits "intellectual theft". Pretension of authorship in a pre-existing work perhaps is more rare than the number of cases where such works, or major parts thereof, have been copied or closely imitated. Though it is often quite obvious that a certain idea pertaining to the compositional structure, instrumentation or studio production of one work occurs in another work published later, it will, in general, be difficult to prove in court that unlawful use of a work in which copyright subsists has been done on purpose.

According to German law, liability in regard to copyright infringements does not depend on intention on the side of the infringer. The UrhG includes regulations pertaining to civil law as well as such based in criminal law (§§ 97-105, 106-111, respectively); while liability with respect to civil law follows from negligence (and more clearly so if there is evidence for culpable negligence), persecution based on criminal law in general requires direct intent or eventual intent (*dolus directus* or *dolus eventualis*), respectively.

With respect to pop and rock music, most lawsuits are in the civil law domain. Plaintiffs typically plead in court that the defendant has violated Authors' rights in making unlawful use of a work protected by the UrhG. Such unlawful use most often means that parts of a song or instrumental have been incorporated into another piece of music that is regarded, in this respect, as an adaptation or rearrangement[9]. According to §§ 23, 24 UrhG, publication and marketing of adaptations and rearrangements require prior consent of the author (or the Rights owner) of the original work. If such consent is missing, infringement of Authors' rights is assumed because significant 'similarity' between two works (or parts thereof) is regarded quite unlikely given that music can be created in very different ways. An assumption reiterated in jurisdiction and comments even is that musical creation could result in an "infinite" number of different works. The idea behind "infinite musical creation" basically seems to be that one could form, out of a scale of twelve pitches as available in "western" music, a huge number of tone rows that could serve as the basis for whatever musical texture. Theoretically, the number of

---

[8] One major idea that had interested both Sweelinck and Bull, and also other composers of the time (e.g., W. Byrd, P. Philips), was how far chromaticism could be explored within the compass of meantone temperament.

[9] Details of plagiarism, adaptation, arrangement, free use, and quotation are discussed, from a legal point of view, in Nordemann/Vinck/Hertin 1994, 1998 as well as in the doctoral dissertations of Schlingloff 1990, Jörger 1992, Dieth 2000, Riekert 2003, Schunke 2008. As to legal and illegal cover-versions under the U.S. Copyright Act, see Pendzich 2004, ch. 2.

tone rows that can be derived from a set of 12 pitches would be n! (n factorial). For n = 12, 479.001.600 tone rows exist, of which 469.015.680 however are transformations, leaving 9.985.920 unique twelve tone rows (Whittall 2008, 24). This number at times has been taken as a proof that twelve tone and serial composition would be "inexhaustible".

The assumption of "infinite possibilities" that might relate to calculations such as mentioned, does not really hold in regard to tonal music based on relations of tones, distinct intervals, motives, melodic contours, chords and other sonorities, metrical and rhythmic patterns that, in total, result in meaningful structures (see Cogan & Escot 1976, Albersheim 1980). There are various constraints imposed on music many of which stem from perception and cognition. Due to a multitude of empirical observations, perception of musical textures follows rules many of which are Gestalt-like (Bregman 1990). It is telling that works of new music written in twelve-tone or serial technique, respectively, often will be perceived according to quasi-tonal patterns rather than in line with the rules of the systems they are believed to represent (for examples, see Warner 1969, Lerdahl 1988).

The number of musically *meaningful* formations (*Gebilde*) in western tonal music certainly is very large yet not "infinite". The range of formations that are "acceptable" within a certain genre or style depends, most of all, on characteristic features that experts as well as ordinary listeners expect to occur. Of course there are always works in which a composer transgresses a given rule, or in which some novelty occurs in regard to musical texture or form. One should, though, not underestimate the amount of "typical" melodic phrases and harmony patterns that are found in western tonal music (and in particular in Baroque as well as in classical music up to ca. 1800).

With respect to pop and rock music styles, there are "typical" metrical and rhythmical structures as well as chord progressions and melodic phrases. Staying within the bounds of musical and stylistic typicality is one condition necessary to ensure that a song or instrumental may become popular. If a song or instrumental is too far off the mainstream, it will hardly meet the expectations of the buying public. To be of interest to consumers, though, pop and rock productions must also include some features or ingredients that appear to be "new", or "surprising" in some way. Therefore, proven strategy for selling pop and (mainstream) rock rests on a concept according to which a marketable production combines known features "typical" of a certain genre and/or style with a modest amount of novelty in regard to textures and sound. Also, some gags in the production (unusual sound samples, moving sound sources in the mix, etc.) or comical phrases in the lyrics (see, e.g., "The Bad Touch" by The Bloodhound Gang) will help to arouse public interest, and possibly to increase sales.

One may argue that, in regard to melodic and chord structures, meter, rhythm, agogics and dynamics, tempo, instrumentation and overall sound balancing, etc., the number of meaningful formations possible in tonal music (including pop and rock genres) is still very large. The reason is that, with a range of variables (e.g., tempo) and parameters (e.g., tonality) available to composers and producers of music, one can form a huge number of musical formations resulting from combinations of parameters and variables as well as from individual settings. Viewed statistically, the likelihood for two works to have identical combinations of m parameters and settings on n variables, will be very small[10]. While 'identity' thereby appears a rare case, there is a certain probability that formations embedded in musical works will exhibit a degree of 'similarity' (which is a scalable property). In fact, the similarity of pairs of tunes can be measured in an objective way by means of special computer algorithms, and the results thus obtained can be checked against data from behavioural experiments (see Müllensiefen & Frieler 2004, Frieler 2008). Comparison of tunes based on algorithms no doubt is of relevance for determining whether or not similarity between melodies may indicate plagiarism (see Müllensiefen & Pendzich 2009).

Plagiarism, in a strict sense, requires intent (see above). Since such intent is difficult to demonstrate in a lawsuit (where the plaintiff must furnish evidence), a high degree of similarity found between two works, or certain parts thereof, is often interpreted so that "unconscious borrowing" (German: *unbewusste Entlehnung*) is assumed. This is kind of a "neutral" expression insofar it does not exempt the defendant from liability, on the one hand, as the fact of "borrowing" alone can be a violation of §§ 23, 24 UrhG. On the other, it does not attest a deliberate act of infringement of rights. Implicitly, some degree of negligence might be assumed, depending on how "unconscious" borrowing actually was. Negligence would be slight if we accept one view according to which, in particular in music, a melody or some chord progression one has heard previously "sinks down to the half-light of unconsciousness, from where it might resurface, on occasion, as [his or her] alleged own intuition" (Hertin 1989, 160).

The psychological process alluded to here has been labelled *cryptomnesia* (meaning "hidden memory", *verborgene Erinnerung*), and has been studied by the psychiatrist and psychologist Carl Gustav Jung (1902, 1905 [1966]). Since cryptomnesia has been introduced, by making reference to Jung, to legal commentaries to account for phenomena of "unconscious borrowing" (see Vinck in Nordemann, Vinck, Hertin 1994, § 24 UrhG, Anhang, Rz. 7), it seems apt to elaborate a bit on Jung's concept. Jung (1905) was of the opinion that every sensation leads to a "memory trace" (*Gedächtnisspur*) which, ultimately, corresponds to some neurophysiological representation in the brain. As our brain contains millions of such memory traces, one can not be aware of them all at a given time. Rather, many if not most of these traces left from sensations (*Eindrücke*), soon sink into the subconscious. From a Jungian perspective (1902, 110ff. = 1966, 90ff., 1905), this means that such traces are only loosely connected with what he calls the *Ich-Komplex* (basically, there are several layers of consciousness and self-consciousness which combine into the *Ich-Komplex*).

Jung (1905) held, in line with many psychologists of the time, that there are several ways to access memory, namely actively by our capacity for remembering, and also when a certain sensation from outside "triggers" memory so that a certain memory trace is activated, and a certain content of the memory reappears. Depending on the kind of sensation, and the way it is processed, the subject will experience either a direct or a more indirect recollection.

---

10 In regard to contemporary („casual") art of the 1960s and 1970s, ‚statistical singularity' has been proposed as a criterion sufficient for the protection of a work under Authors' rights, by jurists (see Gierth 1974). This view emphasised that certain objects can be statistically singular with regard to certain features without being original in the sense of a ‚work'.

An addition to direct and indirect recollection triggered by sensations, cryptomnesia by Jung is viewed as a third type of memory access where the trigger is psychic and intra-personal instead of coming from outside. Cryptomnesia means that a memory trace that had been unavailable for some time is activated by, for example, strong emotions or other psychic forces, to the effect that an image appears in consciousness without necessarily being identified as an image hitherto preserved in memory[11]. According to Jung (1966, 90) it is characteristic of cryptomnesia that the re-emerging image does not show features of an image from memory (*Gedächtnisbild*) since it is barely connected with the self of the person. The reason why an image is so weakly recognized may be that the subject was inattentive at the moment he or she sensed the object (which nonetheless was stored in memory), or that the object contained in the image was of little if any interest at the time a person sensed it.

Because of the weak or almost lacking associations between the object stored as a *Gedächtnisspur* in memory, on the one hand, and the self (the *Ich-Komplex*), on the other, an image may come "with surprise" to a subject that is largely unable to clarify, at that moment, from where the image may have originated.

It is the unclear origin of images "resurfacing" from the unconscious spheres of memory that makes subjects believe they have a sudden notion or idea (*Einfall*). As Jung (1966, 104) states: *Das Wort 'Einfall' drückt das anscheinend Zufällige und Grundlose dieser Erscheinung aus*. (The term 'idea' expresses the casual and causeless nature of such a phenomenon.) Since ideas that come to one's mind via cryptomnesia (that is, involuntary but with a certain "automatism" as Jung emphasizes), they are difficult to pinpoint. This is the reason why a writer or composer may believe that the image he or she suddenly becomes aware of is his or her creative idea (whereas in fact it may be the *Gedächtnisspur* of something perceived long ago). In effect, cryptomnesia thus may mislead artists, writers, and composers who may be accused of plagiarism, yet who are the "victims" of their sub-consciousness that feeds them with images of things they have sensed more or less incidentally.

Because of this casual information pick-up (to express it in Gibsonian terminology) and the automated storage process of information in long term memory (LTM), we all have a broad range of material at our disposal. Material such as colours or shapes are likely to become available, in the course of creative work, to a painter because cryptomnesia happens to retrieve relevant information stored in LTM "automatically". Hence, there is a constant flow of images, some of which are directly connected by their content to the task under way while others may be remote, or even irrelevant. The artist or composer can make use of any of the material, and can combine small elements to form building blocks for his work. Jung (1966, 109) argues that artists or writers typically create works by recombining known elements rather than by inventing new ones.

[11] Jung's view is not so much different from modern concepts of (long term) memory. From a connectionist perspective, it has been argued that long-term memories might be stored permanently in so-called associations of neurons. According to this view, we do not "forget" the content of the memory, but rather are unable to activate the recall path (cue) to that memory (cf. Snyder 2000, ch. 6). It is likely that a "hidden memory" can be retrieved if at least part of its original context is reconstructed after actually perceiving or imagining certain features relevant for that context. This may "trigger" the reconstruction process, to the effect that the respective memory crops up after cues have been restored.

So we now have the picture of the ("innocent") artist or writer, driven by the inflow of information that had been long hidden in LTM, putting together rather than creating works. In fact, this is what some artists and also composers (lat. componere = to put together) do[12].

Jung at this point makes some reservations, though, that are of importance with respect to artistic creation vs. plagiarism. He (1902, 110 = 1966, 90) argues that any information an artist, writer or composer considers important, will remain connected to his or her consciousness because he or she will reflect the content of this information from time to time. Consequently, such information cannot fall into oblivion, or into the "half-shade of unconsciousness". Therefore, if a work produced by an artist, writer, or composer contains one or several striking or otherwise remarkable ideas, one can almost always exclude cryptomnesia as a source. Such ideas must then either be genuine (i.e., belonging to the artist, writer, or composer who includes them into his or her work), or will give rise to questions concerning authorship. If a non-trivial idea known from a preexisting work is reproduced in a later one, Jung states, there is reasonable suspicion that we are dealing with "more or less conscious plagiarism". With this qualification, Jung's point of view is more precise than the summary assertion of "unconscious borrowing" that, it seems, in many cases is as difficult to prove as it is to refute (see BGH GRUR 1971, 266, 268 "Magdalenenarie"; cf. Pimat 2002).

As a prominent case of "unconscious borrowing" in music, George Harrison's song "My sweet Lord" (1971) often is cited, which in fact is highly similar in melodic and chord structure as well as in regard to other features to most of the Chiffons' "He's so fine" (composed by Ronald Mack) that was recorded in 1962, and was a top hit in the U.S. charts in 1963. The single went to no. 12 in the British charts on July 1st, 1963. Harrison who admittedly knew "He's so fine" was accused of plagiarism, and had to pay a huge sum to the copyright holders of "He's so fine" (Bright Tunes Music Corp.) for damages they claimed.

Whether or not "borrowing" is done unintentional, or even "unconsciously" may make a difference with respect to liability as well as public opinion. Therefore, the issue needs to be investigated by the court (with the aid of expert witnesses) in case a defendant's allegation points to either "unconscious borrowing" or to (unintentional, incidental) "double creation" (unbeabsichtigte bzw. zufällige *Doppelschöpfung*). German jurisdiction in regard to these issues is somewhat uneven due to different circumstances that were of relevance in individual cases. On the one hand, jurisdiction of some appellate courts and the BGH has acknowledged that both phenomena may possibly occur, in particular in genres of "light music" (e.g., operetta, Schlager, popsong) where formulaic melodic elements, harmonic schemata, and rhythmic patterns common to many songs abound (BGH GRUR 1971, 266, 268 "Magdalenenarie"; see also BGH GRUR 1988, 810, 811 "Fantasy" and the ruling of the Kammergericht Berlin as appellate court discussed in BGH GRUR 1988, 812, 813f. "Ein bißchen Frieden"). On the other, the BGH insists that numerous possibilities for the creation of individual works exist in music so that it would be extremely unlikely to find substantial convergences between

[12] It is not possible to discuss the notion of ‚creation' and ‚creative process', at this point; see, however, Finke et al. 1992, ch. 2.

two works that indeed result from individual creation (BGH GRUR 1988, 814f. "Ein bißchen Frieden"). This point of view seems plausible if applied to music that has been composed with the intent to create individual works that are unique in some respect, and identifiable as such. In regard to the prevalence of formulaic or even stereotyped elements found in Schlager and mainstream pop, the perspective maintained by the BGH seems unconvincing because it is at odds with facts. Similarly disputable are the rulings of the BGH according to which even the most elementary melodic motives and phrases feasible in a popsong still are treated as if constituting a "personal intellectual creation", and hence are protected as 'works' that merit "Melodienschutz", etc. (cf. BGH GRUR 1981, 267 "Dirlada", BGH GRUR 1991, 531 "Brown Girl I", BGH GRUR 1991, 533 "Brown Girl II").

3.3 Changing values, changing concepts in jurisdiction

A number of rulings of district and appellate courts as well as the BGH from the 1980 and 1990 in regard to such notions as 'work', 'adaptation', 'plagiarism', 'melody' (as relevant for the *Melodienschutz*), in retrospect seem problematic because they reveal, first of all, that the terminology and criteria used for assessment of actual pop productions predominantly stem from aesthetics and music theory of the 19th century. The notion of a 'work' conceived as a "personal intellectual creation" relates to formations that are unique, individual, full-grown, elaborate, and complete. With respect to western art music it was expected of a work "that it offers something new and singular" (Seidel 1987, 39). These criteria were met by works of music that are usually regarded as *opera* (one aspect of the singularity of a work is that an opus number is assigned to it). Of course, there were huge quantities of compositions still in the 19th century which were less demanding, and more modest in regard to musical craft. We know of so many bandmasters who have written mood pieces and similar works, not to forget the genre of *Salonmusik*, etc. Notwithstanding compositional and aesthetic differences that are evident between works acknowledged as opera (or even as *Meisterwerke*), and less inspirational counterparts fabricated by bandmasters, it can be said that both participated, at least in principle, in the idea of the musical and aesthetic work as sketched here.

With respect to craftsmanship, concepts and results the differences between masterworks and bandmaster's pieces are certainly significant, however, they still appear gradual rather than fundamental when compared to much of the Schlager and pop production of the 20th century.

The problems arising from attempts to apply the notion of a 'work' in music to a ditty are very obvious in jurisdiction where even the BGH puzzled about the question if a few trivial notes are enough to constitute a "personal intellectual creation" (see BGH GRUR 1981, 267 "Dirlada", BGH GRUR 1991, 531 "Brown Girl I", BGH GRUR 1991, 533 "Brown Girl II").

The same problems encountered in defining a 'work' as a whole are observed in regard to the notion of 'melody'. A melody can be defined, quite stringently, from a musicological and psychological perspective (see above, section 3.1). Different from opinions expressed in some of the literature on Authors' rights in Germany that have claimed a musicological definition of 'melody' would be lacking, there are such definitions at hand (if one takes the effort to read relevant publications in musicology). For example, one definition, short enough to be used in a legal context, reads: "A melody consists of a number of notes of different pitch arranged so as to constitute a unit as regards (external) form and movement and according to the laws of the functional relations existing between them" (Smits van Waesberghe 1955, 11). By external form, the author understands "the structure of a melody in its various parts, such as arsis and thesis, opening and closing sections, question and answer, and the melodic relationship between the parts". The term movement refers the type of rhythm (free or bound to measure), or to measure alone.

Whereas 'melody' means a musical Gestalt in a more technical sense, the term 'theme' (*Thema*) has been viewed also with respect to aesthetics. Hanslick (1854, 99) held that the theme in musical composition is a self-contained musical idea that cannot be subdivided in regard to aesthetics. Given these qualifications, simple motives of three or four notes as prevalent in many popsongs would hardly have to be reckoned as a 'melody', and even less as a 'theme'. The *Melodienschutz* apparently was introduced into German Authors' rights (LUG 1902, UrhG 1965) in order to protect individual and original melodies that are clearly identifiable because of their Gestalt properties (see above). If, as it happens in jurisdiction, trivial sequences consisting of a small number of notes are subsumed under the notion of 'melody', and are considered worthy of 'Melodienschutz' (§ 24 Abs. 2 UrhG), this is usually done so by referring to the concept of the so-called "small coin" (*Kleine Münze*) which means, in brief, that even very elementary formations (in music, art, writing, etc.) will be protected under Authors' rights as long as a minimum of "personal intellectual creation" is discernible. This minimum meanwhile has reached such a low level (see BGH GRUR 1981, 287 "Dirlada", BGH GRUR 1988, 810 "Fantasy", BGH GRUR 1991, 531 "Brown Girl I", BGH GRUR 1991, 533 "Brown Girl II") that the concept of *Kleine Münze* has come under dispute (see Schulze 1987), and in particular so in regard to ready-made pop music (see Schneider 1992, Schunke 2008, 38ff.). As one author correctly states, the criterion of the level of creation (*Gestaltungshöhe*) in jurisdiction doesn't play a role in regard to defining works of music, to the effect that practically everything qualifies for a 'work', or at least for an 'adaptation' (see Dieth 2000, 111 and also Schneider 1986).

One reason why jurisdiction had a focus on 'melody' of course was the special protection of this musical parameter under the German Melodienschutz. Another reason why traditional musical categories such as melody and harmony have been considered predominantly with respect to "personal intellectual creation" is that these are fundamental to traditional western music. One may in fact find rare melodic formations as well as unique chords and chord progressions (e.g., the famous "Tristan chord" and Wagner's way of combining such chords into sequences). Though complex rhythmic patterns came into play long ago (even before Stravinsky), and are found frequently in jazz and related genres, protection of rhythms wasn't really an issue in Authors' rights and copyright until recently (see Reinfeld 2001). Of even less regard for a long time has been one of the most complex and most fundamental parameters peculiar to music: sound. Discussing music in regard to allegations of plagiarism or "unconscious borrowing" typically meant comparison of notations (see BGH GRUR 1971, 266, 267, "Magdalenenarie", BGH GRUR 1988, 810 "Fantasy", 812 "Ein bißchen Frieden"). To be sure, such notations with respect to pop music almost always are produced ex post, that is, after a piece of music has been recorded in the studio, and has been published by means of duplication on phonogram record, or CD. Notation in this respect is established as a *postscript* whereas, in the times before studio recording technology became available to musicians and composers, notation served as a *prescript* that guided the execution of a work of music in a more or less imperative way. Transcription of musical sound into western staff notation always and inevitably involves interpretation since a person is first listening to a quasi-continuous stream of musical sound and then must seek to break down this stream into a sequence of discrete events and elements expressed as notes written in a staff system (see Schneider 1987). Meanwhile, there are some so-called "automatic transcription" algorithms available as codes. Most of the codes actually can handle only monophonic music (for example, a singer or violin player without accompaniment), and the notation resulting from "automatic transcription" must always be checked for mistakes or missing elements. The transcription thereby is a sketch rather than a detailed notation. In sum, a notation of music from sound recording to this day involves interpretation, a fact that at times has been recognized by the courts when looking closer at the opinions experts had delivered including notations based on transcriptions. In general, however, notations derived from transcriptions have been treated so as if providing objective evidence (cf. see BGH GRUR 1971, 266, 267, "Magdalenenarie", BGH GRUR 1988, 810 "Fantasy", 812 "Ein bißchen Frieden").

It was not before the 1990s that courts fully recognized the importance of 'sound' and production parameters for modern pop music. In a ruling remarkable for the time, the OLG München stated that parameters other than 'melody' can be constitutive for a work of music, and that imitating or even copying the overall sound characteristic of a pop production (not containing a 'melody' protected under whatever regulation of the UrhG) can cause an infringement of Authors' rights (OLG ZUM 1992, 202ff. "Phlegma Madness I"). The judges of the appellate court in this case had based their ruling explicitly on *listening* to the original recording known as "Sadeness I" as well as to the imitation titled "Phlegma Madness I" (the title indicates a parody, see below). The LG München (court of first instance), also after listening to the two recorded songs, had ruled that every element contained in "Sadeness I" probably was preexisting and known, yet that a new musical arrangement of such elements could result in individual sound patterns attributable to the "personality of the author" (LG München 21 O 3095/91, 13.3.1991, p. 8f.). The recording titled "Sadeness I" thereby was ruled to constitute a work of music (§ 2 UrhG), and the other recording apparently imitating "Sadeness" was found to infringe on rights as an unlawful adaptation that had been published without prior consent of the author[s] of the original (§ 23 UrhG). The defendant argued that "Phlegma Madness I" was fabricated as a parody, and that, moreover, it would contain a number of individual elements sufficient to distinguish the two recordings. This defence pleading was neither accepted by the LG München, nor by the OLG München (court of appeal). The appellate court had employed an expert witness who attested that the musical sound of "Sadeness I", based on sampled Gregorian Chant (from another CD, probably an infringement on ancillary rights as defined in §§ 73ff., 85, 86 UrhG), synthesizer patterns and reverb effects was highly unique[13]. The work as a whole was found original and protected under Authors' rights mainly because of the expert witnesses' opinion, but also because of the impression it made on the judges actually listening to the recording.

---

[13] The recording is easily accessible via Internet sources.

The two rulings on "Sadeness I" vs. "Phlegma Madness I" are remarkable in that the role of musical sound in regard to pop music is acknowledged at last. Also, the courts involved noticed the change from traditional concepts of musical composition (i.e., working out a piece of music according to 'melody', 'harmony', etc., and putting everything into notation) to studio production that, in the last fifty years or so, has become fundamental in modern rock and pop music (see Schneider 2009).

4. A case study in forensic music psychology

In the following, a lawsuit will be reported in which forensic music psychology has played an important part. The lawsuit in question concerned the well-known rock guitarist, Gary Moore, who (together with his record company, Virgin) was charged for having infringed on authors' rights when including, in his song "Still got the blues" (published 1990, Virgin Records), a major part from an instrumental titled "Nordrach" that was written, in 1973, by members of the German rock group, Jud's Gallery, and subsequently performed regularly in concert. A version of the work was recorded, on March 29th, 1974, by the German public broadcast station, Südwestfunk (SWF), and was actually broadcast on November 27th, 1974, and apparently also on another date (most likely by the end of July or beginning of August, 1974). There was some circumstantial evidence, brought forward by the plaintiff (and discussed in detail by the court of first instance, LG München, in several hearings as well as in the ruling 21 O 23120/00) that Gary Moore who in the summer of 1974 had spent some time in the area of Bonn, Germany, could have heard "Nordrach" either at a live performance of Jud's Gallery, or from broadcast. The defendant denied to have heard "Nordrach" then or on a later date, and claimed further he had not been influenced by "Nordrach" in whatever respect when writing "Still got the blues".

The lawsuit that, already at first instance, took about eight years (from filing the suit in 2000 to the court delivering the judgement on December 3rd, 2008), contains too many aspects to be detailed here (see LG München 21 O 23120/00, pp. 1-17 for the court stating the facts, and pp. 18-47 with the reasons for the judgement). Both the plaintiff and the two defendants (G. Moore, Virgin) employed private expert witnesses[14], and the court has commissioned two opinions from expert witnesses who both were also present in hearings of the court so that the court, the plaintiff, the defendants, and their respective lawyers could put questions, or could be asking for additional assessments on certain issues.

The first expert witness summoned by the court (Professor Hermann Rauhe, Hamburg) had to analyze the musical structure and quality of the part in "Nordrach" that the plaintiff claimed had been copied in "Still got the blues". The expert witness concluded (2003) that the part in question, though based on a chord progression as well as on fairly conventional melodic elements known before, was a genuine creation resulting from the combination and interplay of the elements, on the one hand, and a distinct way of musical execution and sound found in "Nordrach", on the other. In a

[14] The author of the article at hand was the sole private expert witness for a surviving member of Jud's Gallery, Jürgen Winter, who acted as plaintiff in the case reported here. I decided, in 2005, to support the plaintiff as a private expert witness on grounds scientific, and hence worked for free. I have submitted two opinions in writing to the court, in 2006 and 2008, respectively.

hearing, the expert witness maintained that the interplay of all the constructive elements could be adequately perceived and evaluated only by listening to the recording of "Nordrach"[15], and certainly not from notations resulting from ex post-transcriptions (see above) that would neither include actual phrasing and other details resulting from playing an electric guitar nor the 'sound' so important for rock and pop music. The expert witness emphasized that, in popular music, the decisive level for analysis, comparison and evaluation would be perception and listening experience, and certainly not a notation produced ex post.

The court, assenting to the opinion of the expert witness, attested that the part of "Nordrach" being the matter in dispute constitutes a "personal intellectual creation", and hence is protected under law (§ 2 UrhG). Consequently, if a very similar part would be found in "Still got the blues", this would have to be regarded as an adaptation (§ 3 UrhG) that could only have been published and marketed with the prior consent of the original author (§ 23 UrhG). If such consent was lacking, an unlawful adaptation would have been produced and published for which the infringers could be held liable.

Interlude: Typical strategies for defendants in lawsuits on musical plagiarism

If one reviews documents, one finds that, in particularly in lawsuits dealing with musical plagiarism, there are typical strategies for defence (see also Schlingloff 1990, 54ff., Dieth 2000, 198ff.). There is a standard pattern of arguments one is confronted with in such lawsuits that can be summed up thus:

a. deny that the plaintiff's song contains anything that is new or original; then
b. bring some evidence suited to demonstrate that certain elements (e.g., melodic motives, chordal patterns) contained in the plaintiff's work were already used by this or that composer way back (in the Middle Ages, the Renaissance or Baroque era, or by Bach or Mozart or Schubert or...). Flesh out your argument by fabricating notations which make elements of the plaintiff's work look as "similar" as possible to elements found in "classical" or (if that fails) whatever folk and ethnic music. Look for, and possibly adduce a wide range of "parallels", and present them all in synoptic notation. Do not hesitate to include the most remote or obscure specimen in order to prove that, if such elements were preexisting, they could not have been "invented" by the plaintiff (nota bene: the German UrhG does not rest on a principle of priority as is relevant for patent law; see Vinck, in Nordemann et al.1994, § 24 UrhG, Anhang, Rz. 11).
c. In this respect, do not forget to point to "rambling melodies" (*Wandernde Melodien*), another useful tool that has hardly ever been missed in lawsuits involving music plagiarism[16].

[15] The recording made in 1974 of „Nordrach" by the SWF had become publicly available in 2000 when issued on CD (Jud's Gallery. SWF-Sessions Vol. 1, Longhair Records LHC 1).

[16] The topos of „hiking" or „rambling" melodies stems from the booklet of same title written by Wilhelm Tappert (1868, 2nd ed. 1889). Tappert's argument in general is we have to deal with processes of cultural diffusion and adaptation rather than with independent creation. Tappert, a dedicated follower of Darwin, decried the view of „artistic creation" so firmly rooted in 19th century aesthetics, to assume a process of constant alteration (*Umbildung*) of melodic patterns instead. Many of Tappert's claims in regard to diffusion of melodic patterns or, rather, motives are bold, and not substantiated by facts. As

d. Be sure to present elements that, by and large, look "similar" in *notation* (and don't refer to elements found in the same work that look "dissimilar"); adhere to a perspective that is as atomistic as possible, avoid musical context and meaning, present "skeletal melodies" devoid of actual accent patterns or phrasing.
e. In case you cannot find appropriate "parallels", argue that at least certain elements contained in the plaintiff's work, or in parts thereof, belong to the stock of elements or building blocks common to many, if not all composers, musicians, producers, etc.
f. In sum, dispute that the plaintiff's song or instrumental, or a certain portion thereof, can be regarded to constitute an original 'work' or 'adaptation' that falls under legal protection;
g. If nothing else helps, say you did not know the plaintiff's work, and if there is any "borrowing", it has been unintentional and "unconscious".
h. Finally, deny any act of negligence.

To be sure, quite a few of the points assembled here are found in legal practice.

In regard to the lawsuit under review ("Nordrach" vs. "Still got the blues"), the court ruled that, in order to investigate the relation between the two works, listening experience (*Hörerlebnis*) is decisive whereas an analysis based on notations taken from the recordings (that private expert witnesses had produced for the defendants in quantity, along with notations of preexisting works believed to demonstrate historic or contemporary "parallels"; see Interlude, a-d) could only be considered secondarily. Such notations (as derived from transcription) could not substitute for the experience of music conceived as sound.

In sum, the court ruled that "Nordrach" was an original work of authorship at the time it had been created, in 1974, and had not been based on other works known at that time (p. 28 of the ruling). The court further found that the guitar solo part contained in "Nordrach" was taken to be included in "Still got the blues", and that, when listening to the respective parts, one would hear "striking convergences" with respect to musical form, melody, rhythm, tempo, harmony, instrumentation and arrangement (pp. 34, 35 of the judgement). Taking into account (1) the musical convergences as elaborated by the expert witness Prof. Rauhe in his opinion commissioned by the court, (2) the impression of a high degree of similarity the judges had when listening to the parts of the two recordings, plus (3) some circumstantial evidence with respect to Moore's stay in Germany, in 1974, the court was convinced that the defendant had known, and had made use of, "Nordrach" when writing "Still got the blues". The judges saw no evidence for "independent double creation", and did not accept the allegation brought forward by defendant's lawyers which claims that, even if Gary Moore would have heard "Nordrach" on some occasion, in 1974, he could not have remembered the piece from what they (and their expert witnesses) called "ephemeral listening", for a period of sixteen years (i.e., the year 1990 when Moore actually published "Still got the blues").

Nettl (2005, 118) states, „the compositions from which the motifs are taken may in other ways be only slightly or not at all related to each other. The similarity or identity resides in the motifs".

The question of allegedly "ephemeral listening" and of limited vs. sustaining musical memory (as regards the defendant) became a separate and most interesting issue in this lawsuit for which an opinion by yet another expert witness (Professor Eckart Altenmüller, Hanover) was commissioned by the court (since the defendants had demanded such an expert statement after the court had informed both parties that, taking into account the evidence accumulated in the course of several hearings, an infringements of rights could hardly be denied). We will return to the issue of "ephemeral listening" and fugacious memory soon.

To understand some of the factual evidence relevant for the court's ruling according to which a certain part of "Nordrach" was used in "Still got the blues" (as a so-called dependant borrowing, *abhängige Entlehnung*), the structure of that part will be briefly outlined[17]. The instrumental part in dispute occurs in "Nordrach" (a long piece of 12 minutes) twelve times. It consists of a characteristic melody played on electric guitar (with a modestly distorted "British" sound) over a chordal sequence provided by keyboards and bass that goes:

a-minor → d-minor → G-major → C-major → Fmaj7 (open e-string on guitar) → d-minor → E-major → a-minor

The melody and the chordal pattern integrate very well not only in regard to tonal aspects, but also due to the accent structure that, to be sure, is emphasized by means of articulation and phrasing the musicians employ in "Nordrach". In effect, a compact musical formation (*Gebilde*) is established that is recognized as such by listeners.

The instrumental solo part of "Still got the blues" that corresponds to the part of "Nordrach" as recorded in 1974, contains the same chordal pattern as well as almost identical melodic material. Also, there is a striking similarity in accent structure, tempo as well as in other parameters (as noted by the court, see above). The part in question in "Still got the blues" occurs, all in all, for eleven times (eight turns result from repeating the basic chordal sequence unaltered, and three include slight variants).

In order to check the degree of 'similarity' experienced by musically trained listeners, the present author has carried out an experiment in 2005 in which n = 64 subjects took part. Most of the subjects were students in musicology in their first academic year. No information was given concerning the background of the experiment. None of the subjects had ever heard "Nordrach", and none knew about the lawsuit (as I learned from the subjects when asking them after the experiment was finished). The task of the subjects was to rate the melodic and harmonic similarity (variables 1, 2) between

a. Nordrach, portion of the recording starting at 8'15", duration ca. 60 seconds (ending with a fade out to 0 dB level):
b. Still got the blues, portion of the recording starting at 2'28" of the recording, duration ca. 60 seconds (ending with a fade out to 0 dB level)

---

[17] A detailed analysis of „Nordrach" as well as of „Still got the blues" was provided in the present author's first opinion as an expert witness in this lawsuit (July 27th, 2006).

on a scale ranging from 1 = very low degree of similarity to 7 = very high degree of similarity. The statistics for melodic similarity did yield aM = 5.76, SD = 1.02, Median = 6, and for harmonic similarity aM = 5.86, SD = 1.18, Median = 6. Judgements of the subjects (treated as ordinal data what in fact they are) on the two variables correlate significantly at (Spearman's) rho = 0.33 (p [0.01] = 0.293). Correlation thus corroborates that melodic and harmonic structures of both pieces are regarded as highly similar.

A number of subjects, when hearing "Nordrach" for the first time, believed it to be "an older version" of "Still got the blues" (a piece they knew, or had heard on the radio where it was played so often).

The experiment was repeated, in a different set-up, in 2008 and 2009, respectively. Taking into account the defendant's claim that the portion under dispute found in "Nordrach" and "Still got the blues", would be found also in other pieces of popular music, namely in

- Les feuilles mortes (a.k.a. "Autumn Leaves"), written by Joseph Kosma (music) and Jacques Prévert [French lyrics], published in 1947 (music) and 1955 (lyrics);
- Dana, an instrumental written and produced by Dr. Roland Kovac for a film in 1971, published on LP in 1973.
- Hello, a world hit for Lionel Richie in 1984 (Motown Records),

an experiment was designed based on paired comparison (cf. Sixtl 1982, ch. 5). Of "Les feuilles mortes", the version sung by Jacques Brel was used. From the three works mentioned, portions corresponding to the portions from "Nordrach" and "Still got the blues" under dispute were selected and digitally copied so that for each title a sound file of approx. 1'15" (including repetitions of a relevant part) results. Taking five titles/ sound files, a total of 20 pairs would be possible. Making the assumption that a comparison of A to B would yield almost the same result as when B is compared to A (A : B ≈ B : A), the number of pairs can be halved. This was done in this experiment for the simple reason that the attentiveness of subjects in such experiments may weaken with the duration of the experiment. Consequently, there is a trade-off between presenting a full set of paired samples (including inversions) and attention necessary to make similarity judgements adequately. In order to ensure maximum attention of the n = 22 subjects, a set of n x (n –1)/2 pairs of sound files was provided. The pairs were presented in random order via professional equipment (Tandberg 3000, ElectroVoice Sentry 500 x 2) at a SPL of ca. 65-70 dB in a room with appropriate sound absorption on walls. Between a pair of two sound files, a short period of silence lasting 3 seconds was inserted on the CD that contained all pairs of sound files. The break between pairs of sound files was manually set to ca. 30 seconds so that the subjects had sufficient time to make their similarity judgements with respect to three variables: (1) harmony (*Harmonik*), (2) melody (*Melodik*), (3) overall impression (*Gesamteindruck*). Similarity ratings were made on scales of seven degrees each (1 = very low similarity, 7 = very high similarity).

The ten pairs used in the experiment were

1. Les feuilles mortes ./. Dana
2. Hello ./. Nordrach
3. Dana ./. Still got the blues
4. Les feuilles mortes ./. Hello
5. Dana ./. Nordrach
6. Les feuilles mortes ./. Still got the blues
7. Dana ./. Hello
8. Nordrach ./. Still got the blues
9. Hello ./. Still got the blues
10. Les feuilles mortes ./. Nordrach

Consequently, each title/sound file was heard four times. The statistics for the ten pairs (three variables each) are given in table 1:

Table 1: Descriptive Statistics, paired comparison, 10 pairs, 3 variables, n = 22 subjects

	N	Minimum	Maximum	Mittelwert	Standardabweichung
Harm1	22	3,00	7,00	5,0455	1,36198
Melodik1	22	3,00	7,00	4,7727	1,23179
Gesamt1	22	2,00	7,00	4,6818	1,58524
Harm2	22	3,00	7,00	5,3636	1,32900
Melodik2	22	2,00	7,00	5,2273	1,02036
Gesamt2	22	3,00	7,00	5,5000	1,10195
Harm3	22	4,00	7,00	6,0000	,97590
Melodik3	22	4,00	7,00	6,0000	,92582
Gesamt3	22	2,00	7,00	6,0455	1,17422
Harm4	22	2,00	6,00	4,4091	1,33306
Melodik4	22	2,00	6,00	4,0455	1,46311
Gesamt4	22	1,00	7,00	4,0000	1,71825
Harm5	22	3,00	7,00	5,4091	1,22121
Melodik5	22	3,00	7,00	5,2727	,93513
Gesamt5	22	2,00	7,00	5,1818	1,18065
Harm6	22	2,00	7,00	4,8636	1,67034
Melodik6	22	3,00	7,00	4,7727	1,26986
Gesamt6	22	2,00	7,00	4,5000	1,71131
Harm7	22	3,00	7,00	4,8636	1,12527
Melodik7	22	2,00	6,00	4,8182	1,22032
Gesamt7	22	3,00	7,00	4,7273	1,35161
Harm8	22	4,00	7,00	6,0000	,92582
Melodik8	22	4,00	7,00	5,8636	,94089
Gesamt8	22	4,00	7,00	6,1364	,83355
Harm9	22	3,00	7,00	5,5000	1,14434
Melodik9	22	4,00	7,00	5,5909	,95912
Gesamt9	22	4,00	7,00	5,4091	,85407
Harm10	22	3,00	7,00	4,8636	1,39029
Melodik10	22	2,00	7,00	4,9091	1,15095
Gesamt10	22	2,00	7,00	4,4091	1,43623
Gültige Werte (Listenweise)	22				

From the data in table 1 it is evident that similarity ratings (aM) are highest for "Dana" ./. "Still got the blues" as well as for "Nordrach" ./. "Still got the blues". The sum of means from three variables yields 18.0455 for “Dana” ./. “Still got the blues”, and 18.000 for “Nordrach” ./. “Still got the blues”. It should be noted also that the average similarity rating (aM) for overall impression (*Gesamteindruck*) of the latter pair tops everything else.

The experiment was repeated with a small sample of n = 12 subjects, in 2009 giving approximately equivalent results. On the raw data of both experiments, analyses including multiple mean comparisons, multiple correlation and regression, and cluster analysis were performed. The results cannot be discussed here for lack of space available.

For the purpose of the present article it is sufficient to establish a rank order of the ten pairs for the three variables in regard to similarity ratings from experiment 1 (n = 22 subjects):

Table 2: Mean similarity ratings for ten pairs and three variables in ranking order

Variable: harmony

	Pair	**aM**
Rank 1	Dana ./. Still got the blues	6.0000
Rank 1	Nordrach ./. Still got the blues	6.0000
Rank 2	Hello ./. Still got the blues	5.5000
Rank 3	Dana ./. Nordrach	5.4091
Rank 4	Hello ./. Nordrach	5.3636
Rank 5	Les feuilles mortes ./. Dana	5.0455
Rank 6	Les feuilles mortes ./. Still got ...	4.8636
Rank 6	Dana ./. Hello	4.8636
Rank 6	Les feuilles mortes ./. Nordrach	4.8636
Rank 7	Les feuilles mortes ./. Hello	4.4091

Variable: Melody

Rank 1	Dana ./. Still got the blues	6.0000
Rank 2	Nordrach ./. Still got the blues	5.8636
Rank 3	Hello ./. Still got the blues	5.5909
Rank 4	Dana ./. Nordrach	5.2727
Rank 5	Hello ./. Nordrach	5.2273
Rank 6	Les feuilles nortes ./. Nordrach	4.9091
Rank 7	Dana ./. Hello	4.8182
Rank 8	Les feuilles mortes ./. Dana	4.7727
Rank 8	Les feuilles mortes ./. Still got ...	4.7727
Rank 9	Les feuilles mortes ./. Hello	4.0455

Variable: Overall impression

Rank 1	Nordrach ./. Still got the blues	6.1364
Rank 2	Dana ./. Still got the blues	6.0455
Rank 3	Hello ./. Nordrach	5.5000
Rank 4	Hello ./. Still got the blues	5.4091
Rank 5	Dana ./. Nordrach	5.1818
Rank 6	Dana ./. Hello	4.7273
Rank 7	Les feuilles mortes ./. Dana	4.6818
Rank 8	Les feuilles mortes ./. Still got ...	4.5000
Rank 9	Les feuilles mortes ./. Nordrach	4.4091
Rank 10	Les feuilles mortes ./. Hello	4.0000

To interpret the ranking properly in regard to legal aspects, one of course has to observe the temporal order in which the five works were published because only preexisting works can be possibly used in whatever way in a work produced and published later.

It should be mentioned, at this point, that the author and the publisher of "Dana" had filed a lawsuit against the Gary Moore and Virgin, already in 1996, where they had claimed that a portion of "Still got the blues" (in fact, the same portion disputed also in the "Nordrach" vs. "Still got the blues" case) would be an unlawful adoption (*urheberrechtsverletzende Übernahme*) of the first 16 measures of "Dana". The defendant had rejected this view with the aid of a private expert witness saying that "Dana" was based on motives known from Baroque music, and also on a chord sequence pattern known from that period. Hence the elements contained in "Dana" would belong to the public domain.

The court commissioned an opinion from a lawyer working as an expert witness in matters musical (!) who remarked, in the same line, that "Dana" was based on a chord structure known from Baroque music. This chord pattern, moreover, would be found, in identical manner, in "Autumn leaves" (*Les feuilles mortes*). The plaintiff challenged the qualification of that lawyer to act as an expert witness (without avail), and further criticised that the expert had based his opinion on the sheet music (piano score!) of "Dana", and not on the recording of the instrumental. The court ruled that the first 16 measures of "Dana" under dispute would not constitute a "personal intellectual creation" in regard to § 2 UrhG. Hence the court saw no need to determine whether those measures had been adopted in "Still got the blues" (LG München 7 O 11024/91 [1996], "Dana"). If the expert witnesses as well as the court would have based their respective opinions on actually listening to "Dana" (and then to the instrumental solo part of "Still got the blues") instead on looking at a piano score of "Dana", the ruling might have turned out differently. As the data in tables 1 and 2 demonstrate, a group of musically trained subjects, after listening to "Dana" (measures 1-16) and "Still got the blues" (solo instrumental part) judged the similarity of these two pieces on all three variables to be very high.

The lesson to be learned from this finding as well as from other results reported here in brief is that empirical evidence obtained from experiments with samples of musically trained listeners could be used (and should be used), in lawsuits dealing with music plagiarism in order to check allegations and assertions, on the one hand, and to substantiate opinions and judgements, on the other. In supporting the courts (as well as plaintiffs and defendants, depending on the case), music psychology takes a forensic turn.

To end this article, let's return to the issue mentioned above that became very important in the lawsuit "Nordrach" vs. "Still got the blues": did Gary Moore, if he had heard "Nordrach" in 1974, do so by "ephemeral listening" (as his private expert witnesses presumed), to the effect that he would not have established a trace lasting in memory?

a. The setting hypothesized in this question is that the defendant could have listened to "Nordrach", in the summer of 1974 when he visited the area of Bonn, Germany, where he stayed for a while (how long exactly is a point disputed in the lawsuit). Since he has published "Still got the blues" in 1990, the gap between possibly listening once or twice to "Nordrach" and producing his own song would have amounted to almost 16 years if one assumes that Gary Moore, after having heard "Nordrach", did never try to play the portion of "Nordrach" that appears, in very similar structure and execution, in "Still got the blues", nor did he take any notes. In sum, it is hypothesized that the defendant, provided he could have listed to "Nordrach" in 1974, and in fact did hear the piece, received or maintained no further information in regard to "Nordrach" in the period lasting to 1990 when he produced "Still got the blues". It is under these premises that the model of cryptomnesia (see above, section 3.2) might be applied to attest the defendant "unconscious borrowing".
b. In case one would hypothesize an attitude of "ephemeral listening" that would exclude per se that a (neurally encoded)'representation' of a piece of music, or portions thereof, is stored in long term memory (LTM), no trace would be left. This would even exclude cryptomnesia (ex nihilo nihil!), and hence "unconscious borrowing". The assumptions private expert witnesses made in defence of Gary Moore in fact repeatedly emphasized that, had he ever listened to "Nordrach", it would have been "ephemeral listening" (with a faint and fast fading memory trace that would have been totally erased after a short period of time).
c. Since the defendant's representatives insisted, in repeated yet somewhat limp assertions, on "ephemeral listening" including a fast fading memory and loss of relevant musical information in regard to their client's possible exposure to "Nordrach", the court commissioned an opinion by an expert witness. The issue to be examined by the expert concerned the question whether a person, having heard a piece of music, could have stored this in his memory for a period of about 16 years. The expert witness appointed by the court was Professor Eckart Altenmüller, a well-known neurologist (and also a trained musician) working in the field of music perception and cognition.

   The expert witness, after discussing findings pertaining to memory for music as reported in publications on psychology and neurology, and after evaluating the evidence from the case at hand as well as the pro and con arguments, came to the conclusion that it seems more likely than not that the defendant, when writing "Still got the blues", had been influenced by what he apparently had heard some 16 years ago. The expert argued further that the features found in

"Nordrach" on the one hand, and the capacity for music that must be attested to the defendant, on the other, make it reasonable to assume that a storage of "Nordrach" in the LTM of the defendant could have happened (opinion of the expert witness, 30th of January, 2008). The defendant's legal representatives refused to accept this assessment (and also disputed the scientific qualification of the expert witness – to no avail).

d. Though there have been many studies addressing memory for music in certain respects (an extensive review covering relevant publications and theories is found in Müllensiefen 2004; see also Snyder 2000), systematic studies on LTM are scarce. There is, however, a broad range of information found in biographies of composers and performing artists that clearly demonstrates extraordinary LTM capacity. One of the musicians and composers "famous for his incredible feats of memory is the well-known conductor Arturo Toscanini who is reported to have known every note of every instrument of 250 symphonic scores, 100 operas and numerous other musical works" (Kapardis 2003, 27; the information was taken from Gray 1999, 330). There are more such accounts: the German organist and harpsichordist, Helmut Walcha (1907-1991) who went blind at the age of nineteen[18], has recorded the complete organ works of J.S. Bach twice (as well as works of Buxtehude, Bruhns, and other composers) he knew by heart.
   Basically, every performing artist (violinist, pianist, singer, etc.) must develop a very voluminous, and at the same time very reliable LTM for music. This of course requires a long phase of training as well as a steady backup achieved by playing a work repeatedly.
   In principle, a performing artist in rock music is in no other position than an artist who plays so-called "classical" music except that the rock musician starts to develop his capability for memorizing music from the very beginning because notation is of little if any practical significance in rock. Young rock musicians lern to "hear out" musical structure as well as a bunch of "guitar licks" or other phrases when listening to the recordings of their heroes. So training in fact rests on "aural tradition": you hear something, you find it interesting or even brilliant, and you pick up your guitar, bass or whatever instrument and try to play what you've been listening to. This is how one acquires the tools of the trade, in rock music. Another useful method is watching what performers actually do on stage. As Achim Reichel, founding member of The Rattles (Hamburg, one of Germany's pioneer beat groups), recalls from his musical beginnings, in about 1960, it was like this: "None of us had really learned his instrument ... but when English groups like 'The Beatles', 'Rory Storm & The Hurricanes', Tony Sheridan (and others) played the 'Top Ten', the 'Kaiserkeller', or the 'Indra' [at Hamburg], we used to sit right in front of the stage and kept an eye on their fingers"[19]. This way, they saw how the intro to some of Chuck Berry's songs was played on guitar.

e. If even beginners can learn a peculiar intro or the changes (chord progression) from keeping an eye on someone's fingers, it will be much easier to do so for a professional who also will be capable to understand the structure of a song or of

[18] See http://de.wikipedia.org/wiki/Helmut_Walcha

[19] Achim Reichel (narrating from his life), brochure, WEA Musik GmbH, Hamburg 1988.

a certain part (Intro, solo instrumental, bridge) at once when listening to it. In case the defendant did hear that characteristic part of "Nordrach" to which the solo instrumental part in "Still got the blues" corresponds to closely, it is very likely that he, as a professional rock music already then, would have achieved an immediate understanding of the harmonic, melodic, and rhythmic patterns used for building blocks in "Nordrach", and also of the interplay of these patterns. As was mentioned above, the characteristic sequence of chords, which occurs in "Nordrach" together with melodic phrases played on electric guitar, is repeated over and over again. This would be enough, for a professional musician playing electric guitar (as the defendant does), to attract his attention, to listen to that part, grasp what the constructive features are, and memorize the distinctive features as well as the "profile" of how they interact and fit together.

f. The private expert witnesses working on behalf of the defendant presumed two things which, to the present author, are totally unconvincing: first, that the defendant, a professional rock guitarist already in 1974, would have exercised "ephemeral listening", if exposed to, and actually listening to that characteristic part of "Nordrach" in which the lead guitar is featured prominently. Second, that no trace lasting in memory for more than a short period would have resulted from such an exposure to "Nordrach".
   The first assertion is highly unrealistic since professional rock musicians will find it difficult, I dare to conclude drawing from my own experience and from talks with other rock musicians, to practice "ephemeral listening" when exposed to music that is close in genre, playing style and sound to one's own work.
   The second assertion simply is untenable in the light of evidence from clinical neurology which proves that subjects (not necessarily professional musicians) keep music they have been exposed to in their LTM for a very long time, probably for the rest of their lives (for reports of relevant cases, see Sacks 1987, pt. 3, 2007, ch. 5; Cuddy et al. 2009).

5. Conclusion

In the present paper, some areas in criminal law and civil law have been addressed where music psychology might be usefully employed to examine and to clarify certain aspects as well as to support courts seeking to gain specific information relevant for assessments. Music psychology (as part of systematic musicology) in this regard can be of help to the court because it commands a wide range of empirical-experimental methodology that can be adapted to "forensic" investigations if needed. To illustrate problems and complexities one is confronted with in this field, a recent lawsuit concerning Authors' rights, and in particular aspects of "unconscious borrowing" as well as memory for music has been reported in some detail. This case, if anything, demonstrates that forensic music psychology can contribute substantially to the work of the court.

References

a) books and articles

Albersheim, Gerhard (1980), *Die Tonsprache*. Tutzing: H. Schneider

Bregman, Albert (1990), *Auditory scene analysis*. Cambridge, MA: MIT Pr.

Blackburn, Ronald (1996), What is forensic psychology? *Legal and Criminological Psychology* 1, 3-16.

Cogan, Robert, Pozzi Escot (1976), *Sonic design: the nature of sound and music*, Englewood Cliffs, N.J.: Prentice – Hall.

Cuddy, Lola 2009. Musical Memory in normal and disordered aging. Paper, 7th Triennial Conference of ESCOM, Jyväskylä, Finland, August 2009 (see Conference Abstracts, p. 76).

Dieth, Matthias (2000), *Musikwerk und Musikplagiat im deutschen Urheberrecht*, Baden-Baden: Nomos.

Dornbuch, Chritian, Jan Raabe (eds.) (2002), *RechtsRock: Bestandsaufnahme und Gegenstrategien*, Münster: Unrast-Verlag.

Ehrenfels, Christian von (1890), Ueber Gestaltqualitäten, *Vierteljahrsschrift für wissenschaftliche Philosophie* 14, 249-292.

Fenner, Lucie (2005), *Erinnerung und Entlehnung im Werk von Charles Ives*, Tutzing: H. Schneider.

Finke, Ronald, Thomas Ward, Steven Smith (1992), *Creative Cognition. Theory, research, and applications.* Cambridge, MA: MIT Pr.

Frieler, Klaus (2008), *Mathematik und kognitive Melodieforschung. Grundlagen für quantitative Modelle*. Hamburg: Verlag Dr. Kovac.

Fromm/Nordemann, see Nordemann/Vinck/Hertin.

Fuchs, Eberhard (1983), *Urheberrechtsgedanke und –verletzung in der Geschichte des Plagiats unter besonderer Berücksichtigung der Musik*, Jur. Diss. Univ. Tübingen.

Gierth, Peter (1974), *Individualität und Zufall im Urheberrecht.* Berlin: J. Schweitzer.

Gray, Peter (1999), *Psychology*. 3rd ed. New York: Worth Publ.

Hanslick, Eduard (1854), *Vom Musikalisch-Schönen. Ein Beitrag zur Revision der Ästhetik der Tonkunst*. Leipzig (Repr, Darmstadt: Wiss. Buchges. 1965).

Hertin, Paul (1989), Das Musikzitat.im deutschen Urheberrecht. *GRUR* 91, 159-167.

Isaacs, Alyssa (2008), Parents and critical media literacy: Music. http://criticalmedialiteracy.umn.edu/music.htm.

Jörger, Thomas (1992), *Das Plagiat in der Popularmusik*. Baden-Baden: Nomos.

Jung, Carl Gustav (1902/1966), *Zur Psychologie und Pathologie sogenannter okkulter Phänomene.* Diss. Univ. Zürich. Repr. in C.G. Jung. *Gesammelte Werke*, Bd 1, Zürich, Stuttgart: Rascher 1966, 1-98.

Jung, Carl Gustav (1905), Kryptomnesie. *Die Zukunft* 13, 325-334, Repr. in C.G. Jung. *Gesammelte Werke*, Bd 1, Zürich, Stuttgart: Rascher 1966, 103-115.

Kapardis, Andreas (2003), *Psychology and Law: a critical introduction*, 2nd ed., Cambridge: Cambridge Univ. Pr.

Koffka, Kurt (1936), *Principles of Gestalt psychology*, London: Routledge & Kegan Paul (4th imprint 1955).

Krumhansl, Carol (1997), Effects of perceptual organization and musical form on melodic expectancies. In Marc Leman (ed.). *Music Gestalt, and Computing. Studies in Cognitive and Systematic Musicology*, Berlin, New York: Springer, 294-320.

Lerdahl,. Fred (1988), Cognitive Constraints on compositional systems, In J. Sloboda (ed.), *Generative Processes in music. The Psychology of performance, improvisation, and composition*, Oxford: Clarendon Pr., 231-259.

Meconi, Honey (ed.) (2004), *Early Musical Borrowing*, New York: Routledge.

Miehling, Klaus (2006), *Gewaltmusik – Musikgewalt: populäre Musik und die Folgen*, Würzburg: Königshausen & Neumann.

Müllensiefen, Daniel 2004. *Variabilität und Konstanz von Melodien in der Erinnerung. Ein Beitrag zur musikalischen Gedächtnisforschung*, Phil. Diss. Univ. Hamburg.

Müllensiefen, Daniel, Klaus Frieler (2004), Cognitive Adequacy in the measurement of melodic similarity: algorithms vs. human judgments, *Computing in Musicology* 13, 147-176.

Müllensiefen, Daniel, Marc Pendzich (2009), Court decisions on music plagiarism and the predictive value of similarity algorithms, *Musicae Scientiae Discussion Forum* 48, 257-292.

Nettl, Bruno (2005), *The Study of ethnomusicology. Thirty-one issues and concepts*, 2nd ed. Urbana, Chicago: Univ. of Illinois Pr.

Nordemann, Wilhelm, Kai Vinck, Paul Hertin (1994, 1998), *Urheberrecht*, Stuttgart: Kohlhammer (8th ed. 1994, 9th ed. 1998).

Pendzich, Marc (2004), *Von der Coverversion zum Hit-Recycling. Historische, ökonomische und rechtliche Aspekte eines zentralen Phänomens der Pop- und Rockmusik*, Münster, Hamburg: LIT-Verlag.

Pimat, Manfred (2002), *Beweisprobleme der (angeblich) unbewussten Entlehnung in der Musik*, Osnabrück: Universitätsverlag Rasch.

Reinfeld, Tim (2001), *Der Schutz von Rhythmen im Urheberrecht*, Göttingen: Vandenhoek & Ruprecht.

Riekert, Stephan (2003), *Der Schutz des Musikurhebers bei Coverversionen*, Berlin: BWV.

Sacks, Oliver (1987), *The Man who mistook his wife for a hat*, New York: Harper & Row.

Sacks, Oliver (2007), *Musicophilia. Tales of music and the brain*, New York: A. Knopf (German translation: *Der einarmige Pianist. Über Musik und das Gehirn*, Reinbek: Rowohlt 2008).

Scheel, Karen, Westefeld, John (1999), Heavy metal music and adolescent suicidality: an empirical investigation, *Adolescence* 34, 253-273.

Schlingloff, Jochen (1990), *Unfreie Benutzung und Zitierfreiheit bei urheberrechtlich geschützten Werken der Musik*, Frankfurt/M: Peter Lang.

Schneider, Albrecht (1986), GEMA-Vermutung, Werkbegriff und das Problem sogenannter "GEMA-freier Musik", *GRUR* 88, 657-663.

Schneider, Albrecht (1987), Musik, Sound, Sprache, Schrift: Transkription und Notation in der Vergleichenden Musikwissenschaft und Musikethnologie, *Zeitschrift für Semiotik* 9, 317-343.

Schneider, Albrecht (1992), Traditional, Entlehnung, Werkbegriff: Anmerkungen zu den Entscheidungen "Brown Girl I"/"Brown girl II" des BGH, *GRUR* 94, 82-85.

Schneider, Albrecht (2009), Komposition und Produktion von ‚U-Musik' unter dem Einfluss technischer Medien, In Holger Schramm (Hrsg.), *Handbuch Medien und Musik,* Konstanz: UVK (Universitätsverlag Konstanz), 495-530.

Schulze, Gernot (1987), Der Schutz der kleinen Münze im Urheberrecht, *GRUR* 89, 400-408.

Schunke, Sebastian (2008), *Das Bearbeitungsrecht in der Musik und dessen Wahrnehmung durch die GEMA*. Berlin: de Gruyter.

Seidel, Wilhelm (1987), *Werk und Werkbegriff in der Musikgeschichte*, Darmstadt: Wissenschaftl. Buchges.

Sixtl, Friedrich (1982), *Meßmethoden der Psychologie. Theoretische Grundlagen und Probleme*, 2nd ed., Weinheim, Basel: Beltz.

Smits van Waesberghe, Joseph (1955), *A Textbook of melody. A course in functional melodic analysis*, American Institute of Musicology.

Stack, Steve, James Gundlach, J. Reeves (1994), The heavy metal subculture and suicide, *Suicide and life-threatening Behavior* 24, 15-23.

Stadler, Michael (ed.) (1998), *Handbook of implicit learning*, Thousand Oaks, Cal.: Sage.

Snyder, Bob (2000), *Music and Memory. An Introduction*, Cambridge, MA.: MIT Pr.

Tappert, Wilhelm (1889), *Wandernde Melodien*, 2. Aufl., Berlin: Brachvogel & Ranft.

Tödt, Daniel (2007), Gewalt und Umruhen in den französischen Banlieus: Was hat der Rap damit zu tun?, *EthnoScripts* [Hamburg] Vol. 9, 106-128.

Took, Kevin, David Weiss (1994), The Relationship between heavy metal and rap music on adolescent turmoil: real or artefact?, *Adolescence* 29, 1994, 613-621.

Villani, Susan (2001), Impact of media on children and adolescents: a 10-year review of the research. *Journal of the American Academy of Child and Adolescent Psychiatry* 40, 392-401.

Warner, Theodor (1969), *Das Undurchhörbare. Beiträge zur Hörpsychologie und Didaktik der Moderne*, Baden-Baden, Stuttgart: Agis-Verlag, Kohlhammer.

Wellek, Albert (1963), *Musikpsychologie und Musikästhetik*, Frankfurt/M.: Akad. Verlagsanstalt.

Whittall, Arnold (2008), *The Cambridge Introduction to serialism*, New York, Cambridge: Cambridge U. Pr.

b) German jurisdiction (court of first instance, appellate court, high court rulings)

LG = Landgericht; OLG = Oberlandesgericht; BGH = Bundesgerichtshof, BVerfGE = Entscheidungen des Bundesverfassungsgerichts (Collected rulings).

LG München 7 O 11024/91 (1996), "Dana"

LG München 21 O 23120/00 (2008) "Nordrach" (not final)

LG München 21 O 3095/91 (1991) "Phlegma Madness I"

OLG München ZUM 36, 1992, 202 "Phlegma Madness I"

BGH GRUR 1968, 321 "Haselnuß"

BGH GRUR 1971, 266 "Magdalenenarie"

BGH GRUR 1981, 267 "Dirlada"

BGH GRUR 1988, 810 "Fantasy"

BGH GRUR 1988, 812 "Ein bißchen Frieden"

BGH GRUR 1991, 531 "Brown Girl I"

BGH GRUR 1991, 533 "Brown Girl II"

BGH GRUR 1998, 376 "Coverversion"

BVerfGE 31, 229 "Schulbuchprivileg"

GRUR = Gewerblicher Rechtsschutz und Urheberrecht

ZUM = Zeitschrift für Urheber- und Medienrecht

Rolf Bader
Marie-Kristin Dietz
Paul Elvers
Miriam Elias
Niels Tolkien

Foundation of a Syllogistic Music Theory

Introduction

Music theory tries to explain music by analysing bars, phrases, or forms, the structure of musical textures or the musical thought of the piece. In the history of music and in different cultures many music theories have been proposed. Still not all are governed by a strict formal or logical system. Those music theories derived from musical experience often restrict themselves by this experience, where rules for certain historical, social, or geographical contexts are given which on the other hand do not hold for other kinds of music. The present paper on the other hand tries to follow another approach. It starts with a system of thought in general, and tries to explain music through it, the musical thought (*Musikalischer Gedanke*). The system used is the syllogistic of Aristoteles which is the foundation of modern logic.

Although the logical rules of syllogism are more or less those of our present logic, the syllogism is different in a way crucial for musical analysis. In the writings of Aristoteles, his logic is intrinsically based within the phaenomenon present and cannot be formulated without it. His logic is not restricted to the domain of logic itself as a discipline which is the case with modern logic. Modern systems normally only deal with statement variables, which may be 'true' and 'false' ect.. So when a statement is represented by a variable p, this p means a true or false statement. The statement itself is no longer present and could be any true or false one. Aristoteles did not at all think of such a case when constructing his syllogism. His system is deeply rooted in the things discussed. What makes it the foundation of modern logic is the idea to express the thinking of man and the reasonings and connections of 'nature' (a modern term) in general through variables which are terms or phrases rather than whole statements. This system he presents in terms of statements is straightforward and reasonable and so his syllogism has never been altered and everybody easily agrees with its ease and its beauty[1].

So trying to formulate music in terms of syllogism is – as we will see – stating that musical events are connected in space and time and then looking for the inner reasonings and thoughts, connections, needs, and tensions within music. We do concentrate here upon the development of a tonal system as this is of much interest in music theory. We

[1] We discuss the syllogism of statements, the assertoric logic rather than the modal syllogism which indeed was discussed and found not be be completely closed. Also the three-fold logic which was known to Aristoteles, too is not discussed. We show, that the syllogism of statements is so wide that music theory can be expressed by it to a very large extend.

will also discuss the foundation of Riemann's musical theory as he was one of the last to base his system on philosophy rather than on personal stylistic experience.

Syllogistic in the context of Aristoteles' writings

The formal 'way of thinking' which syllogistic describes is deeply embedded in the philosophy of Aristoteles (Maier 1896, 1900). Although he connects ontology and logic in a non-trivial way, our discussion about music theory lies within the field of logic alone. Nevertheless, syllogistic is also capable to describe ontological relations which need to be discussed when using this method in the field of musical semantics and meaning of music or in general when dealing with the connection between inner- and outer-musical topi.

The formal syllogistic system and all rules of it is described in the writing 'analytikon proteron' and can therefore be considered as the foundation of logic in the western philosophical tradition (Aristoteles 1998). The second or 'later' book 'analytikon ysteron' deals with general problems of reasonings, predictions, assumptions and the like. The writing 'katagorias' (Aristoteles 2005) with its famous 10 categories beginning with the 'ousia', the 'hypokeimenon' which later was called 'substantia' leads to the writing of the 'metaphysics' (Aristoteles 1989, 1991). Here the ontological realm is reached where on the other side the analytic and so the syllogistic writings are in the field of subjective logic. So the katagoria is one point of connection between these two worlds. This connection need not to consider us here as we deal with the logic of music within us only. A further writing which may be of interest within our frame is the 'Topic' (Aristoteles 1995), a very voluminous book about possible situations of reasoning when discussion topics with a person and how to deal with them. As musical logic can be derived by the syllogistic system, a further interesting investigation of situations of 'musical reasoning' in terms of the topi discussed by Aristoteles may be of interest in future work. The rethoric of Aristoteles (Aristoteles 1994) is more about general behaviour in discussions and how to interact with persons and crowds as well as methods and ideas of how to convince people.

The syllogistic has been established as a desperate need of philosophers fighting for the truth in philosophy, social cases, political struggle, educational problems or of course music and against reasonings motivated more by earning money, winning a legal case, succeeding as a politician or winning a discussion for gaining reputation as a philosopher or sophist. 'Catching' a sophist, as Platon puts it in his writing 'sophistes' (Platon 2007), and showing that he is wrong needs a clear definition of the world, of people and all aspects of being showing the relations and therefore the reasonings between things and telling which reasoning lead to a conclusion and which does not. So the Topic deals with very many unserious reasonings where a conclusion is claimed but not really there and tries to show serious reasonings as alternatives. So we may say today that the foundation of science is meant to act against corruption, simplification and the seek for pure personal gains which motivated to develop and establish systems and ways of reasonings based on honest and free logic.

As syllogism is not very well known in Musicology we want to describe the basic formalism here briefly. We are only dealing with the assertoric syllogism here which has never been in serious debate concerning its formality or consistency and holds as one of the straightest systems known (see Dressler 2005). Aristoteles also formulated a modal syllogistic based on possibilities and necessarities still under debate today and not discussed or used here (see Schmidt 2002).

The syllogistic system

The formal system of syllogistic reasoning is simple and consists of three levels, word (phrase), sentence and syllogism. A sentence connects two words like 'Socrates is a man.' or 'All men are human beings.' A syllogism connects two sentences and so leads to a conclusion, like 'Socrates is a human being.' Modern statement-logic only starts from the sentence which is true of false and so avoids the content of the sentences only dealing with the formal aspects. Syllogistic on the other side uses the content and is therefore very useful in music theory as we are dealing with a content there, the music.

1) Word or phrase

The tone c or the tone g can be a word as well as 'cadence', 'tonality', 'musical phrase', 'Rock'n' Roll', 'the tone color which comes after the first introduction of the theme' ect. Here a subject is meant. But the word can also be an adjective like 'bright', 'dense', 'touching', 'never heard before' ect.[2]

2) Sentence

Syllogistic then knows four possible connections between the two words (or phrases), they can be affirmative or negative and they can be complete or partial, leading to four possible statements:

ontologic statement	logic statement	formal statement
A is B	or A is said to be B	or A -> a B (a like in affirmo)
A is not B	or A is said not to be B	or A -> e B (e like in negatio)
A is partly B	or A is said to be partly B	or A -> i B (i like in affirmo)
A is partly not B	or A is said to be partly not B	or A -> o B (e like in negatio)

Note, that in our modern understanding of grammer, the subject is B and not A. So the sentence 'Socrates is a man.' need to be formulated like ''Being a man' is said from Sokrates.' and so in the affirmative sentence A -> a B, A is 'being a man' and B is

[2] Logic did always start from sound, i.e. Petrus Hispanus (2006) in his widely used late scholastic treatise 'Tractatus / Summulae Logicales' defined the words used in logic as phonations consisting of sounds. Also in his foundation of Psychology (De anima), Aristoteles derives the understanding (nous) from the sensing in a two-stage process - which today is called perception - and therefore from a phaenomenon.

‘Socrates’. So if we say ‘An octave consists of twelve half-tones.’ we have to formalize ‘’Twelve half-tones’ -> a ‘Octave’.

The just example is also very good to show another important aspect of syllogistic. This sentence must not to be reversed. So we cannot say ‘Octave’ -> a ‘Twelve half-tones’ because twelve half-tones may also be a row in Schönberg’s sense melodically and harmonically combined. So the sentence ‘Twelve half-tones are an octave.’ is not strictly true, it would need to be ‘Twelve half-tones *can also* be arranged in a way to form an octave.’, syllogistically written correctly as A -> i B. So twelve half-tones can be said to be many things and so partly they are also able to form an octave. To make the thing even a bit more difficult we need to mention that the fact, that an octave may also be more than twelve half-tones, i.e. a perfect harmony or a ratio of 2:1. This does not effect the affirmative connection in the first example because twelve half-tones completely form an octave and so are completely within the frame of any octave. So formally here the fact is represented that everybody would agree with the sentence ‘An octave always consists of twelve half-tones.’ but will disagree with ‘Twelve half-tones always are an octave.’ claiming that they may be but also may be not.

Formally, the turning around of words in a sentence is called a conversion and each statement has a conversion like

statement	conversio
A -> a B	B -> i A
A -> e B	B -> e A
A -> i B	B -> i A
A -> o B	B -> o A

So only the complete affirmative statement need to be changed to a partly affirmative statement when converting the phrases. Still we need to be careful with the partly affirmative conversions. In a hierarchical structure it may be that here the conversion fails. If B is under A together with C, D, … then converting B -> i A to A -> i B may be wrong as i.e. ‘Ska’ may be said to be under the category of ‘Rock’, but then ‘Rock’ is a category and we cannot say as a conversion that it is under the category of ‘Ska’ then.

3) Syllogism

When combining two sentences we formulate a syllogism which lead or lead not to a conclusion. So in the syllogism mentioned above A is ‘human being’, B is ‘man’ and C is ‘Socrates’.

Every man is a human being		A -> a B
Socrates is a man.		B -> a C
So Socrates is a human being.	=>	A -> a C

So now we have a connection between two formally unconnected phrases, ‘Socrates’ and ‘human being’. Note again, that the subject is always at the right side of the formal sentence.

This syllogism is a case called ‘Barbara’ or academic conclusion as it is the only affirmative (‘a’) conclusion (or way of thought to reach ‘a’) possible. As mentioned above, the aim of the syllogistic effort is to tell true reasonings from false ones. So with this formal description of a reasoning we are now able to write down all possible reasonings by combining the A, B, C and the a, e, i, and o and then to take a close look at all of them to determine which syllogism leads to which conclusion and which syllogisms lead to no conclusion at all.

To do this we need to consider all possible variations to combine A, B, and C. We will find three figures possible. In all cases the conclusion is between the formally non-connected phrases A (major) and C (minor) via a middle or connecting phrase B (medium).

Figure 1:	A -> x B	
	B -> x C	=> A -> x C
Figure 2:	B -> x A	
	B -> x C	=> A -> x C
Figure 3:	A -> x B	
	C -> x B	=> A-> x C

Here ‘x’ stands as a variable for a, e, i, or o. Theoretically a forth figure can be constructed but can be shown to be redundant and included in the first three ones. Now when including the a, e, i, and o into the ‘x’ for each figure we have 16 possible syllogisms and so a total of 48 syllogisms. From these only 14 do lead to a conclusion and so are valid syllogisms. All others do lead to more than one conclusion and so give a first selection if our reasoning may lead to a clear point or not. We need to show all valid syllogisms here as we will name them later on according to their traditional artificial names. Although we cannot go into details here these names include the reasoning, the figure and even its proof.

Figure I:

1.	A -> a B	B -> a C	=>	A -> a C	Barbara
2.	A -> e B	B -> a C	=>	A -> e C	Celarent
3.	A -> a B	B -> i C	=>	A -> i C	Darii
4.	A -> e B	B -> i C	=>	A -> o C	Ferio

Figure II:

1.	B -> e A	B -> a C	=>	A -> e C	Cesare
2.	B -> a A	B -> e C	=>	A -> e C	Camestres
3.	B -> e A	B -> i C	=>	A -> o C	Festino
4.	B -> a A	B -> o C	=>	A -> o C	Baroco

Figure III:

1.	A -> a B	C -> a B	=>	A -> i C	Darapti
2.	A -> e B	C -> a B	=>	A -> o C	Felapton
3.	A -> i B	C -> a B	=>	A -> i C	Disamis
4.	A -> a B	C -> i B	=>	A -> i C	Datisi
5.	A -> o B	C -> a B	=>	A -> o C	Bocardo
6.	A -> e B	C -> i B	=>	A -> o C	Ferison

Figure I leads to all possible conclusions, figure II can only conclude in negatio, complete or partially and figure III can only conclude partially and not completely. Again it is beyond the scope of this paper to discuss the proofs why only these syllogisms lead to precisely these conclusions (see Drechsler 2006 for details here). Aristoteles knows three kinds of proofs which have been used up to the 20th century. Here Lukasiewicz (Lukasiewicz 1993), who also discussed the three-fold logic of Aristoteles, started discussions of how to write these proofs in terms of modern logic[3]. Many others have tried this too with their formulation of logic, still none of them has any doubt about the syllogism proofs being correct, and as this discussion is solely in the field of formal logic we can neglect this discussion here.

Musical logic as syllogistic

'Logos', meaning 'speech', is often used by Aristoteles as 'legetai', 'is sayed from' to form the sentences of the syllogisms. Also an ontological formulation using the gammer terms 'is them' or 'is theirs' or simply 'is' is used by him. So 'Socrates is a human being.' and 'Socrates is said to be a human being.' are equivalent in terms of syllogism. When it comes to music we have to find rules for dealing with the connections between the phaenomena present here. As we discuss music theory in this paper we deal with notes or tones of certain pitches. And as music theory is based on reasoning which again is based of logic, the connection between tones can be temporal, spatial or content-based. So before we define the music theory we first have to write down the syllogisms of events following in time, notes being spaced next one to another and the connection of content between them.

[3] Aristoteles himself – as correctly argued by Lukasiewicz – knows sentences which are true, false and true and false. The examples of Aristoteles can be interpreted using the partly affirmative or negative terms as they are not only true (or false). Part of this discussion is also in his writings unfortunately called 'Metaphysics' where he tries to tell the kernel of things (hypokeimenon, later sub-stare or substania, substance) and the things occasionally present (symbebekon, later accidentia, accidential) as the kernel cannot contradict while attributes may. This is not 'unlogical' as this contradiction appears between attributes on different levels, kernel and surrounding.

Some of the reasonings we will find are dialectic in the sense of Hegel. Riemann, following Moritz Hauptmann in this way, argues with Hegelian dialectics when dealing with the cadence which will be discussed below. So first we have to write down this Hegelian dialectics of thesis, antithesis, and synthesis in terms of syllogism at least between the thesis and antithesis. This has already and widely be discussed in the 19th century by Georg Prantl in his famous 'Geschichte der Logik' (history of logic, Prantl 1885) and others to try to cope with the Hegelian Logic in terms of traditional logic. For Hegel, a thing A is only becoming this thing if it first leaves itself (the A) in a *Nichtung* (none-being, vanishing) to again come to A while then finding A and the *Nichtung* of A being the same. This seems unlogical as the none-being of A (what A is not) is then found to be A which contradicts the first finding. Here Prantl argues that this is not a contradiction as Hegel's *Nichtung* refers to a definition of the None-A. So the other thing next to A, the B is syllogistically not written as B -> e A (negative). The syllogism need to be None-A -> a B (affirmative). In this second case we do have an affirmation 'a' of the None-A which only seems to be a contradiction but can easily be written down as a syllogism and so does not contradict classical logic.

1) temporal

With this in mind we can write down the syllogism of the temporal aspect of events in general and of tone events in our case as

Time point B -> a Time point None-A	(maj) : statement of existence
Time point None-A -> a Time point A	(min) : statement of essence

=> Time point B -> Time point A (syll.) : statement of temporal following of time point B after time point A: A – B

(maj) states that a time point B only exists as a time point None-A as A is there and B only follows up. So this statement is about the pure existence of time point B.

(min) states that a time point of A need necessarily be there if a time point None-A shall be there as both cover the 'A'. So as the A is the essence of both time points this statement is of essence, i.e. if a social group comes into existence because it is protesting against another social group both share the same topic the discussion is all about.

(syll.) then only states, that the subject time point A is followed by a time point B because of the medium phrase time point None-A. In other words, if a time point follows a previous one it does so because is shares the same essence to the first one as 'the other' or 'the next' referring to the first one. This is very important to mention as we now have a clearer picture about events following each other not only in a random manner but as referring one to another. Only if time points do so we are able to see connections between them and so music can come into place.

The property of time as a 'one way street' can be explained here, too as the order is not random. If we talk in terms of Husserl's writings about time (Husserl 1928) and use the term retention for remembering a recent time event and protention of forecasting a near one we can write

Time point B -> a Time point A	(syll. from above) : retention, echotic memory
Time point A -> i Time point B	(conversion of syll. from above) : protention, expectation

Here we see that to look back is a complete affirmative case which have needed to be so and looking forward is a partly affirmative case where other possibilities of progression may there be, too.

2) spatial

The reasoning for the spacing of tones is likewise to the temporal case. Here we can write

Tone B -> a Tone None-A	(maj) : statement of existence
Tone None-A -> a Tone A	(min) : statement of essence
=> Tone B -> a Tone A	(syll.) : statement of the special following of tone B next to tone A

(maj) states that a tone B only exists as a spatial point of None-A or the other of A. This statement is of pure existence.

(min) states that tone A and tone None-A have the same essence, namely the 'A' which is common to both. In other words the tone next to A as referring to tone A is only heard as in its context and otherwise it would not exist as tone B if not as tone None-A.

(syll.) then only states that tone B is next to tone A. Note that this only holds if these two tones are taken alone. We will discuss this when forming a tonal system. Still the fact that tone B is heard as 'something else' than tone A can only be if it refers to tone A and is 'the other'.

Again it is interesting to see the conversion.

Tone B -> a Tone A	(syll. from above) : progression, tonal system
Tone A -> i Tone B	(conversio of syll. from above) : open space, musical freedom

The progression in tonal space from tone A to tone B is a necessary one to form a tonal system as we will see below. The conversio on the other hand states, that for tone A, tone B is only one possibility of a spatial neighbour. There may be others and so tone A is free to progress and refer to others. So here a fundamental difference shows up between tones in a tonal system depending upon a fundamental tone. Also this discussion is necessary to get into the foundation of referring of tones in a system and so in the reasoning or logic of it. Again, if there would be no reference between the tones there could not be any tonal system, there could be no music at all as music can only exist because of the temporal and spatial differences and references.

3) content

When formulating a tonal system we look at tones of arbitrary timber and the like. So on this quite abstract basis the references of content between tones cannot be on a level of adjectives like 'bright' or 'exiting'. Only two contents are needed here to construct the system. The first is already known from the previous discussion as the Hegelian *Nichtung* and the second is the old philosophical 'sentence of identity'.

So as already explained above, the *Nichtung* in terms of syllogism is

None-A -> a B

stating that A and B do refer one to another in terms of their existence where B is defined in a complete affirmative way as None-A.

The sentence of identity is 'A is A' or 'A = A' and reads as syllogism

A -> a A.

The important thing about this sentence is that although A is the same on both sides, it appears twice and so differs in terms of appearing at different sides. So we better write

A' -> a A

meaning that the essence is the same while still the place of existence (but not any other property) is different. In terms of a tonal system a simple example is the octave where the same pitch class appears at different points in tonal space.

It shall be noted here that one of the advantages of the use of syllogistic in many cases turned out to be a very efficient tool for deeply understanding musical phaenomena. This is caused by the strict formalism of the syllogistic method. So i.e. to find the above one needed to start the syllogism at its end. Because a complete affirmative conclusion was needed, only one syllogism and so only one way of thinking was possible to lead to useful conclusions. Again, was the members of the conclusion (the A and the C) have also been known the only task was to find a medium term B connecting these two. So the syllogism forces one to strictly and deeply go into the reasoning of musical thought and clearly define its terms and methods.

Riemann's reasoning of tonal cadence

Riemann is following the ideas of Moritz Hauptmann, and his Harmonik und Metrik (Hauptmann 1853), but he is especially focussed on the time-dependent succession of chords, witch is necessary for every piece of music. Therefore he argues against a static perception of music, such as Hauptmann followed, when he determined the cadence only through a complete presentation of the tonal material, all tones of the scale, through sub dominant and dominant, and a closing tonic on the end, disregarding a time-dependency. Riemann follows the ideas of his teacher Hermann Lotze about *economical hearing* (Ökonomie des Hörens). In his *Katechismus der Musikästhetik* (Riemann 1890) he explains this as follows. If a tone or chord is appearing as a first musical event it is taken as a base for all following events. So if a second tone or chord appears it is always perceived *in relation* to the first event. This is more economic than perceiving the second event as a completely new one. Lotze and Riemann argue that tonality is caused by this temporal succession of tones with economic perception. So tonality can only be explained by temporal development of tones as they create a relation between them where the first tone is taken as reference. This is what makes Riemann´s musical logic a striking innovation and determines his understanding of the cadence.

Riemann argues that in every piece each chord can be reduced to its function as thesis, antithesis or synthesis. The notions of thesis, antithesis and synthesis can be led back to the thought of G.W.F Hegel and his *Logik der Wissenschaft* (logic of science) (Hegel 1813, 1832). It can be reduced to the coherence of being (A) and nonentity (None-A), and their genesis. He postulates that the being and the nonentity is the same. The pure being is undefined, it is nothing else than just pure being. The first thought is always the simplest, the most non determined thought. Regarding the content of nothing else than pure being, we can only say that it is nothing, it is nonentity. So therefore, pure being and nonentity is the same. Only the genesis, which Riemann takes as time here, is determining them as different. But they are still the same because the genesis is the unity of both, of being and nonentity. It is a process, wherein both are defined and separated, but still combined. The pure being as the first thought is the pure abstraction, therefore the absolute negative, strictly speaking the nonentity. The nonentity is the straight equality with itself, and therefore it is the same as the being. The moment of truth is the unity of being and nonentity. This unity is the process of genesis.

Riemann`s musical logic follows the same structure of thinking. The thesis is at first a pure being, as well as the antithesis. They are pure beings as well as they are nonentities. Only because of the synthesis or the process of genesis they can be defined as what they are. This final definition is felt as satisfaction at the closing chord of a cadence.

Based on the fact that there is a fixed coherence of musical material within cadences, the theory is explaining every kind of music from the smallest unit to bigger forms and arrangements. He is explaining two different types of cadences, which comply the same logical functions. They are divided into a big cadence (with the middle C major chord in Figure 1) and a small cadence (without this chord).

Thesis	Antithesis		Synthesis	
C	F	C	G	C
I	IV	I	V	I

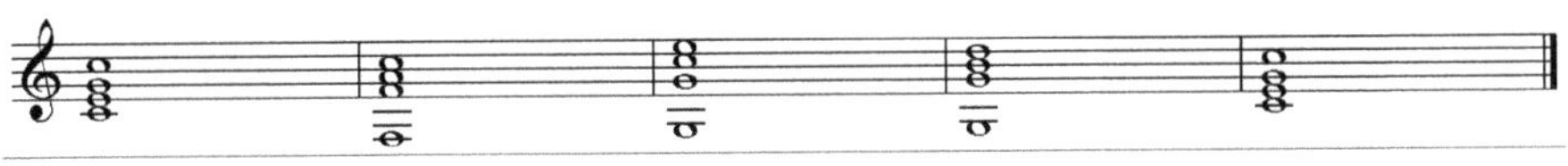

Figure 1: Riemann's cadence structured in thesis, antithesis and synthesis. His reasoning uses the tone c^1 and its succession through the chords.

The reasoning uses the tone c^1 and its development through the cadence. This tone is thetic in C as it is the fundamental tone of this chord. It becomes anthithetic in F as it is the 5th of this chord. Because of the economic hearing the tone, known as fundamental is now 'displaced' into a 5th position and therefore contradicting itself, becoming 'anti', a 'None-A'. This contradiction is hold on during the following C major chord as this chord uses its 5th as bass tone. Then in the following G major chord the tone is replaced by h (english b) which according to Riemann is synthesizing the two contradicting notions of c^1. Riemann writes that the final c^1 is imperiously demanded ("gebieterisch gefordert") by the h (english b). The ending chord C major is presenting this synthesis and therefore closing the cadence. According to Riemann the whole antithesis is the F – C progression and the whole synthesis is the G – C progression. So when using only the F major or talking only about the G major chord they are not the complete form but still have these properties and are therefore called antithetic and synthetic respectively.

For Riemann this reasoning is the cause of the difference between the cadence Tonic-Subdominant-Tonic and Tonic-Dominant-Tonic. According to Riemann the first one sounds skinny and cold, the second including the dominant sounds full and satisfying. Therefore the plagal cadence is not aspiring to close within the musical arrangement, whereas the cadence Tonic-Dominant-Tonic is. The deceptive cadence where the closing Tonic is replaced by another chord is a non closing cadence which is not following the logical coherence of the chords within tonality. The leading tone compels the reappearance of the tonic, but instead of it, there is a non logical chord following, and the desired synthesis is deceptive and false.

Riemann´s musical logic is even going further discussing the chords built upon the second, third, sixth and seventh scale tone. These chords can substitute the main degrees of a cadence. They are ordered in a hierarchic structure according to their relation to the main functions which are thetic, antithetic, and synthetic. The relation is based on the similarity of tone material. That means that the thetic moment, typically built by the tonic, can be substituted by the Tonic-parallel and Dominant-parallel. The antithetic moment can be substituted by the Subdominant-parallel. Finally the subtonic and the mediant can substitute the dominant and build up a synthetic moment within a cadence.

In terms of syllogism Riemann's reasoning can be written as

None-c^1 -> a c^1 : antithetic, c^1 is fundamental in C-major and 5th in F-major

c^1 -> e h : synthetic G – major, c^1 is replaced by h and therefore not h

=> c^1 -> a h : synthetic C – major, c^1 is 'imperiously demanded' by h

This is not a valid syllogism as the form A -> a B, B -> e C does not lead to any conclusion. Even if we were rearranging the order of the sentences like

c^1 -> e h : synthetic G – major, c^1 is replaced by h and therefore not h

None-c^1 -> a c^1 : antithetic, c^1 is fundamental in C-major and 5th in F-major

=> c^1 -> a h : synthetic C – major, c^1 is 'imperiously demanded' by h

and therefore get the form A -> e B, B -> a C which is Celarent the conclusion would be A -> e B and not A -> a B meaning that h is not demanding but rejecting c^1. So although the first part of Riemann's reasoning is clear the synthesis part is not logical. This is also strongly felt within his reasoning when only looking at it without formalization. Why should the tone h demand the tone c^1 because of the antithetic nature of c^1 appearing before? Indeed we could write down a valid syllogism like

c^1 -> a None - c^1 : synthetic G – major, c^1 is replaced by h and therefore not h

None-c^1 -> a h : antithetic, c^1 is fundamental in C-major and 5th in F-major

=> c^1 -> a h : synthetic C – major, c^1 is 'imperiously demanded' by h

Formally this is Barbara and so correct. The problem is the definition of None- c^1 here. If there are more than one None- c^1, then we are not allowed to use a complete affirmative case 'a' but only a partial affirmative case 'i'. But when doing so this syllogism again does not exist and no conclusion can be drawn.

The fact that we feel a strong demand of the h to return to the c^1 must be explained differently. Therefore we try a new definition of the cadence in a syllogistic way. We try to be as general here as possible to include all possible musics may they be western or not.

Musical cadence written as syllogism

From the discussion above we can now quite easily find a reasoning for the cadence. If a cadence closes necessarily (and in german the word 'Schluß' means both, the cadence and a logical conclusion), we at least know the syllogism we have to use as there is only one concluding with a complete affirmation, the Barbara syllogism. We can argue then the

Definition of cadence

None-A -> a A	: existential and temporal, an event following A is None-A
A -> a A'	: essential, the event is the same but different in time, sentence of identity
=> None-A -> a A'	: tension, a coming A is heard within the None-A

After A is sounding another event different from A is perceived as None-A. Now temporally, A is left behind. Still another event A' being the same as the first one but temporally different need to be written according to the sentence of identity. We must conclude then, that within the None-A a coming – not jet present – event A' is *there*. This *there* is a tension because although it is there it is not sounding jet. Riemann's notion of 'demanding' is present here. Still we do not – and cannot – reason that A' *will* appear. We only find the tension present and so when A' appears the tension closes. This is an important fact because if we would have had the conclusion that A' -> a None-A : temporally, then A' must appear after None-A as a kind of natural law, strictly speaking no one would ever be able not to play A' after None-A. This is not true of course and this is not concluded here. The difference to the definition of Riemann is the avoidance of another chord or tone progression before the Dominant. Here the thesis is the identity of two same events different in time, the antithesis is the appearance of another event and the synthesis is the 'demanding' of yet the first event again. Maybe Riemann felt the synthesis case quite equal as he did not state the last C-major chord as synthesis alone but the G – C chord progression including the G-major chord. Note that his is only possible as a time dependent form and not without it as the event A and A' need to be the same and the ' denotes only the temporal difference.

Our definition of a cadence is as wide as possible. We have not only defined a cadence for the classical chord progression of a cadence, it holds for melodies, it also holds for sounds without any pitch, it even can be expanded to time dependent processes not in the field of music. Still, the main progression of 'home – away from home – home again' is logically felt here. Examples may be manifold here, Flamenco music i.e. has a very wide notion of cadence where everything leaving the fundamental chord part of the cadence, melisms in Arabian classical and improvised music know and use this notion, classical Indian music follows this reasoning, too. Indeed we claim this as the fundamental reasoning of all musical events and so also follow Riemann when he states that all music can be decomposed into cadential sections.

Note, that None-A need to be all that is not A. So cadence in this most general form means that everything played after the event A is 'away from home' and so None-A. Also we need to remember A and so feel everything not A as away from it. So we need to enlarge our definition to tonal systems now to learn more about the logic when differentiating None-A.

If we deal with more than two events a fundamental difference appears. In the case of two i.e. tones we could use the content sentence Nicht-A -> a B and the spatial sentence B -> a A. If we enlarge the system to only three tones A, B, and C the spatial relation changes to B -> i A, because the tone C is also around. If we now try to draw conclusions according to the spatial relations only we have the following possible sentences

1. A -> a B : spatial
2. A -> e C : spatial
3. B -> i A : spatial
4. B -> i A : spatial
5. B -> i C : spatial
6. C -> e A : spatial
7. C -> a B : spatial

If we construct syllogisms with these sentences i.e. Ferison

A -> e C : spatial
B -> i C : spatial
=> A -> o B : spatial, there is another note next to B besides the note A

we only arrive at conclusions we have already put into the system and so are tautologies. If we do this with only the temporal development the picture is similar. So i.e. when we play A, then B and then C we can write a Darapti like

C -> a B : temporal
B -> a A : temporal
=> C -> a A : temporal, the note C is played after, although not right after the note A.

Again a trivial conclusion. *So we cannot derive a tonal system based on more than random temporal or spatial neighbourhood on spatial or temporal evolution alone. To come close to the real relations between the tones in a tonal system we have to combine temporal and spatial relations.*

To show the inner relations we can write a Felapton for the sequence of tones A – B – C like

C -> e A : spatial, tone C is not next to tone A
B -> a A : temporal, tone B follows tone A
=> C -> o B : content, after the sounding of A and B another tone C is possible and distinguished from B

Here the temporal evolution splits the None-A - which is B and C together - if a third tone C is not next to A. Note that the conclusion is a negation (partial) and not an affirmation.

If we have the sequence of tones A – C – B then we can write a Disamis like

B -> i A : spatial, one of the tones next to B is A
C -> a A : temporal, tone C sounds after tone A
=> B -> i C : content, after the sounding of A and C another tone B is possible and similar to C

These two last syllogisms show the difference between an extension and a differentiation of the tonal system. In the case of an extension where the new tone lies beyond both previous tones the conclusion is a negaio (o), in the case of an differentiation where the next tone lies within the previous two tones the conclusion is an affirmation (i). This shows that the tonal system is not arbitrarily extended but follows a logic of 'new' (negation) and 'old' (affirmation) depending on the melody line.

It is also important to note that all conclusions which can be drawn from only two tones (or generally events) are only partial ones. We never reached at any complete conclusion like 'a' or 'e' anymore. This does not change when the system is enlarged to more than three notes. So logically we can have a qualitative difference between

System with 1 event:	unlogical
System with 2 events:	complete affirmatio and complete negaio reasoning, cadence
System with 3 or more events:	partially affirmation and partially negaio reasoning, tonal system

Musical consonance written as syllogism

Most tonal systems show harmonic relations of the intervals. These relations are expressed in numbers. The logical content of numbers has been discussed by Frege when founding modern logic. Although Frege in the end failed to find a logical basis for numbers we nevertheless will treat them syllogistically. This is possible by using again a definition of Aristoteles concerning numbers. For him 'one' was not a number. The greek word *amonia* (number) is composed out of *monas*, unity and the prefix *a-* as an a-copulativium meaning 'the togther of unities' or units. So '1' cannot be *amonas* as yet nothing has come together there. So if we distinguish between the number one ('1') and the counter one (*monas*) we can write.

1 -> a *monas* : number 1 is unit (counter, *monas*)
monas -> a 2 : number 2 is contructed out of and only out of units
=> 1 -> a 2 : number 2 is constructed out of number 1

Although the conclusion seems trivial, the reasoning is not. If we assume two tones, say c' as 1 und c'' as two then we find that c'' is *constructed out of* c'. *So the most trivial relationship between 1 : 2 is now logically loaded with the term construction.* This construction means that what is c' is also c'' and musically this is the *Tonigkeit* or pitch class. In other words pitch class logically speaking is a reasoning derived from one unit and one derivative.

We can derive now syllogistically the following intervals starting with the octave already explained and using the definition of *Tonigkeit* from then on[4].

Octave Barbara

1 -> a *monas*
monas -> a 2
=> (syll.) 1 - > a 2 : *Tonigkeit,* Octave

Fifth (2 : 3) Darii

Tonigkeit -> a 1 : 2 : content, 1:2 is definition of *Tonigkeit*
1 : 2 -> i 2 : 3 : content, number 2 present in both intervals
=> (syll.) *Tonigkeit* -> i 2:3 : *Tonigkeit* partially present

Fourth (3 : 4) Ferison

3 : 4 -> e 2 : 3 : content, as 3:4 plus 2:3 form an octave, 3:4 is the other of 2:3
Tonigkeit -> i 2 : 3 : content, shown above
=> (syll.) 3 : 4 -> o *Tonigkeit*: *Tonigkeit* not partially present

Major Third (4 : 5) Darii

Tonigkeit -> a 1 : 4 : content, 1:4 is definition of *Tonigkeit*
1:4 -> i 4 : 5 : content, number 4 present in both intervals
=> (syll.) *Tonigkeit* -> i 4:5 : *Tonigkeit* partially present

Minor Third (5 : 6) Ferison

5 : 6 -> e 4 : 5 : content, as 5:6 plus 4:5 form an fifth, 5:6 is the other of 4:5
Tonigkeit -> i 4 : 5 : content, shown above
=> (syll.) 5 : 6 -> o *Tonigkeit* : *Tonigkeit* not partially present

---

[4] 1:3 is not Tonigkeit as it holds that *monas* -> i 3 (and not *monas* -> a 3), where 3 not only consists of units but also consists also of the '2'. On the other hand, 1:4 and 1:8 are octaves again and therefore *Tonigkeit* holds for them.

Major Second (8 : 9) Darii

Tonigkeit -> a 1 : 8 : content, 1:8 is definition of *Tonigkeit*
1:8 -> i 8 : 9 : content, number 8 present in both intervals
=> (syll.) *Tonigkeit* -> i 8:9 : *Tonigkeit* partially present

Minor Second (9 : 10) Ferison

9 : 10 -> e 8 : 9 : content, as 9:10 plus 8:9 form an third, 9:10 is the other of 8:9
Tonigkeit -> i 8 : 9 : content, shown above
=> (syll.) 9 : 10 -> o *Tonigkeit*: *Tonigkeit* not partially present

All intervals explained are derived from the octave. All except the octave have a partial affirmatio or negatio to pitch class or *Tonigkeit*. The reasoning for partial affirmatio *Tonigkeit* is that the lower tone of the interval is an octave (*monas*) and therefore has *Tonigkeit*. The partial negatio intervals are defined as 'the others' to the related ratios having one number in common and forming one larger interval together. They are always the smaller interval compard to their companion. The following table may make this more clear.

Interval	**kind of *Tonigkeit***	**is lower tone an octave?**	**syllogism**
Octave	complete affirmatio	yes, 1:2	Barbara
Fifth	partial affirmatio	yes, 1:2	Darii
Fourth	partial negatio	no	Ferison
Major Third	partial affirmatio	yes, 1:4	Darii
Minor Third	partial negatio	no	Ferison
Major Second	partial affirmatio	yes, 1:8	Darii
Minor Second	partial negatio	no	Ferison

It clearly appears that we have found a reasoning for musical consonance. Musical consonance is the amount of *Tonigkeit* or pitch class present in a musical interval. The amount is greater with affirmatio than with negaio and it is greater the smaller the octave interval. So we do not only have an ordering of consonant intervals where smaller intervals are less consonant. The logical reasoning also finds a categorical difference between intervals directly connected to *Tonigkeit*, like Fifth, Major Third and Major

Second with more consonance and the ones indirectly connected through larger intervals where they fit with their larger counterparts like Forth, Minor Third and Minor Second.

So the octave as the only complete affirmatio, is perfect consonance. The fifth is directly connected to the octave and therefore still very consonant although as only partially and not completely affirmatio less consonant than the octave. The Forth is negatio in terms of Tonigkeit and as 'the other' of the Fifth when fitting into the octave interval also less consonant. This may explain that in traditional counterpoint (i.e. Fux) the Forth interval was treated as dissonant. This cannot be explained if we only look at the ratios where the Forth is still larger than the Major Third which again is considered to be a consonant interval although then only an incomplete one.

The same schema as with the Fifth and Forth appears with Major Third/Minor Third and Major Second/Minor Second. Again the larger interval is directly connected to the octave where the smaller one is again 'the other' to its bigger part fitting in a larger interval. Fux considers the Thirds as incomplete consonant and the Seconds as dissonant. This can be explained as an arbitrary, yet reasonable decision as the distance between the Major Second and the fundamental (1:8 to 1:4) is twice as large as the distance between the Major Third and the Fifth (1:4 to 1:2).

This theory of consonance is also easily explained in terms of the overtone structure. It claims that the intervals where the lower tone is an octave to the fundamental is heard as consonant as there must exist a residual tone at the fundamental. So the Fifth 2 : 3, the Major Third 4 : 5 and the Major Second 8 : 9 all have a residual tone which is one, two and three octaves below the fundamental sound respectively. This is not true for the Forth, Minor Third and Minor Second interval. They also have a residual of course but in a more complicated harmonic relation to the lower interval tone and only the Forth as a relation to an octave but here with the higher interval sound. So we can also define consonance in terms of partials. *Consonance is the fitting of the residual of two tones to the octave of the lower of the two tones in terms of pitch class or Tonigkeit.*

Note, that if stated in such a way, a theory of consonance would just be a trial theory. We on the other hand have derived this logical relationship between the intervals and found this theory of consonance by using logical reasoning. So we do not only see a physical reason for these consonances we are also able to understand the musical logic it is founded on.

Conclusions

The syllogism is capable to derive a tonal theory based on logic. In this paper only the beginning of such a theory can be presented and more differentiated systems for different musical styles need to be developed later. The advantages of such a system are manifold.

1. It is derived from a kind of logic deeply connected to ontology. So we are sure our understanding is a real understanding of music and do not need to fear that our understanding is restricted to a musical style or a social or historical age.

2. The reasoning is given by a formal syllogism which at the same time can be written as a meaningful sentence clearly and precisely expressing the musical thought. So the syllogism is always its interpretation in 'normal' musical terms.
3. As ontology is present in this formulation of logic many other fields of music may be treated with such a system, i.e. musical meaning, musical performance, sociological systems ect. All of them can be treated syllogistically.
4. The understanding of music itself is very much enhanced by this method. As shown above many non-trivial questions like cadence or consonance can be derived and understood within the global framework of music in its relations. Here new explanations and musical thoughts are derived. So the syllogism is a powerful method as it was constructed for describing thoughts in general and the connections between things.
5. The syllogism is also able to make large-scale conclusions within one syllogism. Often formal methods need many steps to derive results and so are quite lengthy sometimes. This is not true here as the input of the syllogism, the sentences can be as wide as needed.
6. As the syllogism also knows partly true (and false) sentences and conclusions it follows musical language very closely and intuitively as music is not only governed by laws but also by possibilities. In this formulation only the assertoric syllogistic was used, the modal one dealing with possibilities and necessities is still waiting to be explored for musical thought.

So as until now the syllogism was capable of going deeply into musical language and clearing up many points with its precise and clear laws we hope to enlarge the system of syllogistic music theory in terms of tonal theory but also in terms of other fields concerned with music.

References

Aristoteles (1989), *Metaphysik. Bücher I-VI*, Hermann Bonitz (Neubearbeitung der Übersetzung von Wilhelm Christ), griechisch-deutsch, Hamburg: Felix Meiner Verlag.

Aristoteles (1991), *Metaphysik. Bücher VII-XIV*, Hermann Bonitz (Neubearbeitung der Übersetzung von Wilhelm Christ), griechisch-deutsch, Hamburg: Felix Meiner Verlag.

Aristoteles (1998), *Über die Seele*, Horst Seidl (Neubearbeitung der Übersetzung von Willy Theiler), griechisch-deutsch, Hamburg: Felix Meiner Verlag.

Aristoteles (2005), *Die Kategorien*, griechisch-deutsch, Hamburg: Felix Meiner Verlag.

Aristoteles (1998), *Erste Analytik, Zweite Analytik*, Organon Band 3/4, Hans Günther Zekl (Hrsg. und Übers.), griechisch-deutsch, Hamburg: Felix Meiner Verlag.

Aristoteles (1996), *Topic, Book I and VIII*, Transl. with commentary by Robin Smith, Oxford: Claredon.

Aristoteles (1994), *The 'art' of rhetoric*, John Hendry Freese (transl.), Vol 22, Aristoteles in twenty-three volumens, Cambridge: Harvard Univ. Press.

Drechsler, M. (2005), Interpretationen der Beweismethoden in der Syllogistik des Aristoteles, In: *Prismata. Beiträge zur Altertumswissenschaft*, Franfurt am Main: Peter Lang Verlag.

Hauptmann, Moritz (1853), *Die Natur der Harmonik und der Metrik: zur Theorie der Musik*, Leipzig: Breitkopf & Härtel.

Hegel, G.W.F. (1999), *Wissenschaft der Logik*, Hamburg: Felix Meiner Verlag (Original: 1813, 1832).

Husserl, Edmund (1928), Vorlesungen zur Phänomenologie des inneren Zeitbewusstseins, Herausgegeben von Martin Heidegger, *Jahrbuch für Philosophie und phänomenologische Forschung*, 9, Halle a. d. S., viii-ix; 367-498.

Lukasiewicz, Jan (1993), *Über den Satz des Widerspruchs bei Aristoteles*, Jacek Barski (Übersetzung aus dem polnischen), Hildesheim: Georg Olms Verlag.

Maier, Heinrich (1896), *Die Syllogistik des Aristoteles I: Die logische Theorie des Urteils bei Aristoteles*, Tübingen: Verlag der H. Laupp'schen Buchhandlung (im Nachdruck der 2. Aufl. Leipzig 1936, Georg Olms Verlag, Hildesheim, New York 1969).

Maier, Heinich (1900), *Die Syllogistik des Aristoteles. Die logische Theorie des Syllogismus und die Entstehung der Aristotelischen Logik 2 Bd: II,1: Formenlehre und Technik des Syllogismus. II, 2: Die Entstehung der Aristotelischen Logik*, Tübingen: Verlag der H. Laupp'schen Buchhandlung (im Nachdruck Tübingen 1900, Georg Olms Verlag, Hildesheim, New York 1970).

Patzig, Günther (1969), *Die aristotelische Syllogistik. Logisch-philologische Untersuchungen über das Buch A der „Ersten Analytiken"*, Göttingen: Vandenhoek & Ruprecht.

Petrus Hispanus (2006), *Logische Abhandlungen. Tractatus / Summulae Logicales*, Wolfang Degen, Bernhard Pabst (Hrsg. und Übersetzer), München: Philosophia Verlag.

Platon (2007), *Sophistes*, griechisch-deutsch, Frankfurt a. M.: Suhrkamp.

Prantl, Carl (1885), *Geschichte der Logik im Abendlande Bd. 1*, Nachdruck Wissenschaftliche Buchgesellschaft Darmstadt 1957.

Riemann, Hugo (1895 u. 1900), *Präludien und Studien: gesammelte Aufsätze zur Aesthetik, Theorie und Geschichte der Musik*, Nendeln, Kraus 1976.

Riemann, Hugo (1890), *Katechismus der Musikästhetik. „Wie hören wir Musik?"*, Leipzig: Hesse.

Schmidt, Klaus J. (2002), *Die modale Syllogistik des Aristoteles. Eine modal-prädikatenlogische Interpretation*, Mentis-Verlag.

Ulrich Morgenstern

Die Sackpfeifenmelodien der russischen Harmonikaspieler[1]

1. Problemstellung

In der Erstauflage des *Erklärenden Wörterbuches der lebendigen großrussischen Sprache* führt Vladimir I. Dal' (1801–1871) im Artikel *duda* ein bei Hirten, Kindern und Bettlern gebräuchliches Blasaerophon an, das „selten mit den übrigen Volksmusikinstrumenten (*gudok* [Streichlaute], Balalaika, *volynka*, *rožok* [bei Dal': die Hirtenklarinette *žalejka*]) zusammen gebraucht wird."[2] Wie wir sehen, erwähnt der große Lexikograph die russische Sackpfeife (*volynka*) als eines der gewöhnlichen Volksmusikinstrumente. Im *Enzyklopädischen Lexikon* von Adol'f A. Pljušar heißt es über den Bauerntanz *Golubec*, daß dieser „zu Balalaika, *gudok*, *volynka* oder zur Melodie irgendeines Liedes"[3] getanzt werde. Verallgemeinernde Erwähnungen der *volynka* im 19. Jahrhundert werden durch Beobachtungen von nicht weniger kompetenten Autoren wie Tučkov[4], Fürst Odoevskij (im selben Lexikon, Bd. 2, Stichwort *volynka*), Tereščenko[5], und anderen bestätigt. All diese Quellen bezeugen, daß bis zur Einführung der Handharmonika seit den Dreißigerjahren des 19. Jahrhunderts die *volynka* in Rußland weit verbreitet war.[6]

Unglücklicherweise ist heute keine einzige russische Sackpfeife erhalten. Ebenso fehlen Fotos, genauere Beschreibungen oder Tonaufzeichnungen russischer Sackpfeifenmusik. Das Instrument hat jedoch seine Spuren in der rezenten Musikpraxis hinterlassen. An anderer Stelle habe ich versucht, einige Besonderheiten im Vortragsstil von Harmonikaspielern im Gebiet Pskov herauszuarbeiten, die wahrscheinlich mit der Spieltechnik der Sackpfeife verbunden sind.[7] Gleichwohl beziehen sich die hier aufgeführten „Sackpfeifenmerkmale" nur auf Einzelbeobachtungen und erlauben es nicht, konkrete Harmonikastücke, sondern allenfalls bestimmte Gattungen der *volynka* zuzuordnen. Zu-

[1] Bei der vorliegenden Abhandlung handelt es sich um die deutsche Version meiner ursprünglich in russischer Sprache verfaßten Arbeit „Volynočnye naigryši russkich garmonistov", die für den in Vorbereitung befindlichen Sammelband *Severobelorusskij sbornik* (St. Petersburg, Institut der Geschichte der Künste) vorgesehen ist. Ich danke dem Herausgeber Dr. Aleksandr Romodin für die freundliche Erlaubnis für den deutschen Vorabdruck.

[2] Vladimir Ivanovič Dal', *Tolkovyj slovar' živoga velikorusskago jazyka*, Bd. I, Moskva 1861.

[3] *Ėnciklopedičeskij leksikon*, T. XIV, S. 422, Sankt-Peterburg, 1838.

[4] S. A. Tučkov, *Zapiski Sergeja Alekseeviča Tučkova* (Aufzeichnungen von S.A. T. [1780-1809]), Sankt-Peterburg 1906, zit. nach: Konstantin Aleksandrovič Vertkov, *Russkie narodnye muzykal'nye instrumenty* (Die russischen Volksmusikinstrumente), Moskva: Muzyka 1975, S. 265.

[5] A.V. Tereščenko, *Byt russkago naroda* (Leben des russischen Volkes), Sankt-Peterburg 1848, S. 485.

[6] Vgl. auch Vertkov, *Russkie narodnye*, S. 138–139. Noch nicht alle der dem 18.–20. Jahrhundert entstammenden Quellen zur russischen Sackpfeife (auch außerhalb des weißrussischen Grenzgebiets) sind bislang in der Musikwissenschaft bekannt oder ausgewertet. Dies ist Thema einer für die Zukunft geplanten Arbeit des Verfassers.

[7] Ul'rich Morgenštern, K voprosu o kornjach sovremennoj tradicii igry na balalajke i na garmonike v Rossii (Zur Frage nach den Wurzeln der heutigen Tradition des Balalaika- und des Harmonikaspiels in Rußland), in: *Sud'by tradicionnoj kultury. Sbornik statej i materialov pamjati Larisy Ivlevoj*. Sankt-Peterburg: Dmitrij Bulanin, 1998, S. 164–186.

dem sind die vermutlichen Relikte des Sackpfeifenspiels nur entfernt mit historischen oder ethnographischen Zeugnissen zu diesem Instrument in Beziehung zu setzen. In der vorliegenden Arbeit soll ein in beiderlei Hinsicht konsistenteres Bild gezeichnet werden.

Die hier verwendeten Schriftquellen und Feldforschungsdaten beziehen sich auf die Gebiete (*oblast'*) von Pskov, Tver' und Smolensk – wobei die ethnographischen Zeugnisse hauptsächlich dem Grenzgebiet zu Weißrußland entstammen, das dialektologisch dem nordweißrussischen Sprachraum zugehört.[8] Vor diesem Hintergrund mag der Leser mit Recht eine Erklärung für den ethnonymischen Titel der vorliegenden Arbeit verlangen. Bekanntermaßen gehört der Südrand des Gebiets von Pskov und der Südwesten des Gebiets von Tver' auch nach ethnomusikologischen Befunden zur weißrussischen Tradition. Wäre es also nicht angemessen, die Zeugnisse zur Sackpfeife in diesem Gebiet vor allem der berühmten weißrussischen *duda* zuzuordnen? Einwände dieser Art (die mir gegenüber persönlich auch von seiten meiner geschätzten weißrussischen Kollegen vorgebracht wurden) sind nicht unbegründet. Und dennoch erlauben es meiner Meinung nach zwei Umstände, die Frage etwas anders zu stellen: zum einen ist dies die *ethnische Unbestimmtheit* des Instruments und zum anderen die *ethnische Bestimmtheit* des Repertoires, auf das wir auf der Suche nach Relikten des Sackpfeifenspiels stoßen.

Die weißrussische Sackpfeife *duda* ist bislang nur unzureichend erforscht. Ungeachtet einer soliden Zahl von Quellen[9] sind die in Museen (in und außerhalb Weißrußlands) befindlichen Instrumente sowie Fotografien und ethnographischen Zeugnisse noch nicht zusammenfassend dokumentiert und untersucht worden. Ebenso fehlen typologische Karten. Nichtsdestoweniger steht bereits zum gegenwärtigen Zeitpunkt der polyethnische Charakter der in Rede stehenden Instrumententypen fest. In Weißrußland waren sehr hochentwickelte Sackpfeifen im Gebrauch, mit stark gebogenen, gedrechselten Holzstürzen (*rogaven'*) an Bordun- (*huk*) und Melodiepfeife (*perabor*).[10] Hinsichtlich der Kompliziertheit der Konstruktion ist dieser Sackpfeifentyp in Nord- und Osteuropa allenfalls mit der westslawischen und der karpatischen Gruppe vergleichbar (Abb. 1). Bemerkenswert ist auch der von dem Volkskundler Evdokim Romanov erwähnte Umstand, daß die Anfertigung der Stürzen (gedrechselt aus der kostbaren Karelischen Birke) „im Dorf sehr schwierig ist."[11] Analoge Sackpfeifen sind auch aus

[8] Zur ethnokulturellen Situation des Grenzgebiets s. mein Beitrag im vorhergehenden Jahrbuch und die dort aufgeführte russischsprachige Literatur.

[9] Inna Dmitr'evna Nazina, *Belorusskie narodnye muzykal'nye instrumenty. Samozvučaščie, udarnye, duchovye* (Die weißrussischen Volksmusikinstrumente. Selbstklingende, Schlaginstrumente, Blasinstrumente), Minsk: Nauka i technika 1979; Al'bina Vasileŭna Skorabahatčanka, *Belaruskija narodnyja muzyčnyja instrumenty XX stahoddzja* (Die weißrussischen Volksmusikinstrumente des 20. Jahrhunderts), Minsk 2001.

[10] Evdokim Romanovič Romanov, Vymirajuščij [laut Inhaltsverzeichnis: Izchezajuščij] instrument (Ein aussterbendes [verschwindendes]) Instrument, in: *Vilenskij kalendar' na 1910*, Vil'no 1909, S. 124 (eine Kopie dieser in Deutschland nicht zugänglichen Quelle verdanke ich Herrn Ernst Eugen Schmidt. Sie wurde seinerzeit unter Mitwirkung von Lev Kopelev ins Deutsche übersetzt); Nazina, *Belorusskie narodnye*, S. 108. Nazina erwähnt unter Verweis auf Romanovs Arbeit Zeugnisse über eine *duda* „mit drei Bordunen aus dem Kreis Disna des Gouvernement Wilna" (ebd., S. 110). Hier handelt es sich leider um eine Verwechslung. Das bei Romanov abgebildete Instrument (Vymirajuščij instrument, S. 126) mit dem ungewöhnlichen Tripelbordun stammt nicht aus dem slawischsprachigen Disna, sondern aus dem Kreis Ljucin (deutsch: Ludsen) des Gouvernements Vitebsk (heute das lettische Ludza) und ist eher mit der livländisch-lettgallischen Tradition als mit Weißrußland verbunden.

[11] Ebd., S. 125.

dem benachbarten Litauen als *Labanoro dūda* bekannt – nach dem Namen des Städtchens aus dem Kreis Svencioniu des damaligen Gouvernements Vilno (heute: Vilnius), wo das Sackpfeifengewerbe sich konzentrierte (Abb. 2).[12] Gleichzeitig ergibt sich aus dem von Al'bina Skorabahatčanka mitgeteilten ikonographischen Material[13] (hier: Abb. 3, 4), daß im 19. Jahrhundert in Weißrußland nicht nur Sackpfeifen mit gekrümmtem Schallbecher an Melodie- und Bordunpfeife vorkamen, sondern auch Instrumente mit geradem und auch ohne Schallbecher. Instrumente des letzteren Typs waren weit verbreitet in Lettland, Estland und in Schweden.[14] Wir finden sie zudem in der nördlichen Ukraine[15], in Moldavien[16] und ebenso auf (nicht sehr detaillierten) Holzschnitten des 18. Jahrhunderts, die Moskauer Spielleute zeigen.[17] Der östlichste Vertreter dieses Sackpfeifentyps ist nach den derzeit vorliegenden Quellen der tschuwaschische *sarnaj*.[18] Balgbearbeitung (vernähtes Leder), einfache, gedrechselte Melodiepfeife und Bordun nähern den *sarnaj* dem nordeuropäischen Sackpfeifentyp an und grenzen ihn klar von den übrigen Typen des Wolgabeckens ab, die eher orientalisches und mediterranes Gepräge aufweisen.[19] Ohne jeden Zweifel ist die Verbreitung dieses Sackpfeifentyps in Osteuropa von erstrangiger Bedeutung für die Erforschung der russischen *volynka*.

Aber ebenso offensichtlich ist die begrenzte Reichweite einer Untersuchung der Sackpfeife in diesem Teil Europas aus nationaler Perspektive.[20] Dies gilt zumindest für die organographisch-morphologische Untersuchung, also ohne Berücksichtigung von Repertoire und musikalischer Stilistik. Auf der Ebene des Repertoires dagegen tritt die

---

[12] Rimantas Sliužinskas, The Bagpipe in Lithuanian Traditional Instrumental Music (Referat auf dem Symposium des Internationalen Dudelsackpfeifer-Festivals in Strakonice 1995), S. 2. Diese Quelle, wie auch das unten erwähnte Referat von Johannes Taul stellte mir freundlicherweise Dr. Fritz Schneider aus Krefeld zur Verfügung.

[13] Skorabahatčanka, *Belaruskija narodnyja muzyčnyja instrumenty*, Abb. 15, 17. In dieser Publikation über die weißrussischen Volksinstrumente ist zudem auf Foto 10 ein „Dudelsackpfeifer aus dem Gouvernement Vilno (Wilna), Ende 19. – Anfang 20. Jh)" mit der typischen *Labanoro dūda zu* sehen, wobei auf eine Archivquelle verwiesen wird. Tatsächlich handelt es sich um eine Aufnahme von M. Kuściński aus der Warschauer Zeitschrift *Naokoło Świata* (1905, 3, 24, S. 369) mit der eindeutigen Bildunterschrift „Kobziarz litewski" – „Litauischer Dudelsackpfeifer" (Zbigniew Jerzy Przerembski, *Dudy, Instrument mało znany polskim ludoznawcom*,Warszawa: Inst. Sztuki Polskiej Akademii Nauk, Towarzystwo Naukowe Warszawskie 2007, Abb. 10). In Skorabahatčankas Arbeit ist ein manipuliertes Foto wiedergeben, da der Bildrand mit der ethnischen Zuschreibung weggeschnitten ist.

[14] Per-Ulf Allmo, *Säckpipan i Norden*, Stockholm, Uppsala: AllWin 1990.

[15] Konstantin Aleksandrovič Vertkov / Georgij Ivanovič Blagodatov / Ėl'za Ėduardovna Jazovickaja, *Atlas muzykal'nych instrumentov narodov SSSR* (Atlas der Musikinstrumente der Völker der UdSSR), Moskva: Muzyka 1975, № 128.

[16] Ebd. № 163.

[17] *Russkija narodnyja kartinki. Sobral i opisal D.A. Rovinskij. Posmertnyj trud, pečatan pod nabljudeniem N.P. Sobka,* I., Sankt-Peterburg: R. Golike, 1900, S. 352 f., № 262–11, 13, 16). Die erste Abbildung zeigt eine Sackpfeife mit zwei Bordunen. Ein weiterer Holzschnitt des 18. Jahrhunderts wird unter der Bezeichnung „Ensemble von Skomorochen" von Vertkov angeführt (*Russkie narodnye*, nach S. 96), leider ohne Quellenangabe. Er zeigt eine ähnliche Sackpfeife mit einem Bordun.

[18] Vertkov, Blagodatov, Jazovickaja, *Atlas*, № 187.

[19] Vgl. u.a.: Anthony Baines, *Bagpipes* (Dritte Auflage: Occasional Papers on Technology, 9), Oxford: Pitt Rivers Museum 1995, S. 46–49.

[20] Eine Korrelation des ethnischen Elements mit der organographischen Charakteristik in Mittel- und Ostmitteleuropa ist lediglich bei der westslawischen Gruppe der Sackpfeife zu verzeichnen (vgl. u.a.: *Režný*, Josef, *Der sorbische Dudelsack*, Bautzen: Haus für *sorbische* Volkskultur 1993 und 1997).

ethnische Spezifik mitunter deutlicher hervor. Wie wir noch sehen werden, sind die Instrumentalformen, die Rückschlüsse auf die Sackpfeife erlauben, in dem uns interessierenden Areal (ungeachtet seiner ethnischen Heterogenität) hauptsächlich der russischen Tradition zuzuordnen. Sie gehören zur nicht-tanzgebundenen Gattung *pod pesni*[21], die in Weißrußland den Norden des Gebiet Vicebsk (Vitebsk) erfaßt, und zum Repertoire der russischen Einzeltänze (*pljaska*) wie *Russkogo* oder *Kamarinskaja*, das in Weißrußland ebenfalls vor allem im russischen Grenzgebiet ausgeprägt ist.

2. Zur Sackpfeife in der Region Pskov, Tver' und Smolensk

Wenden wir uns zunächst den historischen und ethnographischen Quellen aus der untersuchten Region zu, also aus den Gebieten von Pskov, Tver' und Smolensk. Im frühen 17. Jahrhundert führte der Lübecker Kaufmannsgehilfe Tönnies Fonne (der übrigens kein gesteigertes ethnographisches Interesse zeigte, sondern sein niederdeutsch-nordwestrussisches Gesprächswörterbuch für praktische Zwecke der hanseatischen Kaufleute verfaßte) einige Musikinstrumentennamen auf, darunter *duda* für die „sackpype."[22] Als Ensembleinstrument wird die Sackpfeife in dem erstaunlichen Bericht des dänischen Gesandten Just Julel aus dem Jahr 1709 erwähnt, bei dessen Besuch (immerhin 71 Jahre nach dem Verbot des Spielmannswesens durch Zar Alexej Michajlovič) der Stadtkommandant von Tver' 16 Gaukler, Tänzer und Musiker auftreten ließ: „Sie hatten sich solche Castanetten gemacht, aus Holzlöffeln mit Schellen an den Enden, und schlugen mit ihnen, sie spielten Sackpfeifen und Geigen, tanzten."[23]

Im Gouvernement Pskov belegt Vladimir Dal' den Terminus *guk*[24] als lokale Bezeichnung der Sackpfeife (im Weißrussischen und Slowakischen bezeichnet *huk* die Bordunpfeife). Zusätzlich steht *duda* nach Dal'[25] (unter anderem) für die Sackpfeife in den Gouvernements Pskov und Tver' und ebenso in Weißrußland (im „Westland", *Zapadnij kraj*, nach damaliger Sprachregelung). In Klyment Kvitkas unveröffentlichtem Aufsatz *Volynka*[26] sind kurze Erwähnungen von Dudelsackpfeifern in den Städten Veli-

---

[21] Wörtl.: „Zu Liedern". Diese Gattung umfaßt vor allem instrumental-vokale Formen, die mit den Umzügen der Dorfjugend verbunden waren.

[22] Louis Leonor Hammerich / Roman Jakobson, *Tönnies Fenne's Low German Manual of Spoken Russian, Pskov 1607*, Bd. II: Transliteration and Translation, Kopenhagen: Det Kongelige Danske Videnskabernes Selskab/Munksgaard 1970, S. 39 [1607, S. 57].

[23] „De hafde og giort sig Castanetter af træskeer med bielder paa enden, som de sloge med, spillede paa seckepiiber og giger, dantzede og hafde", *En Rejse til Rusland under Tsar Peter. Dagbogsoptegnelser af Vice- Admiral Jus Juel, Dansk Gesandt i Rusland, 1709-1711*, København: Gyldendalske Boghandels Forlag 1893, S. 130. Diese Quelle habe ich auf der 17ten Tagung der Study group on Historical Sources des ICTM (Stockholm 21.–25. Mai 2008) in meinem Referat „Western Sources on Russian Folk Instrumental Music and Dance in 17-th.-19-th. Centuries and its Reception in Russian Ethnomusicology" vorgestellt, das derzeit zur Veröffetlichung vorbereitet wird.

[24] *Tolkovyj slovar'* (s. Anm. 2), Stichwort *gukat'*.

[25] Ebd., Stichwort *duda*.

[26] Das handschriftliche Manuskript (F. 275, № 234) befindet sich im Glinka-Museum für Musikkultur in Moskau. Ich danke den Mitarbeiterinnen des Handschriftenarchivs für die freundliche Erlaubnis zur Einsichtnahme. Da meine Aufzeichnungen später einem Diebstahl in der Prager U-Bahn zum Opfer fielen und das Archiv derzeit geschlossen ist, bin ich gezwungen, Kvitkas Angaben aus dem Gedächtnis zu zitieren. Der zwischen 1941 und 1943 abgefaßten Handschrift Kvitkas sind maschinenschriftliche Auszüge aus Romanovs Arbeit von 1909 (mit Quellenangabe) beigefügt. Vermutlich aus

kie Luki (19. Jh.) und Nelidovo (1930er Jahre) und ebenso im Dorf Il'ino (heute Gebiet Tver'), also bei den Smolensker Weißrussen, zu finden.[27] Relativ ausführliche Zeugnisse über die *duda* liegen ab Mitte des 19. Jahrhunderts vor, sie entstammen größtenteils dem Territorium der weißrussischen Mundarten. Der Pädagoge und Philologe Pëtr Dmitrievič Šestakov (1826–1889) teilt in einer Untersuchung über die Smolensker Mundarten – hauptsächlich aus dem Kreis (*uezd*) Duchovščina – wertvolle Beobachtungen über die Stellung der *duda* im bäuerlichen Leben mit:

> „Der Smolensker Bauer liebt sein Vieh und hegt es, er ist der Arbeit nicht abgeneigt, aber auch nicht allzu erpicht auf sie, das Vergnügen und sein hausgemachtes Musikinstrument [*muzyka*], die *duda* aber, liebt er über alles. So spricht etwa ein Bäuerlein von seinem Sohn: Ich habe so einen Sohn: der Bart bis zum Gürtel – und auf der *duda* keinen Piepser (d.h. er spielt nichts). Und das sagt er mit großem Verdruß: Man sieht, daß der Makel des Sohnes ihn an der Seele rührt. Ein Smolensker Sprichwort zeigt noch klarer, wie die Smolensker ihr hausgemachtes, nationales Instrument lieben: sie geben ihm den Vorzug vor dem exotischen [*zamorskij*][28] Instrument – der Geige.
>
> | | |
> |---|---|
> | Skripka igraet, da deneg ne maet; | Die Geige spielt und hat kein Geld; |
> | A duda gudet da dengi beret. | Aber die *duda* dröhnt und nimmt Geld. |
>
> [Anmerkung im Original: *gudet* und *beret* mit Betonung auf der letzten Silbe.]
>
> Also, obwohl die Geige gut ist, obwohl sie spielt, hört ihr doch niemand zu; aber die *duda* ist einfach, dröhnt, aber sie nimmt Geld".[29]

Derselbe Autor beschreibt ein Fest zur Heuernte mit Schaukeln[30],

> „auf denen bei den Festen die Erntehelfer sich schwingen; die Mädchen schaukeln mit Liedern, und der Hausherr, gestützt auf einen Stock, schaut zu, wie sich die Jugend vergnügt. Zur Aufmunterung der Gesellschaft erscheint ein hausgemachter Künstler mit der *duda*, dem Lieblingsinstrument des Volkes. Um ihn herum sammelt sich die Menge und ein Lied wird angestimmt wie etwa dieses:
>
> | | |
> |---|---|
> | Aj du du, dudu, dudu! | Aj du du, dudu, dudu! |
> | Sidit žorov na dubu | Der Kranich sitzt auf der Eiche |
> | I igraet na dudu. | Und spielt auf der *duda*. |
> | U (v) žulejku (instrument v rode dudočki) | Auf der *žulejka* (eine Art Pfeife) |
> | Po malenku | Ein wenig[.] |
> | Priletela sinica, | Es kam eine Elster geflogen |

---

diesem Grund werden in Nazinas Arbeit (*Belorusskie narodnye*, S. 111) einige Beobachtungen Romanovs irrtümlich Kvitka zugeschrieben, der selbst jedoch keine entsprechenden Feldforschungen durchgeführt hat.

[27] Über den Doppelflötenspieler E.I. Kljuev schreibt Kvitka in der erwähnten Handschrift, daß dieser auch den Dudelsack zu spielen verstand und ein solches Instrument kaufen würde „wenn sich dazu eine Gelegenheit böte".

[28] Soweit nicht gerade von Südfrüchten die Rede ist, hat *zamorskij* (wörtl.: „von Übersee") im Russischen meist eine ironische oder abschätzige Färbung.

[29] Pëtr Dmitrievič Šestakov, Smolenskij govor [Smolensker Mundart], in: *Smolenskija Gubernskija Vedomosti, čast' neofficial'naja*, 1853 (14. Nov.) № 46, S. 368.

[30] *Krugovye kacheli* bezeichnet eine Vorform des Riesenrads.

Žuravova sestrica,
Udarila po nosu:
Pošlo gore po lesu i.t.d.[31]

Die Schwester des Kranichs,
Und schlug ihm auf die Nase.
Unglück kam in den Wald[,] usw".

Die Beobachtungen Šestakovs zeigen die soziale Bedeutsamkeit der Sackpfeifenkunst, die sich auch in der Bezahlung des Musikers („nimmt Geld") ausdrückt. Interessanterweise war aber auch außerhalb der professionellen Musikausübung das Sackpfeifenspiel als ästhetisches Vergnügen von Bedeutung – wie auch als Attribut eines repräsentationsfähigen erwachsenen Mannes. Dies zeigt die von Šestakov besonders hervorgehobene Enttäuschung des alten Bauern über seinen Sohn, von dem er offensichtlich eine Befähigung im Sackpfeifenspiel erwarten durfte – ebenso wie bis in die heutige Zeit auch der nichtprofessionelle Dorfmusiker nicht selten gleichzeitig ein wirtschaftlich erfolgreicher, im Handwerk bewanderter und allgemein angesehener Mann ist.[32] Ausgesprochen erhellend sind die Mitteilungen über das Verhältnis der Smolensker Bauern zur Geige, wovon noch zu sprechen sein wird.

Aus dem Dorf Bobrovki im Kreis Ržev, Gouvernement Tver' (heute: Bobrovka, Kreis Olenino) liegen knappe Beschreibungen des Erzpriesters S. Razumichin über den Gebrauch der Sackpfeife bei Tanzabenden der Jugend zur Weihnachtszeit (*svjatočnye večerinki*) vor: „Nach dem Lied beginnen gewöhnliche russisch Tänze [*pljaski*]. Man tanzt die *Baryn'ka*, den sogenannten *Golubec* und den *Russischen kazačok* zu Harmonika, Geige, Balalaika, Flöte [*svirel'*] und nicht selten zur *duda*."[33] Die unglückliche Beschreibung Razumichins, der die Funktion von Anblasrohr und Spielpfeife gleichsetzte, forderte schon den Spott Vertkovs heraus.[34] Wir finden diese Verwechslung häufig auch in der Ikonographie – und auch Razumichins Abbildung zeigt einen Ledersack mit lediglich *einer* verkehrt eingesteckten Spielpfeife (Abb. 5). Dennoch schließen die offensichtlichen Mängel der Beschreibung nicht eine richtige Darstellung einer geraden Spielpfeife ohne Schallbecher aus. Hätte die Sackpfeife von Bobrovki einen solchen Trichter (sei er aus Holz oder aus einem Rinderhorn) gehabt, so wäre er auch dem flüchtigen Blick eines organologisch wenig bewanderten Priesters kaum entgangen. Über eine Bordunpfeife wird von Razumchin nichts mitgeteilt, obgleich eine bordunlose Sackpfeife mit Lederbalg in dieser Region im 19. Jahrhundert sehr ungewöhnlich wäre.

Nikolaj Privalov erwähnt eine Fotografie, die in den Jahren 1902–1905 von A.V. Komarov auf einem Volksfest im Kreis Porech'e (heute: Demidov) des Smolensker

[31] Šestakov, Smolenskij govor, ebd.

[32] Persönlichkeitstypen der nordweißrussischen Musiker hat Aleksandr Romodin vielfach beschrieben, s. unter anderem: „Tradicionnye rasskazy o severobelorusskich muzykantach" (Traditionelle Erzählungen über norrdweißrussische Musiker), in: *Muzykant v kul'ture: koncepcija i dejatel'nost'* (Hg. Alisa A. Timošenko), Sankt-Peterburg: Rossijskij institut istoriii iskusstv 2005, S. 102-120. Hier sind auch die zahlreichen Fälle zu erwähnen, in denen die Eltern des angehenden Musikers nicht geringe Mittel aufzuwenden hatten. Dies ist besonders in solchen Traditionen bemerkenswert, in dem ein wirtschaftlicher Nutzen einer musikalischen Betätigung nicht zu erwarten war, vgl. meine Arbeit *Die Musik der Skobari. Studien zu lokalen Traditionen instrumentaler Volksmusik im Gebiet Pskov (Nordwestrußland)*, Bd. 1, Göttingen: Cuvillier 2007, S. 177f.

[33] S. Razumichin, Selo Bobrovki i okružnoj ego okolotok. Tverskaja gubernija, Rževskij uezd (Das Dorf Bobrovki und seine Umgebung im Gouvernement Tver', Kreis Ržev), in: *Ėtnografičeskij zbornik izdavaemyj RGO*, vyp. 1, Sankt-Peterburg: RGO, 1853, S. 269.

[34] Vertkov, *Russkie narodnye*, S. 139.

Gouvernements aufgenommen wurde. Sie zeigt einen Dudelsackpfeifer, der ein Instrument mit einfacher Bordunpfeife spielt.[35]

Außerordentlich wertvolle Hinweise auf die Sackpfeife (*duda*) im Kreis Žarkovskij, im Südwesten des Gebiets von Tver', teilte mir großzügigerweise der Musiker und Forscher Sergej Starostin mit, der sie zweifellos für eine eigene Untersuchung hätte verwenden können, auf die man dennoch hoffen möchte. Gestützt auf die Empfehlung des Moskauer Kollegen konnte ich in den Ortschaften Kaščënki und Zelen'kovo Erkundungen durchführen, die später auf die Kreise Duchovščina und Demidov im Gebiet Smolensk ausgedehnt wurden.[36]

In Kaščënki erinnert man sich bis heute an die Sackpfeife oder an Erzählungen über das Instrument:

> Das war noch vor unserer Zeit, ich war noch nicht auf der Welt (…) Meine Mutter erzählt, daß sie zur *duda* tanzten [*guljali*]. (…) Und man sagt, er [sic!] ist aufgeblasen, drückt sie unter dem Arm. Und da ist wohl so eine Pfeife [*piščik*], die pfiff [*piščal*]. Und man sagte, sie tanzten, man bat den, auf den Festen [zu spielen], der die *duda* spielte. (Anna Vasil'evna Miščenkova, *1928 in Zaluž'e, Interview vom 21.10.2005)

Wie dieser Zeugin so ist auch anderen aus Zaluž'e stammenden Gewährsleuten der Dudelsackpfeifer Semën in Erinnerung, der noch vor dem Krieg in hohem Alter starb. Varvara Valentinovna (*1922, Familienname unbekannt) sah diesen Musiker noch selbst: „Bei uns spielte so ein alter Mann. (...) Wir erinnern uns, er ist zu unserer Zeit gestorben (...) Sie lebten sehr arm“. Semën lebte seinerzeit auf einem Einzelgehöft (*chutor*). Aber die Sackpfeife wurde auch von fahrenden Musikern gespielt:

> Es gab so welche, man lud sie auf Hochzeiten ein, nun, wohl auch auf irgendeinen Tanzabend. Aber ... wir haben sie nicht gesehen (...) sie hatten auch so verschiedene Instrumente. Und diese *duda*. (...) Die Eltern erzählten es, wie sie umherzogen, was man ihnen gab. Auf Hochzeiten lud man sie ein. Damals gab es nicht so viele Musiker. Das sind jetzt erst viele geworden. (Interview vom 22.10.2005)

Auf die Frage, ob die Fahrenden ihre eigene Wirtschaft führten, antwortete die Gewährsfrau entschieden: „Ach was! Nichts hatten die. So sind sie umhergezogen.“

Höchstwahrscheinlich auf Semën beziehen sich die Erinnerungen von Anatolij Evdokimovič Krylov (*1933) aus dem Dorf Zelen'kovo. Ende der Dreißigerjahre „zog so ein Mann umher, der spielte und sich damit ein Stück Brot verdiente. (...) Er zog über die

---

[35] Nikolaj Ivanovič Privalov, *Opisanie narodnych muzykal'nych instrumentov kollekcii im. N. I. Privalova* (Beschreibung der russischen Volksmusikinstrumente der Sammlung N. I. Privalov). Handschriftliches Ms. der Petersburger Instrumentensammlung, 1912, S. 57. Bislang ist es mir leider nicht gelungen, diese Fotografie ausfindig zu machen.

[36] Die Feldforschungsdaten für die vorliegende Untersuchung wurden in den Jahren 2005 und 2006 erhoben im Rahmen des von der Deutschen Forschungsgemeinschaft geförderten und am Musikwissenschaftlichen Institut der Universität Hamburg angesiedelten Projekts „‚Ich spiele nicht auf eure Art.‘ Stil und Repertoire der traditionellen russischen Instrumentalmusik im Grenzgebiet zu Weißrußland als Faktor für die Bildung von ethnischer und Generationenidentität“. An den Unternehmungen waren auch meine Kollegen vom Rußländischen Institut für die Geschichte der Künste (St. Petersburg) Dr. Aleksandr Romodin und Valentin Vinogradov beteiligt sowie der Journalist Igor' Stesev, die Musiklehrerin Elizaveta Borodulina und der Student des Konservatoriums und Fotograf Rostislav Koševel.

Dörfer. Der eine gibt dies, der andere das. Er spielt auf dieser *duda*". Auf die Frage, ob zum Spiel dieses Musikers auch getanzt wurde, antwortete Krylov: „Man tanzte, aber natürlich (...). Und er selbst sang Lieder zu dieser *duda.*" (Interview vom 24.10.2005)

Den Balg „machte man aus Schafshaut[37]. Irgendwie vernähte man sie, blies sie auf, setzte ein Rohr ein und blies". Die Melodiepfeife sah aus „wie ein *piščik*, den es früher gab. (...) Gerade war sie. Aus Esche machte man den *piščik. Piščik* ist eine lokale Bezeichnung für das Rohr der Hirtenklarinette (*rog*) ohne den Schallbecher".[38] Die Verwendung eines Schallbechers aus Horn verneinte Krylov für die *duda*. Als ich ihm zur Kontrolle schematische Zeichnungen einer Melodiepfeife mit Kuhhorn, mit gebogenem Schallbecher wie bei der litauisch-weißrussischen Labanoro dūda sowie ohne Schallbecher vorlegte, bezeugte der Gewährsmann die letztere Version. Als vormals allgemein gebräuchliches Instrument auf Tanzabenden wird die Sackpfeife in Erzählungen erwähnt, die Anna Ivanovna Antonova (*1929) aus Afonino, einem Nachbardorf Zelen'kovos, mitteilte:

> Sie war ... so ein großes Rohr hatte man gemacht (...) und so einen Balg, der wird aufgeblasen. Und dann fängt man an zu singen, zu spielen. Und da wird der Balg aufgeblasen. Was er [der Spieler] ausspricht[39], das spricht der Balg aus. (…) Früher gab es ja keine solchen Instrumente, alle waren selbstgemacht. (…) Ich habe sie nicht gesehen. Ich weiß nur, daß man es so erzählte. (…) Noch meine Mutter und mein Vater sagten, daß sie auf dem Tanzabend waren – man spielte die *duda* und das war alles. (Interview vom 25.10.2005)

Die Erwähnung des „großen Rohres" dürfte auf eine Bordunpfeife verweisen.

Erinnerungen an die Sackpfeife konnten auch im Kreis Demidovo aufgezeichnet werden, von wo das bei Privalov erwähnte Foto stammt. Wie der Harmonikaspieler Anatolij Nikolaevič Antonov (*1937) berichtete, spielte das Instrument im Dorf Kozjuli Danil Gerasimov (gestorben ungefähr 1990 im Alter von 90 Jahren). Als Luftsack diente eine Ochsenblase, die Spielpfeife war mit Bast umwickelt, das Anblasrohr „gebogen zum Mund wie ein Saxophon ungefähr. (…) Er spielte und drückte dieses ... diesen Ballon mit den Knien und gleichzeitig spielte er". Antonov verdanken wir auch Einzelheiten über das Repertoire dieses Musikers:

> Die *Kamarinskaja* spielte er wohl, wie ich mich erinnere. Ein wenig. Denn sie war schon kaputt. Sie gab gerade noch ein paar Töne von sich, als ich sie sah. Aber ich erinnere mich, daß er die *Kamarinskaja* anspielte, ein wenig. (Interview vom 7.11.2005)

---

[37] Etwas später entsann sich der Gewährsmann, daß eine Ziegenhaut verwendet wurde.

[38] Variante A der *žalejka* im oberen Wolgagebiet nach der Typologie Starostins: Sergej Nikolaevič Starostin, O nekotorych raznovidnostjach pastuš'ich muzykal'nych instrumentov verchnej Volgi" (Über einige Arten von Hirtenmusikinstrumenten der oberen Wolga), in: *Metody muzykal'no-fol'klorističeskogo issledovanija. Sbornik naučnych trudov*, Moskva: Moskovskaja gosudarstvennyj dvaždyj ordena Lenina Konservatoija imeni Petra Il'iča Čajkovskogo 1989 (Hg. Vjačeslav Michajlovič Ščurov), S. 64-81. Michail Michailovič Goržkov (*Ol'chovyj rog*, Moskva: Gosudarstvennyj centr russkogo fol'klora, 2002) verwendet für analoge Instrumente mit Rindentrichter und geringer Zahl von Grifflöchern, die er am Oberlauf der Düna dokumentieren konnte, die Bezeichnung „Eschenhorn".

[39] *Vygovarivat'* ist eine häufige metaphorische Wendung für ein gelungenes, ausdrucksstarkes Instrumentalspiel.

Wirkungsbereich des Dudelsackpfeifers war – wie so oft – auch hier das Hochzeitszeremoniell: „Er sagte, daß er auf Hochzeiten gespielt hätte" (im Dorf Kozjuli). Die Verwendung einer Ochsenblase nähert das Instrument Gerasimovs in gewisser Weise an die russischen Sackpfeifen an.[40] Auch die Tradition von bastumwickelten Blasaerophonen verweist eher auf Mittel- und Nordrußland als auf den weißrussischen Raum, dem Demidov dialektologisch gesehen angehört.

Aufschlußreich sind auch mittelbare Zeugnisse über die Sackpfeife aus dem Bereich der Phraseologie. Im Pskover Gebietswörterbuch[41] wird der Vergleich einer Zecke mit einem Musikinstrument erwähnt: „die Zecke bläst sich auf wie eine *duda*, dick ist sie" (Kreis Pustoška). Im selben Landkreis belegt ist die idiomatische Wendung „sich aufblasen wie eine *duda*" im Sinne von „ein unzufriedenes Gesicht machen, beleidigt sein".[42] Ungeachtet der Tatsache, daß *duda* im heutigen dörflichen Sprachgebrauch alle möglichen Aerophone bezeichnen kann (besonders, wenn der Terminus vom Feldforscher ins Spiel gebracht wird), so steht hier doch der Vergleich mit dem Balg der Sackpfeife außer Zweifel (ein anderes Aerophon kann sich nicht aufblasen). Im Kreis Žarkovskij, Gebiet Tver', stieß ich auf eine ähnliche Redensart, verbunden mit einem konkreten Dudelsackpfeifer: „Und seitdem nennen wir uns so: ‚Ach, du Semënovs *duda*!' (…) Wenn jemand böse ist, sagen wir: ‚Ah, ach, du Semënovs *duda*!'" (Anna Vasil'evna Miščenkova, *1928 in Zaluž'e, Interview vom 21.10.2005). „Wenn jemand böse ist, (...) schmollend [*tjanuvšis'*, also wörtl.: aufgeblasen] dasitzt: ‚Was sitzt du da wie Semënovs *duda*, aufgeblasen? Warum bist du nicht lustig?' Das (…) sagen wir. Und wie kommt das, daß diese *duda* so unfroh war?" (so eine Dorfnachbarin der genannten Gewährsfrau). Eine Modifikation der Redensart, die offensichtlich durch Unkenntnis der Sackpfeife zu erklären ist, wurde im Dorf Medvëdovo des Kreises Bežanicy, Gebiet Pskov, aufgezeichnet: *natjanut'sja kak dudar'* – sich aufblasen wie ein *dudar'*, was nicht das Instrument, sondern den Musiker, höchstwahrscheinlich einen Flötenspieler bezeichnet. Die historische Verbindung zur Sackpfeife ist jedoch auch hier sehr wahrscheinlich. Weitere Untersuchungen zur Phraseologie der Sackpfeife könnten Genaueres über deren Verbreitung in der Vergangenheit ans Licht bringen. Auf die Sackpfeife bezieht sich sicherlich auch ein in Spicino (nördlicher Kreis Velikie Luki), also bei den *Skobari* und nicht bei den Trägern weißrussischer Mundarten (*Poljaki*) aufgezeichnetes Zeugnis: „Es gab die *duda*, sie ist wie ein Korken, mit dem Atem muß man spielen."[43] Der nicht ganz

---

[40] S. die zusammenfassenden Beobachtungen Jakob von Stählins (Nachrichten von der Musik in Rußland, in: Haigold, M. J. J.: *Beylagen zum Neuveränderten Rußland*, T. 2. Riga und Leipzig 1770, S. 70) und ebenso die von Johann Joachim Bellermann (*Bemerkungen über Rußland in Rücksicht auf Wissenschaft, Kunst, Religion und andere merkwürdige Verhältnisse*, T. 1, Erfurt: Keyser 1788, S. 364 f.). Dieser nennt die „Walnika" „bei großen geselligen Freuden das wichtigste Instrument" und hebt die Nähe der einfach gebauten russischen Sackpfeifen in der Umgebung von St. Petersburg zu dem estnischen „Torupill" hervor. Von der Verwendung einer Ochsenblase zeugt auch die *Byline vom großen Stier*, aus dessen Blase ein Skomoroch eine Sackpfeife verfertigt, s. Zoja Ivanovna Vlasova, K voprosu o tradicii v fol'klore („Starina o bol'šom byke" v svete istoriko-ėtnografičeskich dannych) [Zur Frage nach der Tradition in der Folklore („Die Byline vom großen Stier" im Lichte historisch-ethnographischer Daten)], in: *Russkaja literatura*, 2, 1982, S. 170.

[41] *Pskovskij oblastnoj slovar'*, vyp. 10, Sankt-Peterburg 1994, S. 37.

[42] Ebd.

[43] Ebd. Die Angaben zu dem heute zum Kreis Kun'ja gehörenden Dorf sind der Kartei des Wörterbuches an der Philologischen Fakultät der Universität St. Petersburg entnommen, wo mir freundlicher-

klare Vergleich mit dem Korken dürfte wohl der Assoziation mit einem dicht verschlossenen und möglicherweise unter Druck stehenden Behälter geschuldet sein, wie dies der Luftsack der Sackpfeife ist.

3. Sackpfeife und Geige in Nordweißrußland

In weiten Teilen Europas nahm die Sackpfeife lange Zeit den ersten Platz in der bäuerlichen Musikpraxis ein. Ihre Verdrängung ist in erster Linie mit der massenhaften Verbreitung der Violine und häufig auch mit dem der Barockmusik entsprungenen Ensemblekonzept einer beweglichen, akzentuierten Baßbegleitung zu erklären (später spielten auch die Harmonikainstrumente eine Rolle). Dieser Prozeß begann in Deutschland und Österreich schon im 18. Jahrhundert[44] und erfaßte im 19. Jahrhundert (in höherem oder geringerem Maße) andere europäische Länder wie etwa Ungarn[45], Rumänien[46], Schweden[47], Estland[48] und Lettland[49]. Keine Ausnahme stellen hier auch die weißrussischen Sackpfeifen dar, obgleich der Hauptgrund ihres Verschwindens meines Erachtens noch nicht ausreichend in der ethnoorganologischen Literatur reflektiert worden ist.

Nach den erwähnten Beobachtungen Šestakovs war die Sackpfeife noch Mitte des 19. Jahrhunderts das „Lieblingsinstrument" der Smolensker Bauern. Nur wenig später erscheinen bereits die ersten Zeugnisse über die Verdrängung der Sackpfeife durch Violine und andere Instrumente bei den nordöstlichen Weißrussen. Der Journalist und Schriftsteller V.O. Michnevič führt in seiner Musikgeschichte Rußlands ein wohl auf

weise Einsicht in diese auch in ethnographischer Hinsicht äußerst wertvolle Quellensammlung gewährt wurde.

44 Im deutschsprachigen Raum – das Egerland ausgenommen – verschwand die Sackpfeife im 18. Jahrhundert (John Henry van der Meer, Beitrag zur Typologie der westeuropäischen Sackpfeifen, in: *Studia Instrumentorum Musicae Popularis*, Bd. I, 1969, S. 104). Gerlinde Haid verbindet dies für Österreich einerseits mit der Verbreitung der Violine, andererseits mit der Verdrängung des Bordunprinzips durch das einer harmonisch orientierten Baßbegleitung.

45 In Ungarn waren es im Laufe des 19. Jahrhunderts vor allem die Zigeunerkapellen, die die Sackpfeife verdrängten, die zum Instrument von Amateurmusikern oder Bettelmusikanten absank (Bálint Sárosi, Professionelle und nichtprofessionelle Volksmusikanten in Ungarn, in: *Studia Instrumentorum Musicae Popularis*, VII, 1981, S. 14 f.).

46 Gottfried Habe nicht, Die rumänischen Sackpfeifen, in: *Jahrbuch für Volksliedforschung*, Nr. 19, 1974, S. 123.

47 Jan Ling, *Švedskaja narodnaja muzyka* (Schwedische Volksmusik), Moskva 1981, S. 46.

48 „Die Blütezeit der estnischen Dudelsackmusik dauerte bis zum Anfang des 19. Jahrhunderts, dann begann ihr Rückgang. Der Dudelsack wurde zunächst durch die Geige verdrängt" (Johannes Taul, Der estnische Dudelsack *torupill*, Referat auf dem Symposium des Internationalen Dudelsackpfeifer-Festivals in Strakonice 1979, S. 2. Das Referatsmanuskript verdanke ich Dr. Fritz Schneider aus Krefeld). Auf der traditionellen estnischen Hochzeit ersetzte die Geige die Sackpfeife in der Mitte des 19. Jahrhunderts, s. Igor Tõnurist, Pillimees pulmas // Pillid ja pillimäng eesti külaelus (Instrumente und Instrumentalmusik in estnischen Dörfern), Tallinn: Teaduste Akadeemia Kirjastus 1996, S. 100 (dt. Resümee, S.102-104).

49 „Gradually replacing the old instruments, especially bagpipes, and partly adopting their repertoires, the violin became the most popular folk-music instrument in and after the 1800s" (Valdis Muktupavels, Latvia, in: *The Garland Encyclopedia of World Music*, vol. 8: Europe, Hg. Timothy Rice, James Porter, Chris Goertzen, New York, London: Garland publishing, Inc. 2000., S. 504).

Dal'[50] zurückgehendes, ironisches Sprichwort an, mit dem man sich über die musikliebenden Vicebsker Weißrussen lustig machte: „*Volynka* und *gudok* haben das Haus zusammengehalten, Pflug und Egge haben unsere Häuser verfallen lassen".[51] An gleicher Stelle fragt der Autor:

> Heißt dies, daß die Vicebsker durch Sackpfeifer berühmt waren, ähnlich wie die Jaroslaver durch ihre Sänger, die Vjatischen durch die Pfeifentänze? Heute in jedem Fall sind bekanntlich *volynka* und *gudok* weniger gebräuchlich als früher, sie wurden durch Instrumente neuester Erfindung wie Harmonika, Geige u.ä. verdrängt.

Der große Sammler Pavel Šejn schreibt, daß in Weißrußland „die Dudelsackpfeifer [*dudari*] nun ihren Platz den Geigern und Skomorochen abgetreten"[52] hätten. Ebenso nennt der weißrussische Volkskundler Nikolaj Nikiforovskij im Kapitel „Dudar' i muzýka" (Dudelsackpfeifer und Geiger) seiner Vitebsker Aufzeichnungen ersteren als den „Vorgänger des *muzýka*, also des Geigers".[53] Schließlich bezeugt Romanov die Verdrängung der Sackpfeife durch Bauernensembles unter Beteiligung der Geige:

> Auf einer Hochzeit kann man dagegen den Dudelsackpfeifer nur in dem Fall sehen, wenn sie in seinem Dorf stattfindet. Da ist er der Held des Tages und die neuesten Instrumente [‚*muzyki*'] – Geige und Hackbrett – machen wohl oder übel dem Dudelsackpfeifer Platz.[54]

Über das Hochzeitszeremoniell im Kreis Haradok [russ.: Gorodok] bemerkt derselbe Autor:

> Nach dem Essen führt der Bräutigam die Braut nach Hause; hinter ihnen fahren die Brautwerber und singen Lieder und die Musiker spielen die Sackpfeifen [*na dúdach*] und in letzter Zeit Geige und Schellentrommel [*buben*].[55]

Offenkundig sind die Parallelen der zunächst skeptischen Haltung gegenüber der Geige als Volksmusikinstrument – sowohl von seiten der Bauern als auch der Intelligenz

---

[50] *Tolkovyj slovar' živoga velikorusskago jazyka*, Bd. IV, Moskva 1861, Stichwort „sbirat'". Das Sprichwort spielt ironisch auf die unsicheren Lebensverhältnisse des Musikers an (vgl. ähnliche mündliche Quellen bei Walter Salmen, *Der Spielmann im Mittelalter*, Innsbruck: Helbling 1983, S. 14).

[51] Vladimir Osipovič Michnevič, *Očerk istorii muzyki v Rossii v kul'turno-obščestvennom otnošenii* (Abriß der Musikgeschichte in Rußland in kulturell-gesellschaftlicher Hinsicht), Sankt-Peterburg 1879, S. 84 f.

[52] Pavel V. Šejn, *Materialy dlja izučenija byta i jazyka russkago naselenija Severo-Zapadnogo kraja* (Materialien zur Erforschung von Alltag und Sprache der russischen Bevölkerung der nördlichen Westlande), tom I, čast' 1, Sankt-Peterburg 1887: 530. *Skomoroch* war im nordweißrussischen Sprachraum eine recht häufige Bezeichnung für den Geiger, allerdings auch für andere Musiker. Der Sammler räumt auch eine frühere Verbreitung der Geige ein: „Von den Musikinstrumenten waren in alter Zeit Sackpfeife [*duda*] und Geige in Gebrauch, und zwar vorwiegend erstere", ebd.). Hierbei stützt sich Šejn jedoch auf ein für die historische Organologie so unzuverlässigen Quellentyp wie Liedtexte, während die oben zitierte Einschätzung zweifellos auf eigenen Beobachtungen beruht.

[53] Nikolaj Jakovlevič Nikiforovskij, Očerki vitebskoj Belorussii, 2. Dudar' i Muzýka, in: *Ėtnografičeskoe obozrenie*, 1892, Nr. 2-3, S. 185.

[54] Romanov, Vymirajuščij instrument, S. 127.

[55] Evdokim Romanovič Romanov, *Belorusskij sbornik*. Vyp. 8: Byt' belorusa (Weißrussischer Sammelband. Teil 8: Leben des Weißrussen), Vilno 1912, S. 395.

(Michnevič, Romanov) – zu der Ablehnung der Harmonika nur wenig später, als die Geige bereits eingebürgert war. In einem häufig zitierten Zeitungsartikel heißt es über das Gouvernement Smolensk:

> Auf Dorffesten sind die berüchtigten Harmonikas sehr selten anzutreffen. Die Bauern bevorzugen Geige, Balalaika, *svirel'* [Flöte] und *rožok* [Hirtenklarinette?], in der Meinung, daß die Harmonika allenfalls zum Lärmen in der Schenke taugt, nicht aber für Lieder, das heißt, nicht zur Gesangsbegleitung.[56]

Dessen ungeachtet enthält der Artikel auch Hinweise auf ein spätes Aufkommen der Volksgeige:

> Dank dieser Liebe zur Musik wurde im Gouvernement in letzter Zeit die Geigenkunst [der Geigenbau] als bäuerliches Handwerk [*kustarnyj promysl*] begründet.[57]

Noch Romanov bestätigt folgendes Urteil über die Harmonika:

> Die ordinäre Harmonika ist nach volkstümlichen Anschauungen ein Symbol der Zuchtlosigkeit und Verkommenheit der Jugend, die es schaffte, von den Früchten der berüchtigten Fabrikzivilisation zu kosten, während sie sich während des Wandergewerbes [vom Dorf] entfernte.[58]

Aber welche Karriere machte die „ordinäre Harmonika" innerhalb weniger Jahrzehnte, wenn man die professionelle Schule der berühmten Harmonikavirtuosen bedenkt, die in den Achtzigerjahren des 20. Jahrhunderts Aleksandr Romodin im Kreis Haradok des Gebietes von Vicebsk entdeckt hat.[59]

Wie wir sehen, bestätigen gewichtige Zeugnisse von Erforschern des Vicebsker Weißrußland von der Mitte des 19. bis zum Beginn des 20. Jahrhunderts im Ganzen den Prozeß der allmählichen Verdrängung der Sackpfeife durch die Violine, insbesondere als Ensembleinstrument. Die Gründe hierfür sind einerseits in den Schwierigkeiten zu suchen, die Herstellung und Instandhaltung des Instruments mit sich bringen, jedoch auch in Veränderungen von Vortragsstil und Repertoire.[60] Allerdings beschrieb Romanov auch Fälle, in denen das Publikum der Sackpfeife den Vorzug gegenüber den „neuesten

---

[56] *Charkovskie gubernskie vedomosti*, čast' neofficial'naja, № 233, 11. Sept. 1889, S. 3. Die Ausführungen fassen einen Artikel der Zeitung *Novoe Vremja* zusammnen (leider ohne genauere Quellenangabe), der im Original leider noch nicht Eingang in die ethnoorganologische Forschung gefunden hat.

[57] Ebd.

[58] Romanov, Vymirajuščij instrument, S. 123.

[59] Aleksandr Vadimovič Romodin / Irina Aleksandrovna Romodina: *Tradicionnoe iskusstvo poozer'ja* (Traditionelle Kunst des Poozer'e), Folge 1: *Obrjadovaja muzyka* (Brauchtumsmusik), Langspielplatte, Leningrad: Melodija 1989, S20 29387 000, Folge 2: *Večerinočnaja muzyka* (Musik der Tanzabende), Leningrad: Melodija 1990, S20 30043 002.

[60] Nikiforovskij, Očerki, S. 82, Nazina, *Belorusskie narodnye*, S. 119 f. Nach Romanovs Zeugnis „weigert sich [der Dudelsackpfeifer] beharrlich, eingeschleppte ‚Quadrillen', ‚Polkas', den ‚Len'tej' [der Faulenzer] (Ljans'e) [Lancier] u. ä. zu spielen." (Vymirajuščij instrument, S. 127). Nichtsdestotrotz spielte der berühmte G. K. Slaŭchik (*1836, russ.: Slavčik), dessen Spiel 1931 Evgenij Gippius dokumentieren konnte, neben älteren Tanz- und Hochzeitsmelodien auch eine Polka („*polečka*"), vgl. die Transkription bei Morgenštern, K voprosu o kornjach, № 4, die Titelbezeichnung des Archivs bezieht sich wohl nur auf das erste Stück.

Instrumenten“ gibt (s. o.). Die pragmatischen Erwägungen des Musikers können also auch mit den ästhetischen Vorstellungen seiner Umgebung in Widerspruch geraten. Es darf auch nicht vergessen werden, daß die Sackpfeife in spieltechnischer Hinsicht bedeutende Schwierigkeiten bereitet. Wenn der angehende Geiger (mag er sich auch nach neuerer Auffassung für „das allerschwerste Instrument“[61] entschieden haben) zweifellos nach einige Tagen in der Lage ist, im Staccato oder Portato ein kurzes Motiv zu artikulieren, so ist für den Dudelsackpfeifer die Aneignung der elementaren Artikulationsarten (Portato und Staccato bei gleichbleibendem Luftstrom mit rein ornamentalen Mitteln) nur durch lange anhaltende Übung möglich.

Die frühesten weißrussischen Zeugnisse über ein Instrument mit dem Namen „Geige“ (*skripka* oder „*skripica*/*skrypica*) belaufen sich auf das 16. Jahrhundert.[62] In Polen war bereits im 15. Jahrhundert die *skrzypce* ein Attribut der Unterschichten.[63] Möglicherweise war dies eine Variante der Fiedel, jedenfalls keine Violine, die damals noch nicht existierte.[64] Auch die weißrussischen Quellen des 16. Jahrhunderts dürften kaum eine andere Interpretation erlauben. Auf was für *skrypki* im 17. Jahrhundert weißrussische städtische Skomorochen spielten, ist schwer festzustellen.[65] Zuverlässige Zeugnisse über die Geige im weißrussischen Bauernmilieu liegen erst für das 19. Jahrhundert vor. Dies korrigiert in gewisser Weise weit verbreitete Vorstellungen über die Geige als Trägerin alter Streichinstrumentenpraxis. In Nordostweißrußland waren (im Gegensatz zu den zentralen Landkreisen von Pskov[66]) zum Zeitpunkt der Einführung der Violine Instrumente vom Typ der mittelalterlichen Fiedeln (*gudok*) längst außer Gebrauch gekommen. Die späte Entlehnung der Violine durch die Smolensker und Vicebsker Bauern ist je-

---

[61] Inna Dmitr’evna Nazina: *Belorusskie narodnye muzykal’nye instrumenty. Strunnye* (Die weißrussischen Volksmusikinstrumente. Saiteninstrumente). Minsk: Nauka i technika 1982, S. 96.

[62] Ebd., S. 109.

[63] Geige und Geigenspiel in der polnischen Volksüberlieferung, in: *Die Geige in der europäischen Volksmusik* (Hg. Walter Deutsch, Gerlinde Haid), Wien: Schendl 1975, S. 26.

[64] Ebd., S. 22.

[65] Nazina: Belorusskie narodnye muzykal’nye instrumenty. Strunnye, S. 96.

[66] Vgl. die Zeugnisse über Instrumente ohne seitliche Einbuchtungen aus den Kreisen Opočka bei: Anatolij Michajlovič Mechnecov (Hg.), *Narodnaja tradicionnaja kul’tura Pskovskoj oblasti. Obzor ėkspedicionnych materialov* (Traditionelle Volkskultur des Gebiets Pskov. Übersicht der Feldforschungsmaterialien), tom I, Sankt-Peterburg 2002, S. 458, wie auch aus den Kreisen Loknja und Novosokol’niki (ebd., S. 59) und aus dem Kreis Ostrov (Morgenstern, *Skobari*, Bd. 1, S. 109–110). Diese Angaben korrespondieren mit Privalovs ikonographischem Beleg aus der Mitte des 19. Jahrhunderts über den *gudok* in der Stadt Pskov (Nikolaj Ivanovič Privalov, *Gudok. Drevne-russkij muzykal’nyj instrument, v svjazi s smyckovymi instrumentami drugich stran* (Der Gudok. Ein altrussisches Musikinstrument in bezug auf die Bogeninstrumente anderer Länder, Sankt-Peterburg: Tipografija I. N. Skorochodova 1904: S. 20, 21) und ebenso mit zeitgleichen Zeugnissen aus dem Kreis Ostrov (Morgenstern, *Skobari*, Bd. 1, S. 108–109). Wichtig sind hier auch die zahlreichen Belege über die vertikale Spielhaltung der Violine, die in Weißrußland unbekannt ist, wo die Spielweise des Instruments ausführlich untersucht wurde (Nazina: *Belorusskie narodnye muzykal’nye instrumenty. Strunnye*, S. 93–96). Entsprechendee Beobachtungen Privalovs „in einem entlegenen Winkel des Gouvernements Pskov“ (Privalov, *Gudok*, S. 29) wurden unlängst durch Mechnecovs Feldforschungen in den Kreisen Gdov (*Narodnaja tradicionnaja kul’tura*, tom 1: S. 84), Ostrov, Palkino (ebd., S. 274), Krasnogorodskoe (ebd., S. 458–459), Porchov (ebd., S. 638) sowie Loknja und Novosokol’niki (*Narodnaja tradicionnaja kul’tura, tom.* 2: S. 59) bestätigt. Problematischer ist dagegen die generelle Verknüpfung von Geigen mit ausgehöhltem Korpus sowie der Faktur der russischen Volksgeige mit der Tradition des *gudok* (ausführlicher: Morgenstern, *Skobari*, S. 361–364).

doch in keiner Weise ein mechanischer Prozeß und man kann nur staunen, wie organisch sich letztlich die „exotische Geige“ in den lokalen Traditionen instrumentaler Musikpraxis eingelebt hat.

4. Die Nachwirkungen der Sackpfeife

Kehren wir nun zu den Sackpfeifenelementen im Harmonikaspiel zurück. In meiner oben erwähnten Arbeit habe ich versucht, folgende sackpfeifentypische Merkmale herauszuarbeiten: 1) durchgängiger Bordun für die Dauer einer ganzen Instrumentalperiode oder eines wesentlichen Teils von dieser, 2) ausgeprägte Umspielung vor allem von Repetitionstönen (die der Dudelsackpfeifer zu ihrer Artikulation benötigt), 3) eine deszendente Eingangsphrase. Das letzte Merkmal kann allerdings auch auf die im Pskover Gebiet nachgewiesenen Doppelklarinette (*žalejka*) zurückgehen, die höchstwahrscheinlich einen starken Einfluß auf Vortragsstil der lokaltypischen alten Harmonika von Novoržev (*tal'janka*, *rezucha*) nicht nur in Faktur und harmonischer Struktur, sondern selbst in der Ornamentik ausgeübt hat.[67] Regelmäßig gebraucht eine solche Eingangsphrase auch der erwähnte weißrussische Dudelsackpfeifer G. K. Slaŭčik[68] (Notenbeispiel 1, 2).[69] Slaŭčik markiert gewöhnlich mit einer solchen Phrase den Übergang zwischen den thematischen Abschnitten einer Ausführung.

Überall in den Gebieten von Pskov, Tver' und Smolensk, wo konkrete historische und ethnographische Zeugnisse über die Sackpfeife vorliegen[70], treten mehr oder weniger stark ausgeprägt typologische Merkmale des Sackpfeifenspiels bei den örtlichen Harmonikaspielern auf. Im Repertoire der überwiegenden Mehrheit der Musiker gehören sie zu den lokaltypischen Stücken der Gattung *pod pesni*. In der *Pskovščina* und in den angrenzenden Teilen des Gebiets von Tver' ist dies der *Velikolukische Skobarja*.[71] Ebenso finden sich Spuren des Sackpfeifenspiels in der Tanzform *Kamarinskaja*, in Ostweißrußland (und bei den *poljakí* der südlichen *Pskovščina*) ebenfalls bekannt als *Ljavonicha*.[72] Von allen mir bekannten Harmonikaspielern war es nur Vladimir Prokopievič Jakovlev (1937–2007) aus Manëvo, Bezirk Puchnovo, Kreis Kun'ja, der die Sackpfeifenmelodik und -faktur auch auf spätere Tanzformen (wie *Krakovjak*, *Korobočka* u.a.) ausweitete.

Die tonale Grundlage des *Velikolukischen Skobarja* ist der ionische Hexachord. Die *Kamarinskaja* kann sich auch im Quint- oder Quartrahmen bewegen. Im *Skobarja* er-

---

[67] Morgenstern, *Skobari*, S. 367–369.

[68] S. Anm. 60.

[69] S. das von Nazina (*Belaruskaja narodnaja* № 135) zitierte Stück aus der Sammlung Eval'ds *Pesni belorusskogo Poles'ja* (Moskva 1979: S. 135) sowie die in meinem in Anm. 7 genannten Beitrag beigefügten Transkriptionen.

[70] Lediglich für den Kreis Porchov sind dies indirekte Quellen (*huk* [= Bordunpfeife] als Bezeichnung der Akkordtasten auf der Baßseite der Harmonika).

[71] Über diesen s. mein Beitrag im vorangegangenen Jahrbuch. Die generalisierende Titelbezeichnung ist typologischer Natur. In der Tradition heißt das Stück mit seinen funktionalen Varietäten *Pod pesni*, *Pod draku*, *Skobar'/Skobarja*, *K devkam*, *Ot devok*, im Gebiet Tver' [zeitweise: Kalinin] auch *Po derevne* oder (toponymisch) *Kalinka*.

[72] Vgl. die Anmerkungen in: I. Dz. Nazina, *Belaruskaja narodnaja instrumental'naja. muzyka* (Weißrussische instrumentale Volksmusik), Minsk: Navuka i technika, 1989: S. 639, zu dem Stück *Ljavonicha pa-daŭnejšamu* (Ljavonicha auf alte Art), № 218.

reichen einige Harmonikaspieler auch die Oberoktave (dann meist mit mixolydischer Septime), weit häufiger wird jedoch die Subquarte angespielt. Eine solche hypoionische Organisation liegt zahlreichen westslawischen Sackpfeifen zugrunde, etwa dem böhmischen Bock:

$$g - h - c - d^1 - e^1 - f^1 - g^1 - a^1$$

(hier modellhaft transponiert, tonales Zentrum: c^1). Der unvollständige hypoionische Modus (ohne zweite Stufe) im Rahmen der None korrespondiert mit zahlreichen nordweißrussischen instrumentalen Hochzeitsmelodien.[73] Die von G. K. Slaŭčik gespielte *duda* unterscheidet sich von der Bockstimmung lediglich durch die Moll-Färbung (№ 2), seltener die neutrale Terz (№ 1)[74], und ebenso durch die Erhöhung des sehr selten gebrauchten höchsten Tones um einen Halbton. Die Stimmung der mit der weißrussischen *duda* nahezu baugleichen *Labanoro dūda* entspricht – nach dem Spiel von Juozas Voldemaras aus dem Kreis Svencioniu des Gouvernements Vilno[75] zu urteilen – dem unvollständigen hypomixolydischen Modus (ohne zweite Stufe) in der Oktave.

Bei der Stimmung der litauisch-weißrussischen Sackpfeife fällt ein Widerspruch zwischen phonographischen und schriftlichen Quellen auf: Slaŭčik und Voldemaras spielten Instrumente mit plagalem Modus (hypoäolisch bzw. hypomixolydisch) und einem dem tonalen Zentrum auf der vierten Stufe entsprechenden Bordun. Romanov dagegen nennt als Stimmung der Melodiepfeife „die volle Oktave der Töne der einfachen Leiter“[76], wobei nach seinen Angaben der Bordun eine Oktave unter dem tiefsten Ton der Melodiepfeife erklingt. Rimantas Sliužinskas nennt unter Verweis auf die Diplomarbeit von A. Fokas *Lietuviškas dūdmaišis* ebenfalls die ionische Oktave, also einen authentischen Modus.[77] Auch in den modernen Rekonstruktionen der Sackpfeife in weißrussischen und litauischen Revival-Ensembles scheinen authentische Stimmungen zu dominieren. Diese Rekonstruktionen bedürfen jedoch m. E. einer genaueren Prüfung. Hier wäre auch eine Analyse der weißrussischen Melodien zu den von Romanov mitgeteilten Liedtexten aus dem Repertoire des Dudelsackpfeifers aus dem Kreis Disna hilfreich (sie umfassen ein Ernte-, ein Heische- und mehrere Hochzeitslieder).[78] Höchstwahrscheinlich entsprechen auch diese Lieder plagalen Modi, wie sie so charakteristisch für die nordweißrussischen

---

[73] Elena Nikolaevna Razumovskaja, *Tradicionnaja muzyka russkogo poozer'ja (po materialam ėkspedicij 1971–1992)* (Traditionelle Musik des Poozer'e. Nach Materialien aus Feldforschungen der Jahre 1971 bis 1992 [mit 3 Musikkassetten]). Sankt-Peterburg: Kompozitor 1998, № 64, 65, 67, 68. Vollständig fällt die Leiter des Bocks mit der vokalen Imitation einer Hochzeitsmelodie von Anna Ivanovna Zvereva (*1921) aus dem Dorf Ragozy (Kreis Usvjaty, Gebiet Pskov) zusammen. S. *Gološene nevesty ‚pod jazyk' pri nadele* (Vokalimitation der Brautklage bei der Beschenkung) auf der Schallplatte *Obrjadovaja muzyka* (Brauchtumsmusik), Schallplattenserie: *Tradicionnoe iskusstvo poozer'ja* (Traditionelle Kunst des Poozer'e), Folge 1. (Hg. Aleksandr Romodin / Irina Romodina) Leningrad: Melodija 1989, S20 29387 000. Das Duett ist auch auf der CD *Igraj dudka, – Spiel, Dudelsack* von Aleksandr Romodins Ensemble „Studio St. Petersburg“ zu hören (LuboMusik. LC 7154).

[74] Wie aus der Transkription ersichtlich, stimmt der Musiker sein Instrument speziell für dieses Stück entsprechend um – vermutlich durch Erweiterung des mit Stimmwachs verengten Grifflochs.

[75] Aufnahme von Eduard Wolter 1908 im Berliner Phonogrammarchiv, veröffentlicht und transkribiert in: *Lietuvių etnografines muzikos fonogramos 1908–1942. The Phonograms of Lithuanian etnographic music 1908–1942* (Buch mit Doppel-CD, Hg. Austė Nakienė, Rūta Žarskienė), № 11, 12, Vilnius 2007.

[76] Romanov, Vymirajuščij instrument, S. 127.

[77] Konservatorium Klaipeda 1989.

[78] Ebd., S 128–131.

Brauchtumsgattungen sind. Deren Melodien können mit einer authentisch gestimmten Spielpfeife, also ohne Subquarte und Subsekunde, größtenteils nicht vollständig wiedergegeben werden.

Selbstverständlich können die tonalen Übereinstimmungen der Harmonikamelodien mit der Sackpfeife für sich genommen noch keine historisch-genetischen Schlussfolgerungen erlauben. Hierfür müssen auch Faktur und Melodik (einschließlich Mikromelodik) untersucht werden. Das erste Sackpfeifenmerkmal beim Harmonikaspiel ist der Bordun – entweder als lange anhaltender Baß (oder Akkord) oder als eine einstufige Rhythmussbegleitung, die sich über die ganze Periode erstreckt. Den aus dem Kreis Ostrov des Gebiets von Pskov entstammenden Beispielen für die beschriebenen Borduntechniken[79] entspricht einerseits das von Natal'ja Kotikova und Sergej Slonimskij in Velikie Luki aufgezeichnete *Pod pesni* (Notenbeispiel 3). Hier geraten die ausgedehnten Bässe in Widerspruch zu dem sonst für die Gattung gebräuchlichen Prinzip der binären Funktionsharmonik (stabile Verteilung von Hauptakkord und variablen Nebenakkorden). Besondere Aufmerksamkeit verdienen andererseits einige Versionen lokaler Ostinatomelodien, die der Heimat des Dudelsackpfeifers Semën im Kreis Žarkovskij des Gebiets von Tver' (№ 4, 5) entstammen. Die kognitive Verankerung der Borduntechnik im Bewußtsein der Harmonikaspieler – zumindest im Sinne einer zulässigen Gestaltungsmöglichkeit – wird nicht nur durch das Spiel mehrerer Musiker bestätigt, sondern auch durch die Bemerkungen von Nikolaj Gavrilov (*1929) unmittelbar während des Vortrags der *Morozovskaja* (Notenbeispiel 4). Notenbeispiel 5 stammt von Evdokim Krylov (*1933), der als Kind noch die Sackpfeife hören konnte (s.o.). Der Bordun wird hier nur zum Abschluß der meist sechstaktigen Periode, also beim dritten Durchgang des zweitaktigen Kernmotivs durch eine Kadenzwendung unterbrochen.

Das zweite, in der Harmonikamusik stärker ausgeprägte Merkmal des Sackpfeifenspiels betrifft die Ornamentik. Bekanntlich kann auf Sackpfeifen mit offener Spielpfeife (zu denen die nord- und osteuropäischen Typen gehören) kein Melodieton ohne die Einfügung eines Vorschlages wiederholt werden, da der Luftstrom nicht unterbrochen werden kann. Häufig gehört dieser Ornamentalton nicht zur Leiter der Melodie.[80] Streng genommen handelt es sich auch nicht um eine Verzierung, sondern um eine notwendige Spieltechnik für die klare Artikulation der Melodietöne. Man bezeichnet die Subquarte deswegen zuweilen auch als „Staccatoquinte". Gut ist sie im Spiel Slaŭčiks zu erkennen, wie allgemein auch bei mittel- und südosteuropäischen Dudelsackpfeifern.[81] Grifftechnisch realisiert der Spieler sie durch die bekannte *geschlossene Griffweise*: Er schließt alle Grifflöcher, wobei meist nur das für den jeweiligen Melodieton nötige geöffnet – und vor Anspielung des folgenden Tones geschlossen wird, wodurch der „Vorschlag" in der Subquarte entsteht. Recht häufig tritt dieser aber auch als gerader Sechzehntelwert auf, jedoch wird er auch dann nicht als Teil der Hauptmelodie wahrgenommen, sondern verleiht der vorausgehenden Sechzehntel die Artikulationsart Staccato.[82] Dessen unge-

---

[79] Morgenštern, K voprosu o kornjach, № 1, 2; auch bei Morgenstern, *Skobari*, Bd. 2, № 77, 54.

[80] Bernard Garaj spricht von „melodisch nichtfunktionellen Tönen": *Gajdy a gajdošská tradícia na Slovensku* (Die Sackpfeife und die Tradition der Dudelsackpfeifer in der Slowakei), Bratislava: ASCO 1995, S. 71.

[81] Timothy Rice, *May It Fill Your Soul: Experiencing Bulgarian Music* (Chicago Studies in Ethnomusicology): Chicago: The University of Chicago Press 1994.

[82] Ebd., S. 73,74.

achtet sind die hier beigefügten Transkriptionen mehr an den realen Notenwerten orientiert als an der Wahrnehmung der Artikulationsarten.[83] Dieses Verfahren ist in unserem Kontext besser geeignet, die Analogien zum Harmonikaspiel deutlich zu machen.

In russischen Harmonikastücken ist die Ornamentierung durch die unterste Stufe relativ selten. Boris Smirnov[84] hat im Süden des Gebiets von Archangel'sk ein Stück dokumentiert, bei dem als Vorschlag fast ausschließlich der Grundton der Melodie verwendet wird. Der Harmonikatyp wird von Smirnov nicht eindeutig mitgeteilt, es handelt sich entweder um eine *tal'janka* oder um die lokaltypische sogenannte *volynka*, deren Bezeichnung keineswegs zufällig entstanden sein dürfte.

Die Harmonikaspieler von Pskov verwenden die Subquarte vor allem in schnellen melodischen Wendungen auf den geraden Sechzehnteln (Notenbeispiel 6–8), wie dies auch der Dudelsackpfeifer Slaŭčik in seiner *Ljavonicha* tut (Notenbeispiel 2). Die Subquarte tritt in diesen Stücken als fragmentarischer Bordun auf und hat keinerlei melisches Eigengewicht. Würde man die Subquarte durch Sechzehntelpausen ersetzen, bliebe das Stück zweifellos erkennbar. In der *Ljavonicha* Leonid Kozlovs (*1950, Notenbeispiel 9), die in sehr schnellem Tempo gespielt wird, erscheint die Subquarte gewöhnlich auf den geraden Achteln. Von besonderem Interesse ist ein Harmonikastück von Anatolij Kozlov (*1950) aus dem Kreis Duchovščina des Smolensker Gebiets (Notenbeispiel 10). Die rechte wie auch die linke Hand sind spieltechnisch in besonderer Weise gefordert. Umso schwieriger ist die Koordination von Melodie- und Baßpart. Aber auch die Melodik der Harmonikaspieler der *Pskovščina* ist ungewöhnlich hoch entwickelt.[85] Auf der gleichtönigen zweireihigen Harmonika (*chromka*) ist die ständig eingestreute Subquarte für die Melodiehand ausgesprochen unbequem: Der Zeigefinger ist an den Bordun gebunden, während die drei weniger beweglichen Spielfinger sich an der zumeist sechsstufigen Melodie abarbeiten müssen. Für solche Blasaerophone, deren hypoionische Leiter der diatonische Hexachord mit Subquarte und Subsekunde bildet, ist eine solche Mikromelodik dagegen – geschlossene Griffweise vorausgesetzt – ein ganz natürliches Gestaltungsmittel. Über eine solche Leiter verfügen aber von allen traditionellen osteuropäischen Blasaerophonen nur die Sackpfeifen.

Im heutigen Landkreis Olenino, im Gebiet Tver'(wo der Erzpriester Razumichin seinerzeit seine nicht ganz glückliche Beschreibung der örtlichen Sackpfeife abfaßte), verwenden einige Harmonikaspieler als fragmentarische Bordunelemente nicht die Subquarte, sondern die Tonika (Notenbeispiele 11–13). Die tonale Interpretation ist gleich-

[83] Für die Transkriptionen in meinem russischsprachigen Beitrag wurde die als Kassettenaufnahme vorliegende Tonspur in Originalgeschwindigkeit abgehört, wodurch die Wahrnehmung mehr der Artikulation folgte. Erst das für die vorliegende Arbeit benutzte computergestützte Verfahren der verlangsamten Wiedergabe ließ die häufigen Sechzehntelfolgen erkennen.

[84] Boris Fedorovič Smirnov, *Iskusstvo sel'skich garmonistov* (Die Kunst der dörflichen Harmonikaspieler), Moskva: Sovetskij kompozitor 1962, S. 43 f.

[85] Die besondere Virtuosität der Harmonikaspieler im Kreis Kun'ja betont Aleksandr Romodin. S.: Narodnye muzykanty juga Pskovščiny (Volksmusiker der südlichen Pskovščina), in: *Živaja starina*, 4(24) 1999, S. 8–10. Noch zehn Jahre vor dem Erscheinen dieses Beitrags war es mein Kollege Romodin, der mich als angehenden Feldforscher mit Tonaufnahmen der Harmonikaspieler Kirill Alekseevič Kutuzov (1932–2005) bekanntmachte – und später bei Feldforschungen im Kreis Usvjaty mit Vladimir Prokopievič Jakovlev (1937–2006). Beide Musiker beherrschten in besonderer Weise die Sackpfeifentechniken in Ornamentierung und Mikromelodik.

wohl nicht ganz einfach, da in dem Stück *Po derevne*[86] (*Zavidovka*) das untere g^1 in gewisser Weise mit dem Moll-Zentrum a^1 (jeweils mit dem entsprechenden Baß/Akkord) konkurriert (Notenbeispiel 12). In diesen Instrumentalformen können die Bordunfragmente unmittelbar nicht auf die Sackpfeife, sondern auf die *žalejka* (lokal: *dudka*, *rožok*) zurückgehen, deren Leiter gewöhnlich nicht mit der Subquarte, sondern mit dem Grundton der ionischen (mixolydischen) bzw. der Subsekunde der äolischen Skala beginnt. Für die Verknüpfung mit der Hirtenklarinette sprechen auch Berichte des Harmonikaspielers Nikolaj Kožanov (*1932 im Kreis Belyj, Gebiet Tver'), der noch während des Krieges in das Dorf Lipovka, Kreis Olenino, übersiedelte. Der Musiker erinnert sich genau an das Spiel eines Hirten, der auf der *žalejka* nicht nur Signalmelodien, sondern auch das lokaltypische *Po derevne* spielte. Kožanov versuchte sogar den Hirten zum gemeinsamen Ensemblespiel zu bewegen, was jedoch nicht gelang. Sehr wahrscheinlich, daß in seinem *Po derevne* (Notenbeispiel 12) das häufige Pendeln zwischen dem melodischen Hauptton und dem tiefsten Ton g^1 im dritten Takt[87] auf eine durch geschlossene Griffweise der *žalejka* bedingte Melodik zurückgeht. Ebenfalls wahrscheinlich ist, daß die *žalejka*, wie sie Kožanov beschrieb (mit Schallbecher aus Rinderhorn und sechs Grifflöchern), seinerzeit von Dudelsackpfeifern als Spielpfeife verwendet wurde. Zeugnisse einer solchen Verwendung der *žalejka* konnte Sergej Starostin im äußersten Südosten des Gebiets von Tver' aufzeichnen.[88]

Auf der Diskantseite der Harmonika können Bordunelemente auf der Subquarte außer in Ornamentik und Mikromelodik auch als fragmentarische Zweistimmigkeit auftreten (Notenbeispiel 14-17). Eine solche Bordun-Zweistimmigkeit im Diskant auf der zweireihigen *chromka* ist nur dann möglich, wenn die untere Tonika auf der inneren Tastenreihe liegt (wodurch sich die Subquarte schräg gegenüber auf der Außenreihe befindet), ähnliches gilt für die Subquarte im einstimmigen Spiel. Spieltechnisch wird die simultane Subquarte entweder mit der Zeigefingerkuppe erzeugt (so verfahren die Musiker aus Slavkoviči, s.u.) oder durch eine Art Barré-Griff, bei dem das zweite Zeigefingerglied die entsprechende Taste niederdrückt (diese Technik ist im Kreis Kun'ja besonders verbreitet). Solche Zweiklänge sind auf einer Sackpfeife mit einfacher Spielpfeife natürlich nicht zu erzeugen. Dennoch sind die ständigen Bordunfragmente auf der Diskantseite der Harmonika (vor allem die auf der Subquarte) unmöglich ohne einen Einfluß der Sackpfeifenmelodik zu erklären. Auch nördlich des hier untersuchten Areals, im Dorf Slavkovichi, im Kreis Porchov der mittleren *Pskovščina*, stoßen wir auf solche Sackpfeifenelemente. Evgenij Sergeevich Murov (*1920) gebraucht die simultane Subquarte in der *Kamarinskaja* in manchen Perioden fast ununterbrochen[89], sein Dorfnachbar Ivan Vasil'evich Vasil'ev (1915–1996) verwendet ebenfalls hartnäckig die Subquarte über mehrere Perioden in den Stücken der Gattung *pod pesni*.[90] Somit können wir über die instrumentale Terminologie hinaus[91] auch anhand musikalisch-stilistischer

---

[86] Wörtl.: „Durch das Dorf" – eine vor allem im nordrussischen Raum verbreitete Bezeichnung lokaler Instrumentalformen der Gattung *pod pesni*.

[87] Vgl. auch Kožanovs *Zalivočka* (Notenbeispiel 9 in meinem Beitrag im vorangehenden Jahrbuch).

[88] Mündliche Mitteilung von Sergej Starostin, für die ich mich herzlich bedanke.

[89] Morgenstern, *Skobari*, Bd. 2: № 103.

[90] Ebd., № 32, 33.

[91] Vgl. die Übereinstimmung der Bezeichnung der Akkordtasten der Harmonika im Kreis Porchov/Novoržev und des Borduns der weißrussischen Sackpfeife (Morgenštern, *K voprosu o kornjach*, S. 169).

Merkmale auf Nachwirkungen des Sackpfeifenspiels weiter nördlich der weißrussischen Sprachgrenze schließen.

5. Schluß

Historische, ethnographische und ebenso linguistische Befunde (Musikterminologie, Alltagsphraseologie) bestätigen die Verbreitung der Sackpfeife in dem untersuchten Areal nicht nur auf dem Territorium der weißrussischen Mundarten, sondern auch in der mittleren *Pskovščina*. Im Repertoire der Harmonikaspieler finden sich bis in die Gegenwart regelmäßig Stilmittel der Faktur (episodischer oder durchgängiger Bordun im Baß, fragmentarische Zweistimmigkeit im Diskant) und ebenso der Melodik (Bordunton auf den geraden Sechzehnteln), die ohne den Einfluß der Sackpfeife nicht zu erklären sind. In der *Pskovščina* hatten die Sackpfeifen mit hoher Wahrscheinlichkeit eine unvollständige hypoionische Stimmung (ohne zweite Stufe) im Rahmen der None, analog nicht nur zu den westslawischen Sackpfeifen, sondern wahrscheinlich auch zu dem Instrument des schwedisch-estnischen Dudelsackpfeifers Jaan Pulk.[92] Im nordwestlichen Gebiet von Smolensk kann, ausgehend von den Moll-Melodien Kozlovs und anderer Harmonikaspieler, eine Sackpfeife mit hypoäolischer Stimmung angenommen werden, wie diese auch der Weißrusse G. K. Slaŭčik spielte.

Das Repertoire, in dem Stilelemente der Sackpfeife anzutreffen sind, profiliert sich einigermaßen klar. Vor allem sind dies lokale Instrumentalformen *pod pesni*: der Velikolukische Typ des *Skobarja* und andere. Die *Kamarinskaja* (*Ljavonicha*) ist am Nordostrand des weißrussischen Sprachraums mehrfach als Sackpfeifenstück belegt (Landkreise Haradok, Gebiet Vicebsk, und Demidovo, Gebiet Smolensk). Aber auch eine Verbindung der *Kamarinskaja* mit dem Repertoire der russischen Dudelsackpfeifer, insbesondere der späten Skomorochen, kann als sehr wahrscheinlich gelten. Die Tanzmelodie war im 18. Jahrhundert weit verbreitet (also lange vor dem Verschwinden der *volynka* in Rußland), wovon die Sammlung Kirša Danilovs und andere Belege zeugen.[93] Für ihr höheres Alter spricht ihre Nähe zu dem *raëk*-Vers (*raëšnyj stich*) des 17. Jahrhunderts[94] und ebenso ihre enge Verbindung an die satirischen oder hyperbolischen *skomorošiny*[95], besonders an die grotesk-phantastischen *nebylicy*. Als *Rysk dans* finden wir zwei Varianten der *Kamarinskaja* auch im Repertoire des erwähnten Dudelsackpfeifers Jaan Pulk.[96]

---

[92] Nach freundlicher Mitteilung von Bo Nyberg verfügt die im Stockholmer Musikinstrumentenmuseum befindliche Sackpfeife von Jaan Pulk über sechs vorderständige Grifflöcher und ein Daumenloch. Die hypoionosche Stimmung ergibt sich aus einer summarischen Auswertung der Transkriptionen von O. Anderson, s. Per-Ulf Allmo, *Säckpipan i Norden* (vgl. Anm. 14), S. 278, 469. Der entsprechende Abschnitt wurde von Bo Nyberg bearbeitet.

[93] Viktor Abramovič Cukkerman, *„Kamarinskaja" M. I. Glinki i ee tradicii v russkoj muzyke* (Die *Kamarinskaja* M. I. Glinkas und ihre Traditionen in der russischen Musik), Moskva: MuzGIz: 1957, S. 67 f., 162.

[94] Ebd., S. 75.

[95] Besonders aufschlußreich ist die oben erwähnte *Byline vom großen Stier*, die der Versform der *Kamarinskaja* entspricht.

[96] Allmo, *Säckpipan i Norden*, 1990: 469, № II:21, II:22. Ein anderer von Bo Nyberg mitgeteilter russische Tanz aus dem Repertoire Pulks (ebd., S. 278) erinnert an die im 18. Jahrhundert bekannte Tanzmusikformel 2/4 /TS/TD/, für die ich seinerzeit die typologische Bezeichnung *Golubec* vorgeschlagen habe (*Volksmusikinstrumente und instrumentale Volksmusik in Rußland*, Berlin: Verlag

Wir haben also allen Grund anzunehmen, daß im russisch-weißrussischen Grenzgebiet, aber auch im mittleren Gebiet von Pskov, vor der Einführung der Handharmonika eine nicht unbedeutende Zahl von Dudelsackpfeifern aktiv war. Zu ihrem Repertoire gehörten mit hoher Wahrscheinlichkeit die *Kamarinskaja* und ebenso instrumental-vokale Formen vom Typ des *Velikolukischen Skobarja*, der *Zavidovka* und weitere lokale Ausprägungen der Gattung *pod pesni*. Es ist zu hoffen, daß künftige Untersuchungen, auch in anderen Regionen Rußlands, unsere Kenntnisse über das Erbe der russischen Dudelsackpfeifer noch wesentlich erweitern werden.

Ernst Kuhn 1995, S. 114 f.). Bo Nyberg vermutet estnische oder russische Musiker mit anderen Instrumenten, möglicherweise der weit verbreiteten Petersburger Harmonika (*minorka*) als Vermittler (freundliche Auskunft von Bo Nyberg vom 11.08.2008). Gänzlich ausgeschlossen werden kann eine frühe Bekanntschaft des 1904 bereits über 50jährigen Pulk mit russischen Dudelsackpfeifern freilich nicht, wenn wir Vladimir Dal's Belege aus dem an Estland angrenzenden Gouvernement Pskov berücksichtigen.

6. Abbildungen

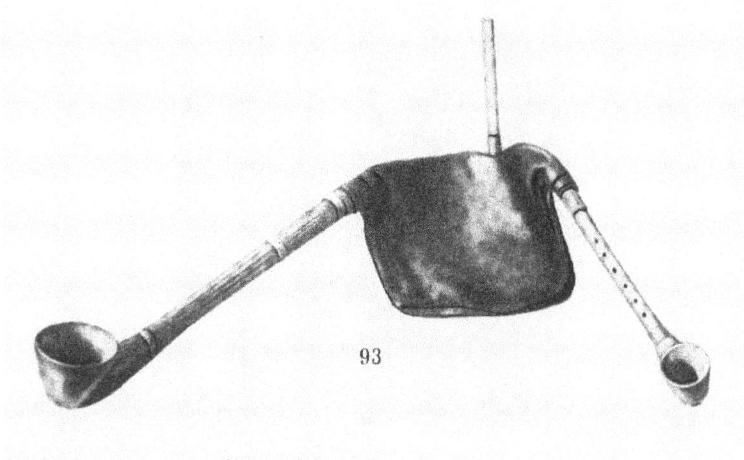

Abb. 1: Konstantin Aleksandrovič Vertkov / Georgij Ivanovič Blagodatov / Ėl'za Ėduardovna Jazovickaja, *Atlas muzykal'nych instrumentov narodov SSSR* (Atlas der Musikinstrumente der Völker der UdSSR), Moskva: Muzyka 1975, № 93, s. Anm. 15).

Abb. 2: Litauischer Dudelsackpfeifer (Nakienė / Žarskienė, *The Phonograms*, S. 62, vgl. Anm. 75).

Abb. 3: „Weißrussischer Dudelsackpfeifer aus dem Gouvernement Vilno". Zeichnung von P Anenskij aus: Spilevskij, PM: Putešestvie po Poles'ju i belorusskomu kraju, Bd III, Minsk 1855 (nach Skorabahatčanka *Belaruskija narodnyja*, Zeichnung 15).

Abb. 4:, „Vor einer weißrussischen Schenke. Tanz zur *duda*". Zeichnung von R. Žukoŭskij aus dem Almanach „Ročnik ljutėrackij" (19. Jh., nach Skorabahatčanka, *Belaruskija narodnyja*, Zeichnung 17).

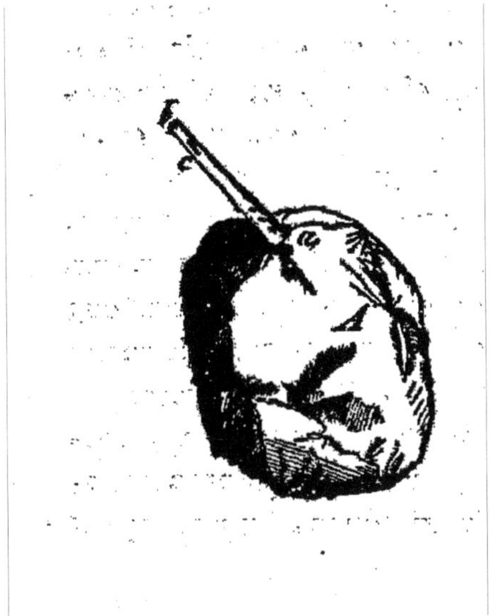

Abb. 5: Mißverständliche Zeichnung einer Sackpfeife aus dem Dorf Bobrovki im Kreis Ržev des Gouvernements Tver' (heute: Bobrovki im Kreis Olenino des Gebiets Tver'), nach: Razumichin, Selo Bobrovki, S. 270, s. Anm. 33.

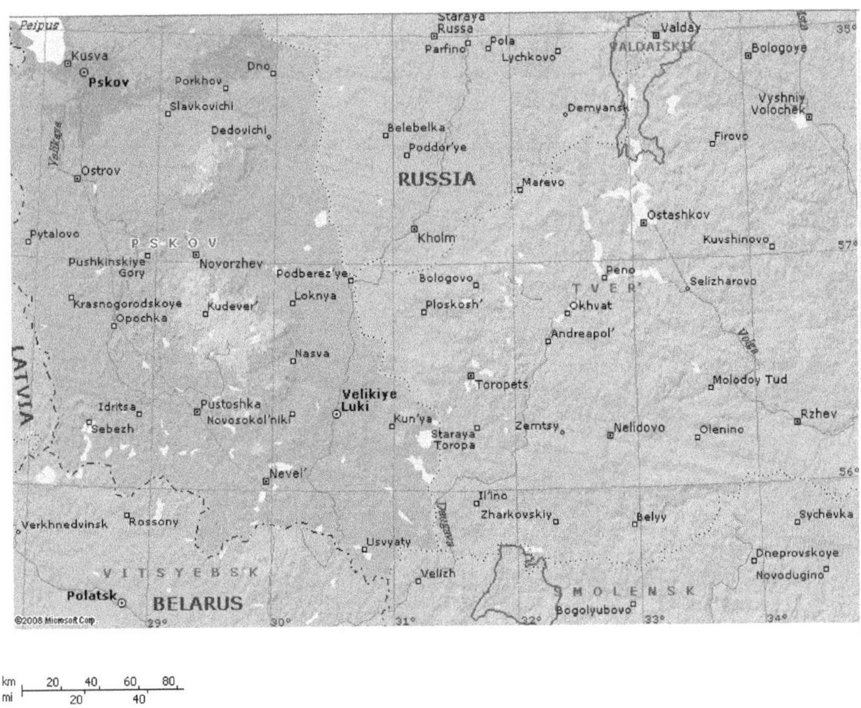

Abb. 6: Karte des Forschungsgebiets Quelle: World Atlas http://encarta.msn.com.

7. Notenbeispiele

1. Russkaja (Russkogo)

G. K. Slaŭčik
(*1836, Gouvernement Vitebsk, Kreis Haradok)

2. Pljasovaja (Ljavonicha)

G. K. Slaūčik
(Gouvernement Vitebsk, Kreis Haradok)

3. Pod pesni (Skobar')

Gebiet Pskov, Velikie Luki

4. Morozovskaja (Po derevne)

N.V. Gavrilov
(*1929, Gebiet Tver', Kreis Žarkovskij)

«Die Morozovskaja. Hier, siehst du – ein Akkord. Ein Finger. Und es macht nichts, daß es nur ein Finger ist.»

«Mit einem Finger kann man es spielen.»

5. Po derevne

A. E. Krylov
(*1933, Gebiet Tver', Kreis Žarkovskij)

6. Pod draku (Skobarja)

I.I. Ryskov (1939-2006, Gebiet Pskov, Kreis Kun'ja)

7. Skobarja

D.T. Podstežonok
(1937–2005, Gebiet Pskov, Kreis Kun'ja)

8. Skobarja (Loknjanskaja)

V.N. Pavlučenkov
(*1934, Gebiet Pskov, Kreis Novosokol'niki)

9. Ljavonicha

L.N. Kozlov
(*1950, Gebiet Pskov, Kreis Kun'ja)

10. Podpripevočnaja

A. A. Kozlov
(*1950, Gebiet Smolensk, Kreis Duchovščina)

11. Po derevne (Oleninskaja)

V.M. Korguzov
(*1932, Gebiet Tver', Kreis Olenino)

12. Po derevne (Zavidovka)

N.I. Kožanov
(*1932, Gebiet Tver', Kreis Olenino)

13. Levonicha

N.I. Kožanov
(*1932, Gebiet Tver', Kreis Olenino)

14. Skobarja

I. N. Nikiforov
(*1929, Gebiet Pskov, Kreis Velikie Luki)

15. K devkam (Skobarja)

V.V. Valuev
(*1942, Gebiet Pskov, Kreis Kun'ja)

16. Pod draku (Skobarja)

P. V. Maksimov
(*1924, Gebiet Pskov, Kreis Kun'ja)

17. Dlinnyj Skobar'

V. M. Grigor'ev
(*1957, Gebiet Pskov, Kreis Kun'ja)

8. Nachweis der Notenbeispiele

Bei den Transkriptionen von Sackpfeife (№ 1, 2) und der Harmonika *tal'janka* (№ 3) wird die originale Tonhöhe beibehalten. Alle anderen Harmonikastücke, die auf der gleichtönigen, halbchromatischen *chromka* gespielt wurden, werden zur besseren Vergleichbarkeit der Spieltechnik in einer einheitlichen Stimmung notiert, die den Diskanttönen $c^1 - c^5$ entspricht.

1. *Pljasovaja [Ljavonicha]*[97]
Sackpfeife (*duda*), gespielt von G.K. Slaŭčik (*1836), Kreis Haradok, Gebiet Vicebsk. Aufgezeichnet am 15.10.1931 von Evgenij Gippius. Phonogrammarchiv des Instituts für russische Literatur und Kunst der Akademie der Wissenschaften in St. Petersburg (Puškin-Haus), FV 3503-2.

2. *Pljasovoj naigryš* (Tanzstück) *Russkaja [Russkogo]*
S. №. 1, FV 3504-3. Die Walzenaufzeichnung ist an zwei Stellen unterbrochen.

3. *Pod pesni* [„Zu Liedern"] (*Skobar'*)
Harmonika (*tal'janka*), unbekannter Musiker, Velikie Luki, Gebiet Pskov. Aufgezeichnet 1963 von Natalija Kotikova und Sergej Slonimskij. Phonogrammarchiv des Instituts für russische Literatur und Kunst, MF2226-01. Dem Petersburger Komponisten Sergej Slonimskij danke ich für die freundliche Erlaubnis zur Verwendung der Tonaufnahmen.

4. *Morozovskaja (Po derevne)* [„Durch das Dorf"]
Harmonika (*chromka*), gespielt von Nikolaj Vasil'evič Gavrilov (*1929), Danilino, Kreis Žarkovskij, Gebiet Tver'. Aufgezeichnet am 25.10.2005 gemeinsam mit Valentin Vinogradov.

5. *Po derevne*
Harmonika (*chromka*), gespielt von Anatolij Evdokimovič Krylov (*1933), Zelen'kovo, Kreis Žarkovskij, Gebiet Tver'. Aufgezeichnet am 24.10.2005.

6. *Pod draku* [„Zur Prügelei"] *(Skobarja)*
Harmonika (*chromka*), gespielt von Ivan Ionovič Ryskov, (*1939 in Kamasy, † 2006), Usmyn', Kreis Kun'ja, Gebiet Pskov. Aufgezeichnet am 23.09.2005 gemeinsam mit Igor' Stesev.

7. *Skobarja*
Harmonika (*chromka*), gespielt von Dmitrij Tarasovič Podstežonok(1937–2005), Andronovo, Kreis Kun'ja, Gebiet Pskov. Aufgezeichnet im August 1988 von einem Freund des Musikers aus Usmyn', dessen Name aus den in Anm. 26 genannten Gründen nicht eruierbar ist.

8. *Skobarja (Loknjanskaja)*
Harmonika (*chromka*), gespielt von Vasilij Nikolaevič Pavlučenkov (*1934 in Černogubovo), Žolobovo, Kreis Novosolol'niki, Gebiet Pskov. Aufgezeichnet am 8.09.2006 gemeinsam mit Igor' Stesev.

[97] In Beispiel 1 und 2 sind die Titel nach den Archivangaben angegeben, in eckigen Klammern die von dem Musiker während der Aufnahme verwendete Bezeichnung.

9. *Ljavonicha*
Harmonika (*chromka*), gespielt von Leonid Nikolaevič Kozlov (*1950), Žigray, Gebiet Pskov, Kreis Kun'ja. Aufgezeichnet am 13.09.2006 gemeinsam mit Igor' Stesev.

10. *Podpripevočnaja*
Harmonika (*chromka*), gespielt von Anatolij Aleksandrovič Kozlov (*1950 in Tula), Kreisstadt Duchovščina, Gebiet Smolensk. Aufgezeichnet am 13.09.2006 gemeinsam mit Valentin Vinogradov.

11. *Po derevne* [„Durch das Dorf"] (*Oleninskaja*)
Harmonika (*chromka*), gespielt von Vitalij Michajlovič Korguzov (*1932 in Kiselëvo), Molodoj Tud, Kreis Olenino, Gebiet Tver'. Aufgezeichnet am 7.07.2006 gemeinsam mit Aleksandr Romodin, Elizaveta Borodulina und Rostislav Koshelev.

12. *Po derevne (Zavidovka)*
Harmonika (*chromka*), gespielt von Nikolaj Ivanovič Kožanov (*1932 in Karskoe, Kreis Bely), Lipovka, Kreis Olenino, Gebiet Tver'. Aufgezeichnet am 12.10.2005 gemeinsam mit Valentin Vinogradov.

13. *Levonicha*
S. №. 12.

14. *Skobarja*
Harmonika (*chromka*), gespielt von Ivan Nikiforovič Nikiforov (*1929 in Rud'kovo), Velikie Luki, Gebiet Pskov. Aufgezeichnet am 11.09.2006 gemeinsam mit Igor' Stesev.

15. *K devkam* [„Zu den Mädels"] *(Skobarja)*
Harmonika (*chromka*), gespielt von Vasilij Vladimirovič Valuev (*1942 in Lužok), Chomušino, Kreis Kun'ja, Gebiet Pskov. Aufgezeichnet am 24.09.2005 gemeinsam mit Igor' Stesev.

16. *Pod draku* [„Zur Prügelei"] *(Skobarja)*
Harmonika (*chromka*), gespielt von Pëtr Vasil'evič Maksimov (*1924 in Kurilkovo), Lipicy, Kreis Kun'ja, Gebiet Pskov. Aufgezeichnet am 10.08.2006 gemeinsam mit Aleksandr Romodin und Elizaveta Borodulina.

17. *Dlinnyj Skobar'*
Harmonika (*chromka*), gespielt von Vladimir Michajlovič Grigor'ev (*1957 in Chlebanicha), Prichaby, Kreis Kun'ja, Gebiet Pskov. Aufgezeichnet am 25.08.2005 gemeinsam mit Igor' Stesev.

Martin Pfleiderer

Stimmen populärer Musik. Vokale Gestaltungsmittel und Aspekte der Rezeption

Die raue Stimme von Louis Armstrong, der Crooning-Gesang von Bing Crosby, die vokalen Ausbrüche des Rhythm'n'Blues-Sängers Howlin' Wolf, der virtuose Scat-Gesang von Ella Fitzgerald, die exzentrischen Rufe des „Godfathers of Soul" James Brown, das weiche Falsett von Curtis Mayfield oder William „Smokey" Robinson, die kunstvolle Melismatik von Soul-Vokalisten wie Aretha Franklin, die gepresste Kopfstimme des Hardrock-Sängers Robert Plant, das gutturale Röcheln von Death Metal-Sängern, der akzentuierte Sprechgesang von Rappern – die Vielfalt der Stimmen und vokalen Gestaltungsmittel in populärer Musik ist fast unüberschaubar groß. Vokale Ausdrucksmittel sind dabei nicht nur charakteristisch für bestimmte Stile und Genres der populären Musik und deren spezifische Ausdruckswelten, sondern verweisen darüber hinaus auf Stimmungen und Emotionen sowie auf vermeintliche Persönlichkeitsdispositionen von Sängerinnen und Sängern, in deren Stimmen zudem biographische Erfahrungen ihre Spuren hinterlassen.

Im Mittelpunkt der populären Musik im 20. und 21. Jahrhundert stehen bekanntlich bestimmte Aufführungen von Stücken, auf der Bühne oder im Studio, die auf verschiedenen Arten von Tonträgern und Speichermedien festgehalten und durch Rundfunk, Film und Fernsehen, seit einiger Zeit auch durch das Internet, verbreitet werden. Dagegen dienen die kompositorischen Vorlagen und Notentexte spätestens seit Mitte des 20. Jahrhunderts fast nur noch als Urheberrechtsbelege zur Sicherung von Tantiemen. Natürlich sind in vielen Bereichen populärer Musik die Songtexte wichtig, und insbesondere wenn der Sänger oder die Sängerin etwas zu erzählen und zu berichten hat, ist Textverständlichkeit ein entscheidendes Kriterium für die Wahl der vokalen Gestaltungsmittel. Bei vielen Aufführungen und Aufnahmen populärer Musik zählen jedoch nicht so sehr die sprachlich vermittelten Inhalte der Songtexte, sondern der vokale Ausdruck selbst: wichtig ist nicht *was*, sondern vielmehr *wie* etwas gesungen wird – oder mit den Worten des britischen Popmusikforschers Simon Frith: „Song words are not about ideas ('content') but about their expression" (Frith, 1996, S. 164). Oft geht es hier vor allem darum, mit stimmlichen Mitteln bestimmten Gefühlen und Stimmungen Ausdruck zu verleihen. Richard Middleton (2000, S. 29f.) unterscheidet folgerichtig zwischen den beiden idealtypischen Extremen eines narrativen Stimmgebrauchs, der sich eng an das Sprechen anlehnt und einen Schwerpunkt auf die erzählten Inhalte legt, und einem gestischen oder klanglichen Stimmgebrauch, bei dem die sprachliche Semantik in den musikalischen Fluss „aufgesogen" („absorbed") wird und die Stimme dazu tendiert, zu einem Musikinstrument zu werden. Zwischen diesen Extremen steht, so Middleton, die in populärer Musik verbreitete Möglichkeit, dass „words merge with melody, resulting in intoned expression" (Middleton 2000, S. 30).[1]

[1] Dies gilt, so Middleton weiter, insbesondere für die markanten Hook-Phrasen eines Songs: „A typical situation, perhaps, is where a vocal ‚hook' works by bundling together the meaning, resonances and sound-shapes of the words together with the melodic, rhythmic, timbral and articulatory dimensions of

Nicht in den Liedkompositionen, sondern in den Stimmen, die diese Lieder singen, steckt somit der Schlüssel zum Verständnis populärer Musik, zum Genuss der Pophörer und womöglich generell zur Popularität der Musik. „Voices not songs hold the key to our pop pleasures", lautet daher nicht von ungefähr eine pointierte These von Simon Frith (1996, S. 201). Wobei es sich hier in der Regel nicht um einen „klassisch" geschulten Gesang handelt, sondern um stimmliche Ausdrucksweisen, die weit näher an der alltäglichen Sprechstimme liegen als am artifiziellen Opern- und Liedgesang und dem Ideal von Klangschönheit der europäischen Kunstmusiktradition. Die Nähe zu alltäglichen Ausdrucksweisen der Stimme bietet einen möglichen Erklärungsansatz für den unmittelbaren Zugang einer breiten Hörerschaft zu populären Stimmen – und zur populären Musik insgesamt. Denn hierdurch wird nicht nur den musikalischen gebildeten Hörern, sondern prinzipiell jedem ein Nachfühlen und Nachahmen der Stimmen ermöglicht – und dadurch ein imaginierendes „Hineinschlüpfen" in die Stimmungs- und Gefühlslage, ja in die vorgestellte Persönlichkeit der jeweiligen Sängerin und des jeweiligen Sängers.

Stimme und vokaler Ausdruck stehen offensichtlich im Zentrum populärer Musik und müssten daher auch ein zentrales Thema der Popularmusikforschung sein – *müssten*, denn tatsächlich werden dort Stimme, Gesang und vokaler Ausdruck bislang eher sporadisch reflektiert und untersucht.[2] Dies liegt nicht nur an der fast unüberschaubar großen Vielfalt der vokalen Ausdrucksweisen in populärer Musik, sondern ist vielmehr, so meine Einschätzung, zu einem großen Teil darauf zurückführen, dass eine allgemein anerkannte deskriptive Terminologie vokaler Äußerungen sowie adäquate Analysemethoden bislang nur in Ansätzen vorhanden sind. Zugleich gestaltet sich das angedeutete Verhältnis von den Stimmen, den Stimmungen und Gefühlen der realen oder imaginierten Person des Vokalisten, den Songtexten, den Aufführungskontexten bzw. der medialen Vermittlung sowie den Prozessen der Rezeption, der Imagination und Identifikation doch recht komplex und bietet Raum für zahlreiche unterschiedliche Untersuchungs- und Interpretationsansätze. Will man den vielen Dimensionen des Phänomens der Stimme in populärer Musik gerecht werden, so ist daher die Kooperation einer Reihe von Forschungsdisziplinen sinnvoll: von der Musikgeschichte und musikalischen Analyse über die musikalische Akustik, die Phonetik und die Gesangspädagogik, die Psychologie der Emotionen und der Persönlichkeit bis hin zu kulturwissenschaftlichen und medienwissenschaftlichen Ansätzen, in denen Stimmphänomene in den

---

their sung performance, encapsulating that particular semantic-affective field which will come to be associated with the song" (Middleton 2000, S. 29).

[2] Ausnahmen bilden die Überlegungen von Bradby/Torode (1984), Cubitt (1984), Frith (1996, S. 183-202), Middleton (2000) und Moore (2001, S. 44ff.), sowie von Potter (1998, S. 87-112 und S. 133-189), Banfield (2000) und Bielefeldt (2006, 2008). David Brackett (1995) thematisiert in seinen ausführlichen Analysen der Aufnahmen von Bing Crosby, Billie Holiday, Hank Williams, James Brown und Elvis Costello auch die stimmlichen Gestaltungsmittel der betreffenden Sängerinnen und Sänger. Im Bereich der Jazzforschung hat sich vor allem William R. Bauer verschiedenen Aspekten der Vokalgestaltung im Jazz, u.a. dem Scat-Gesang, gewidmet (Bauer 1993, 2001/02, s. auch Stewart 1987). Überblicksartige Darstellungen zum Gesang im American Popular Song und Jazz geben Pleasants (1974), Friedwald (1990) und Crowther/Pinfold (1986).

vergangenen Jahren vermehrt diskutiert wurden, ohne dass dabei populäre Musik als wichtiger Themenbereich erkannt worden wäre.[3]

Im Rahmen dieses Beitrags ist es natürlich nicht möglich, die ganze Vielfalt der Gesangstile und vokalen Gestaltungsmittel in populärer Musik auch nur einigermaßen erschöpfend auszuloten und darzustellen. Meine Absicht ist es vielmehr, einige zentrale Dimensionen von Stimme und Gesang in populärer Musik theoretisch zu reflektieren und systematisch zu erschließen, wobei sowohl Aspekte der Produktion und musikalisch-klanglichen Gestaltung als auch der Rezeption berücksichtigt werden sollen. Hierdurch möchte ich einerseits eine Grundlage für musikanalytische Studien zu stimmlichen Gestaltungsmitteln und Ausdrucksweisen von Vokalisten in verschiedenen Genres populärer Musik schaffen, andererseits mögliche empirische Forschungsansätze zu Prozessen der Rezeption populärer Stimmen diskutieren. In der Verknüpfung und gegenseitigen Ergänzung musikhistorisch-analytischer und musikpsychologisch-empirischer Herangehensweisen eröffnen sich, meiner Ansicht nach, vielversprechende Perspektiven zu einer umfassenden wissenschaftlichen Untersuchung von Stimmen populärer Musik.

Am Ausgangspunkt meiner Überlegungen stehen zunächst die folgenden Fragen: Wie lassen sich Stimmen populärer Musik adäquat beschreiben? Wo liegen die Unterschiede zur Sprechstimme einerseits, zum „klassischen" Gesang andererseits? Diesen Fragen widmet sich der erste Teil des Beitrags. Auf der Grundlage einer systematischen Darstellung der vokalen Gestaltungsdimensionen (Tonhöhe, Rhythmus, Stimmqualität) möchte ich dabei auf verschiedene vokale Gestaltungsmittel in populärer Musik hinweisen und dies an Hand der Analyse einer Aufnahme von Ray Charles („Come Back, Baby" 1954) illustrieren. Außerdem werden Aspekte der Gesangstechnik und der technischen Bearbeitung von Stimmen erörtert. Überlegungen zu kulturellen Mustern des vokalen Ausdrucks und zur „persönlichen Stimme" einer Sängerin oder eines Sängers leiten zum zweiten, empirisch orientierten Teil meines Beitrags über. Nach einigen methodologischen Ausführungen möchte ich hier eine Pilotstudie zur Stimmrezeption vorstellen, in der es um die Klärung der Zusammenhänge zwischen stimmlichem Ausdruck und emotionalem Eindruck bzw. Affekt- und Persönlichkeitszuschreibungen durch die Hörer geht: Auf welche akustischen und musikalischen Eigenheiten des stimmlichen Ausdrucks lassen sich die Zuschreibungen von Emotionen und Persönlichkeitsmerkmalen zurückführen und mit welchen Methoden lassen sich die Zusammenhänge von Stimme, musikalischer Gestaltung, Emotion und Persönlichkeit empirisch erforschen?

Dimensionen der menschlichen Sprech- und Singstimme

Unter Stimme versteht man die Gesamtheit lautlicher und lautsprachlicher Produktion und Rezeption. Gemeinhin wird zwischen sprachlichen (linguistischen), sprachbegleitenden (para-linguistischen) und nicht-sprachlichen (extra-linguistischen) Aspekten vokaler Äußerungen unterschieden. In der Phonetik wird nun versucht, die zahlreichen Facetten der Stimme vor allem bei sprachlichen Äußerungen präzise und systematisch zu charakterisieren (vgl. etwa Pétursson/Neppert 2002). Allerdings finden sich auch bei

[3] Vgl. Göttert (1998), Kittler/Weigel/Macho (2002), Epping-Jäger/Linz (2003), Felderer (2004) und Kolesch/Krämer (2006). Bislang haben nur Bielefeldt (2006, 2008) und Middleton (2006) Verbindungslinien zwischen Popmusikforschung und kultur- bzw. medienwissenschaftlichen Theorien der Stimme geknüpft.

sprachlichen Äußerungen sowohl para-linguistische Merkmale (z.B. Sprechrhythmus und Intonation) als auch extralinguistische Eigenheiten (z.B. Atemgeräusche, Näseln), wie sie zum Teil auch bei Tierlauten verbreitet sind (Trojahn 1975, Fischer 2006). Obwohl die menschliche Sprechstimme in erster Linie der Übermittlung von sprachlichen Inhalten dient, kann sie darüber hinaus auch ohne Rückgriff auf Sprache kommunikative Botschaften übermitteln, so etwa durch Schreie, Lachen, Weinen, Jammern usw. Der stimmliche Ausdruck hat bei diesen nicht-sprachlichen Äußerungen den Signalcharakter, der ihnen im Tierreich zukommt, auch für die Menschen bewahrt. So gilt z.B. Brüllen kulturübergreifend als Ausdruck von Wut, Zorn und Aggressivität, Kreischen als Ausdruck von Furcht, Schreien als Ausdruck von Angst und Schmerz, freundliches Grunzen und Lachen als Ausdruck von Wohlbehagen, Schluchzen und Weinen dagegen als Ausdruck von Traurigkeit.

Neben der Unterscheidung in sprachliche, para-linguistische und extra-linguistische Aspekte gibt es weitere relevante Dimensionen, die der Einteilung von Stimmphänomenen dienen können (vgl. König/Brandt 2006). So lässt sich zwischen vorübergehenden (temporären) und permanenten, entweder anatomisch bedingten, oder aber habituellen Eigenschaften einer Stimme unterscheiden. Weiterhin sind vokale Aspekte entweder kontrollierbar, oder aber unwillkürlich und nicht beeinflussbar. Schließlich kann zwischen jenen vokalen Eigenheiten, die über die wechselnden affektiven Zustände, die Persönlichkeit und die soziale oder ethnische Gruppenzugehörigkeit eines Sprechers informieren, und jenen Aspekten, die allein der Übermittlung und Verdeutlichung von kommunikativen Inhalten dienen, unterschieden werden. Zugleich können eine Reihe von Dimensionen unterschieden werden, anhand derer sich bestimmte Stimmen dauerhaft und grundlegend voneinander unterscheiden oder sich auch nur innerhalb einer spezifischen Sprechsituation charakterisieren lassen: die Lautstärke, die Tonlage und das Ausmaß der Abweichungen von einer vorherrschenden Tonhöhe sowie die Sprechgeschwindigkeit und der Sprechrhythmus. Bei der Sprechstimme bildet der Tonhöhenverlauf zusammen mit Akzentuierungen durch Lautstärke und Eigenheiten der zeitlichen Gestaltung das, was als Sprechintonation einer sprachlichen Äußerung bezeichnet wird. Hinzu kommt ein weiteres, zentrales Unterscheidungsmerkmal: die Stimmqualität, die teils mit anatomischen Besonderheiten, teils mit gelernten Eigenheiten der Stimmgebung und der Klangformung zusammenhängt.

Viele der genannten Dimensionen und Aspekte spielen nicht nur bei der Sprechstimme, sondern ebenso im Gesang und Sprechgesang eine wichtige Rolle. Während es sich beim Sprechen und bei nicht-sprachlichen vokalen Äußerungen in der Regel um „natürliche" oder alltägliche Ausdrucksweisen der Stimme handelt, auf die auch in einem künstlerisch-performativen Sprechkontext, z.B. im Theater oder im Film, zurückgegriffen wird, so rückt die Stimme beim Singen in einen ästhetischen Zusammenhang, in dem in der Regel der stilisierten Struktur der Tonhöhen und Tondauern, aber auch Eigenheiten der Stimmqualität und einem klangbetonten Einsatz der Stimme besondere Bedeutung zukommt. Oftmals liegt dem Stimmgebrauch hier eine gewisse Schulung und Übung zugrunde, die alle physiologisch beeinflussbaren vokalen Aspekte umfasst: von der Lautstärke und dem präzisen Treffen von Tonhöhen über eine rhythmisch-metrische Gestaltung der zeitlichen Abfolge bis hin zum gezielten Einsatz von Stimmqualitäten und Klangfarben, wobei generell auch die Orientierung an dem von den Begleitstimmen bzw. Begleitinstrumenten vorgegebenen tonalen, harmonischen, rhythmisch-metrischen und klanglichen Bezugsrahmen eine wichtige Rolle spielt. Bevor ich diese Aspekte vertiefe, möchte ich mit einem kurzen Exkurs zur Physiologie und Funktionsweise des menschlichen Stimmapparats die grundlegenden Eigenschaften der menschlichen Lautproduktion in Erinnerung rufen (vgl. hierzu auch Sundberg 1987, Eckert/Laver 1994, Pétursson/Neppert 2002, Wendler/Seidner/Eysholdt 2005).

Die menschliche Stimme – Physiologie und Funktionsweise

Grundsätzlich lassen sich drei physiologische Bereiche der stimmlichen Lautbildung unterscheiden: die Luft- und Energiezufuhr durch die Atmung (Respiration), die Stimmgebung (Phonation) im Kehlkopf und die Klangformung (Artikulation) im Vokaltrakt, also im Rachen, Mundraum und Nasenbereich.

Der Kehlkopf (Larynx) ist ein hohles, kastenartiges Gebilde, das sich aus mehreren gegeneinander beweglichen Knorpeln zusammensetzt: dem Schildknorpel, dem Ringknorpel, den beiden Ary- oder Stellknorpeln sowie dem Kehldeckel. An Schild- und Stellknorpeln sind die beiden Stimmlippen (auch: Stimmfalten oder Stimmbänder; engl.: vocal folds) befestigt, durch deren Schwingungen die Stimmgebung erfolgt. Die Stimmlippen sind ein Paar weicher Gewebewülste, deren Form und Steifigkeit durch verschiedene kleine Muskeln verändert werden kann. Sie sind vorne am Schildknorpel, hinten an den beiden beweglichen, pyramidenförmigen Stellknorpeln befestigt; der Ringknorpel sitzt unterhalb der Stellknorpel und gibt ihnen hinten durch eine Erhöhung Halt. Die v-förmige Öffnung zwischen den beiden Stimmlippen wird als Stimmritze (Glottis) bezeichnet. Sie ist bei größter Öffnung ca. 2 cm lang und 1 cm breit. Bei der Atmung halten die Stellknorpel die Stimmritze offen. Die Stimmritze kann geschlossen werden (Adduktion), indem die beiden Stellknorpel aneinandergedrückt werden (mediale Spannung). Zusätzlich kann die Längsspannung in den Stimmlippen erhöht werden, indem die Stimmlippen durch ein Abkippen des Schildknorpels etwas verlängert (passive Längsspannung) und die Muskeln in den Stimmlippen angespannt werden (aktive Längsspannung). Bei der Stimmgebung sind die Stimmlippen zunächst geschlossen, sodass sich unterhalb der Glottis ein Luftüberdruck aufbauen kann. Dieser Druck öffnet die Stimmlippen. Die Rückstellkräfte des elastischen Materials, die Spannung der Kehlkopfmuskeln, mit denen die Stimmlippen aneinander gepresst werden, sowie der sog. Bernoulli-Effekt führen dazu, dass sich die Stimmlippen sehr schnell wieder schließen. Der Atemdruck baut sich sodann erneut auf, und der Vorgang wiederholt sich. Auf diese Weise strömen bei der Phonation periodisch Schübe von Luft in den Vokaltrakt ein. Die Häufigkeit oder Frequenz dieser periodischen Luftschübe hängt von der Spannung, Dicke und Länge der Stimmbänder sowie von der Stärke des subglottalen Überdrucks ab. Die Stimmbänder schwingen beim Sprechen mit einer Frequenz von ca. 70-200 Hz (Männer) bzw. ca. 140-400 Hz (Frauen); die tiefere Frequenz der Männer kommt durch die schwereren und längeren Stimmlippen zustande. Die Wellenform der periodischen Luftschübe, die aufgrund der schlagartigen Bewegung der Stimmlippen zustande kommt, prägt das obertonreiche Klangspektrum des Stimmklangs, das durch besondere Phonationsweisen sowie durch die Länge und Form des Vokaltraktes weiter verändert werden kann.

An den Kehlkopf schließt sich die Rachenhöhle (Pharynx) an. Das weiche Gaumensegel (Velum) öffnet bzw. schließt den Vokaltrakt zur Nasenhöhle oder zum Mundraum hin; außerdem wirken auch Schädel und Lunge als Resonanzraum und beeinflussen

Schallabstrahlung und Stimmklang. Die unterschiedlichen Vokale werden durch eine unterschiedliche Formung des Vokaltrakts erzeugt, vor allem durch verschiedene Stellungen der Zunge, wodurch sich besonders ausgeprägte Frequenzbereiche des Klangspektrums, die für die einzelnen Vokale charakteristisch sind, die sog. Formanten, verschieben lassen. Die Vokale können bereits anhand der jeweiligen Lage des ersten und zweiten Formanten (im Bereich 250 Hz bis 1 kHz bzw. 600 Hz bis 2,5 kHz) bestimmt werden, z.B. „u" mit tiefem zweiten Formanten, „i" mit hohem zweiten Formanten, „a" mit hohem ersten Formanten. Die Lage des dritten und des vierten Formanten (ca. 1,7-3,5 kHz bzw. ca. 2,7-3,7 kHz) ist dagegen für die sprecherspezifische individuelle Klangfärbung oder Stimmqualität einer Stimme mitverantwortlich. Allerdings lässt sich auch die Lage dieser höheren Formanten etwa durch die Formung der Lippen und das Senken des Kehlkopfes, wie es im klassischen Gesang üblich ist, verschieben. Neben den Vokalen gibt es mit den zahlreichen Konsonanten weitere Laute, die freilich keiner oder einer nur begleitenden Stimmgebung durch die Stimmlippen bedürfen, sondern vielmehr aufgrund geräuschhafter Wirbel der ausströmenden Luft artikuliert werden.

Gestaltung von Tonhöhe, Tempo und Rhythmus beim Sprechen und Singen

Jede Person wählt beim Sprechen eine charakteristische Tonlage, die vor allem von der Länge der Stimmlippen abhängt. Die typische Tonlage eines Sprechers lässt sich als Mittelwert der verschiedenen Tonhöhen ermitteln, die der Sprecher im Verlauf seiner Äußerungen verwendet. Bei Männern liegt sie in der Regel zwischen 110 und 130 Hz, bei Frauen zwischen 200 und 230 Hz. Kinder haben eine entsprechend höhere Sprechstimme, bei älteren Menschen sinkt die Sprechtonlage. Auch beim Singen werden bekanntlich typische Registerlagen unterschieden, die in der klassischen europäischen Tradition als Bass, Bariton, Tenor, Alt, Sopran usw. bezeichnet werden.[4]

Im Laufe einer Stimmäußerung verändert sich allerdings die Tonhöhe ständig – innerhalb eines Satzes, aber auch innerhalb einzelner Wörter und Vokale (vgl. König/Brandt 2006, S. 117). Das Ausmaß der Variation der Tonhöhe ist beim Sprechen normalerweise nicht sehr groß. Die Abweichungen liegen in der Regel im Bereich von 20 Hz um die Grundfrequenz. Größere Abweichungen gelten bereits als „melodische" Sprechweise, die allerdings für manche Personen und Sprachgruppen charakteristisch ist. Die Tonhöhenabweichungen, die beim Sprechen also nicht mehr als eine Terz ausmachen, werden beim Singen in der Regel um ein Vielfaches überschritten. Hier findet sich ein Tonhöhenwechsel im Bereich von zwei bis drei Oktaven. Eine Begrenzung des Ambitus auf eine Quinte oder weniger sind allerdings bei manchen Vokalisten in populärer Musik nichts Ungewöhnliches, so z.B. beim Singsang eines Bob Dylan oder im Sprechgesang der Rap Music, wo oftmals nur eine, zwei oder drei Tonhöhen verwendet werden, sodass der Ambitus kaum größer ist als beim Sprechen.

Während beim Sprechen die Tonhöhe innerhalb von Wörtern und Sätzen ständig leicht variiert wird, versucht man in einem musikalischen Kontext in der Regel, jeden Vokal und jede Silbe auf einer diskreten, unveränderten Tonhöhe zu singen, wozu eine

[4] Vgl. Fischer (1993), Seidner/Wendner (2004). Sundberg (1987, S. 50f.) problematisiert die mitunter verwirrende Terminologie der Registerbezeichnungen, die sich nicht nur auf Tonlagen, sondern zum Teil auch auf unterschiedliche Phonationsweisen beziehen.

konstante Spannung der Stimmlippen erforderlich ist. Dieser Vorgang, das genaue Ansteuern von Tonhöhen, wird in der Musik als Intonation bezeichnet und bedarf normalerweise einer gewissen Übung. Leichte Tonhöhenschwankungen sowie stufenloses Gleiten zwischen verschiedenen Tonhöhen („portamento") innerhalb von Silben, Wörtern und Phrasen sind allerdings bei vielen Sängerinnen und Sängern in der populären Musik durchaus üblich. Portamenti finden sich überall dort, wo Sprechen und Singen fließend ineinander übergehen, so beispielsweise im Blues-Gesang. Manche Vokalisten verwenden eine kunstvolle melismatische Gestaltung, d.h. bei einzelnen Textsilben bzw. einem einzigen Vokal wird die Tonhöhe mehrfach gleitend gewechselt – so besonders ausgeprägt im Rhythm & Blues (R&B), etwa bei Ray Charles (s. unten), im Soul, etwa bei Aretha Franklin (Putschögl 1998), oder im neueren R&B, z.B. bei D'Angelo oder Lauryn Hill (Pfleiderer 2006b), und inzwischen allgemein bei der durch afroamerikanische Gesangsstile geprägten Interpretation von Pop-Balladen.

Ein weiteres wichtiges Unterscheidungsmerkmal zwischen Sprech- und Gesangsstimme ist die Sprechgeschwindigkeit und die Häufigkeit und Länge der Pausen zwischen Silben, Wörtern und Phrasen. Bei normalem Sprechen wird nur in ca. 50-60 % der Zeit tatsächlich gesprochen, der Rest der Zeit wird pausiert (König/Brandt 2006, S. 118). Die durchschnittliche Sprechgeschwindigkeit liegt bei ca. fünf artikulierten Silben pro Sekunde – natürlich mit großen persönlichen und situationsgebundenen Abweichungen. Beim Singen ist die Geschwindigkeit oftmals, aber nicht immer deutlich langsamer als beim Sprechen – ein weiterer Hinweis darauf, dass es hier nicht unbedingt um das möglichst effektive Übermitteln von sprachlichen Inhalten geht. Ausnahmen bilden vorwiegend narrativ orientierte Genres wie etwa Rap Music oder Singer-Songwriter.

Der Sprechfluss wird in der Regel durch einen regelmäßigen Rhythmus strukturiert, wobei zwischen Sprachen mit regelmäßiger Silbenverteilung (z.B. französisch) und Sprachen mit regelmäßiger Akzentverteilung (z.B. englisch) unterschieden werden kann.[5] Allerdings wird beim Sprechen eine regelmäßige metrische Ordnung nicht so strikt eingehalten, wie dies in einem musikalischen Zusammenhang erforderlich ist, in dem verschiedene Stimmen und Instrumente miteinander koordiniert und daher auf einen gemeinsamen metrischen Rahmen bezogen werden. Bei bestimmten Formen des Rap oder des jamaikanischen Deejay-Sprechgesangs tritt die rhythmisch-metrische Gestaltung – zum Teil mit kunstvollen kreuzrhythmischen Überlagerungen[6] – ganz ins Zentrum der Vokalperformance.

Im populären Gesang kommt es allerdings trotz bzw. vor dem Hintergrund der zumeist strikt durchgehaltenen metrischen Begleitung einerseits zu Akzentuierungen von Tönen, die nicht auf den Grundschlag (Beat) fallen, sog. Offbeat-Akzentuierungen, andererseits zu einer relativ freien, mikrorhythmisch beweglichen oder rubatohaften Ausgestaltung der Melodienlinien. Jeff Todd Titon führt entsprechende Veränderungen der Singweise in der populären Musik seit den 1920er Jahren auf den Einfluss des afroamerikanischen Jazz- und Vaudeville Blues-Gesangs zurück: „(...) in the 1920s,

[5] Nicht die Artikulation von Konsonanten, sondern der Beginn der periodischen Schwingungen der Vokale wird als Anfangspunkt einer gesprochenen oder gesungenen Silbe wahrgenommen; die Konsonanten müssen somit bei einem metrisch exakten Sprechen oder Singen vor den Zeitpunkten des metrischen Rasters artikuliert werden (Sundberg 1994).

[6] So z.B. bei den afroamerikanischen Rappern Chuck D von Public Enemy, vgl. Pfleiderer (2006a, S. 322f.), oder KRS-One, vgl. Krims (2000, S. 58ff.)

aptly called the Jazz Age, Bessie Smith and other African-American jazz and blues singers revolutionized the craft of singing popular music. Their approach was close to the rhythm and tone of ordinary talk, and this natural way of singing caught on“ (Titon 1996, S. 144). Der neue Gesangsstil lässt sich insbesondere an einer neuen mikro-rhythmischen Gestaltung ablesen. So charakterisiert Henry Pleasants den rubatoartigen Stil von zahlreichen Sängern in der Tradition des American Popular Song als eine Möglichkeit der Spannungssteigerung, „(…) a device, now more widely employed by popular than by classical singers, by which time is stolen from one note and given to another producing in the listener a sensation of suspense as he waits for restitution to be accomplished“ (zit. nach Hudson 1994, S. 407). Diese Gestaltungsweise wird auch von Jazzvokalisten verwendet. So sind viele Aufnahmen von Blues- und Tin Pan Alley-Songs durch die Jazzsängerin Billie Holiday Musterbeispiele für eine Gestaltungsweise, die Robert Ashley (2002) als „back phrasing“ bezeichnet hat. Holiday beginnt viele Phrasenanfänge im Vergleich zur Vorlage und zum metrischen Bezugsschema zu spät, „holt“ aber im Verlauf einer Phrase wieder „auf“, sodass sie zum Schluss einer Zwei- oder Viertaktgruppe gemäß oder nur wenig hinter der Melodievorlage singt (vgl. Bauer 1999).

Durch den Einsatz von Tonhöhe, Lautstärke und Silbendauer werden beim Sprechen inhaltliche Aspekte von Sprachäußerungen verdeutlicht. Dies wird unter dem Oberbegriff der Intonation oder Sprechintonation[7] zusammengefasst. Die drei zentralen Dimensionen der Sprechintonation sind der charakteristische Konturverlauf der Tonhöhe innerhalb einer Äußerung (fallend, steigend, steigend und fallend usw.), die Platzierung von Akzenten (Lautstärkebetonungen) innerhalb von Wörtern und Sätzen sowie die Größe von Sinneinheiten in Bezug auf Akzente und Pausen. Die Sprechintonation dient vor allem der Textverständlichkeit – ein abweichender Tonhöhenverlauf und eine falsche Akzentplatzierung kann den Sinn einer Aussage verändern oder die Textverständlichkeit erschweren. Zugleich können aufgrund der Melodik, Rhythmik und Dynamik auf einer paralinguistischen Ebene Informationen über bestimmte Haltungen, Einstellungen und Stimmungen des Sprechers zum Ausdruck kommen, die nicht unbedingt mit dem Text zusammenhängen. In einem musikalischen Kontext können beide Dimensionen der Sprechintonation relevant werden.

Ray Charles „Come Back, Baby“ (1954)

Ich möchte nun einige der genannten melodisch-rhythmischen Gestaltungsdimensionen am Beispiel des Songs „Come Back Baby“ des afroamerikanischen R&B-Sängers Ray Charles illustrieren.[8] Ray Charles (Ray Charles Robinson, 1930-2004) ist ein bekannter und einflussreicher afroamerikanischer Sänger und Pianist, der an die Traditionen des Blues, der Country Music und des Amercian Popular Song anknüpft und darüber hinaus aufgrund seiner Verwendung von Gestaltungsmitteln aus der afroamerikanischen Gospel Music als ein Wegbereiter des Soul der 1960er Jahre gilt. „Come Back, Baby“, auf-

---

[7] Um eine Verwechslung mit „Intonation“ im musikalischen Sinne zu vermeiden, bevorzuge ich den Ausdruck „Sprechintonation“.

[8] Teile der folgenden Analysen wurden erstmals im Rahmen eines Referats präsentiert, das der Autor gemeinsam mit Christian Bielefeldt auf der Arbeitstagung des Arbeitskreises Studium Populäre Musik (ASPM) im Oktober 2006 in Rauischholzhausen gehalten hat.

genommen am 18. November 1954 in Atlanta, wurde ursprünglich als Rückseite einer Single mit dem Titelsong „I've Got a Woman" veröffentlicht, in dem Charles einen Gospel-Song mit einem neuen, weltlichen Text interpretiert.[9]

„Come Back Baby" besteht aus vier achttaktigen Strophen in einem relativ langsamen Tempo (punktierte Viertel = ca. 45 beats per minute); die vierte Strophe hebt sich von den vorhergehenden Strophen durch einen Stop-Chorus ab (Tutti-Akzente auf jedem Taktanfang gefolgt von einer Pause der Begleitband). In harmonischer Hinsicht dominieren Dreiklangs-Akkorde; der plagale Schluss im vorletzten Takt ist typisch für die afroamerikanische Gospel Music:

Eb – – –, Eb – – –, Ab – – –, Cm – Ab A°,
Eb – Cm –, Fm – Bb –, Eb – Ab –, Eb – Bb^+ –.

Zwar knüpft der Text an ein typisches Blues-Thema an. Geschildert wird die emotionale Mischung aus Zorn und Selbstmitleid eines Mannes, der von seinem „Baby" verlassen wurde bzw. verlassen wird, worauf er zunächst mit der Bitte „come back, Baby", in der letzten Strophe jedoch mit der Drohung, seinerseits das Weite zu suchen („and I'll be gone") reagiert. Die eher versöhnlich wirkende Wendung: „let's talk it over one more time" sowie, in der zweiten Strophe, das Zugeständnis „well, I admit, Baby, that I was wrong" sind freilich für Blues-Lyrics eher ungewöhnlich und verweisen auf einen Wandel in der Beziehung zwischen den Geschlechtern, wie er im Soul der 1960er Jahre zum Thema vieler Songs wurde. Ray Charles ergänzt den Text regelmäßig durch Rufe („yeah", „ooh"), die sich zumeist durch eine besondere Stimmqualität auszeichnen (vgl. unten).

Grundlage der analytischen Beschreibung von Eigenheiten der melodischen und rhythmischen Gestaltung ist eine notenschriftliche Transkription der ersten Strophe (Abb. 1).

Abb. 1: Transkription der ersten Gesangsstrophe von Ray Charles „Come Back, Baby" (1954).

[9] Die beiden Songs der Single werden von Peter Shapiro als „one of the two or three most significant singles ever released" (Shapiro 2000, S. 24) bezeichnet. Shapiro bringt zudem den Gesang und die Piano-Figuren von „Come Back Baby" mit dem Gospel-Song „(I Will) Move On Up a Little Higher" in Verbindung. Neben Ray Charles am Piano wirken an der Aufnahme eine Rhythmusgruppe (Wesley Jackson, Gitarre, Jimmy Bell, Bass, und Glenn Brooks, Schlagzeug) sowie zwei Trompeter (Joe Bridgewater und Charles Whitley) und zwei Saxophonisten (Don Wilkerson und David Newman) mit.

Auffällig ist zunächst die bogenförmige Kontur der Melodielinien in „Come back, Baby“, d.h. auf jede aufsteigende Melodiephrase folgt eine Phrase mit einer absteigenden Bewegung zum Phrasenende hin. Der Tonvorrat ist vorwiegend pentatonisch, wobei die Terz oftmals labil zwischen g und ges intoniert wird, in der Transkription jedoch vorwiegend als ges notiert wurde. Charakteristisch ist außerdem die melismatische Ausgestaltung einzelner Silben vor allem am Schluss von Phrasen („yeah“, „know“, „over“, „time“). Für die rhythmische Gestaltung des in einem triolischen Vierer-Metrum ($^{12}/_{8}$-Takt) gehaltenen Stückes ist vor allem das regelmäßige Vorziehen der melodischen und textlichen Betonungen auf den Offbeat vor dem Taktanfang charakteristisch. Diese Offbeat-Akzentuierungen, durch die ein drängender Ausdruck erzeugt wird, sind typisch für einen afroamerikanisch geprägten Gesang (vgl. Moore 2001, S. 46f., Pfleiderer 2006a, S. 215ff.). Im Vergleich zum strikten Metrum der Begleitband ist die rhythmische Feingestaltung des Gesangs zudem recht beweglich. So klingt z.B. die zweite Phrase („oh, mama please don’t go“) beschleunigt und dadurch besonders drängend. Größere Freiheiten in der mikrorhythmischen Gestaltung werden insbesondere bei den Melismen genutzt, die sich daher nur annähernd in diskreten Notendauern darstellen lassen. Sie können jedoch ähnlich wie Feinheiten der Tonhöhengestaltung, z.B. das Gleiten zwischen Tonhöhen innerhalb eines Vokals oder einer Phrase, mit einer Spektraldarstellung visualisiert werden.

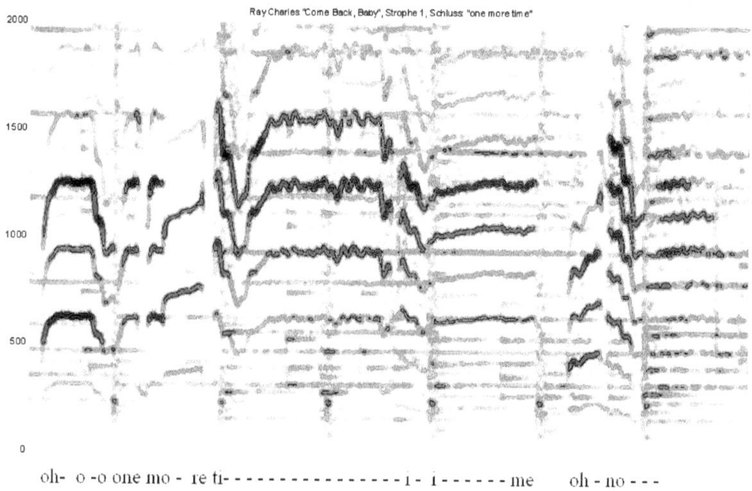

Abb. 2: Spektraldarstellung der Gesangslinie von Ray Charles in „Come Back, Baby“ (1954) am Ende der ersten Strophe.

Abb. 2 zeigt den Schluss der ersten Strophe in einer Spektraldarstellung.[10] Die Gesangslinie hebt sich dabei klar als parallel verlaufende horizontale Linien (Grundton und

[10] Das Spektrogramm wurde mit der Software *Praat 4.6.36* erstellt, die von Paul Boersma and David Weenink am Institute of Phonetic Sciences der Universität Amsterdam entwickelt wurde (s. www.praat.org). Verwendet wurde eine FFT (Fast Fourier Transformation) mit einem Gauß-

Obertöne) gegen die leiseren Klänge der Begleitband ab. Als metrische Orientierung können die regelmäßig auf jedem Grundschlag erklingenden Snare Drum-Schläge dienen, die im Spektrogramm aufgrund ihres breiten Klangspektrums als vertikale Linien erkennbar sind. Die regelmäßigen Schläge können auch als Orientierung für eine detaillierte mikrorhythmische Analyse der Vokalphrasierung von Charles herangezogen werden.

In der Spektraldarstellung wird das melismatische Gleiten zwischen den einzelnen Tonhöhen und das schnelle Vibrato sichtbar, das Charles bei einigen – aber nicht allen – gehaltenen Tönen einsetzt. Außerdem wird erkennbar, wie der Sänger etwa am Schluss der Silbe „time" langsam die Tonhöhe nach oben zieht. All diese Feinheiten gehen in der konventionellen Notenschrift verloren, genau wie Eigenheiten der Stimmqualität und der klangfarblichen Gestaltung.

Stimmqualitäten: Stimmgebung, Artikulation und Klangfarbe

Neben der charakteristischen Sprechtonlage, Intonationsweise und Sprechgeschwindigkeit unterscheiden sich Stimmen vor allem durch eine spezifische Klanglichkeit, die auch als Stimmqualität bezeichnet wird (Laver 1980). Die Stimmqualität resultiert aus physiologisch begründeten Besonderheiten der Stimmgebung, aus der Formung des Vokaltraktes und der Resonanzräume sowie aus habituellen, also gelernten und gewohnheitsmäßig beibehaltenen Dispositionen der Phonation und Artikulation. Eine spezifische Stimmqualität kann daher als das zumeist unwillkürliche Charakteristikum einer individuellen Stimme angesehen werden. Allerdings können Eigenheiten der Stimmqualität in einem gewissen Rahmen auch beeinflusst und unterschiedliche Stimmgebungs- und Artikulationseigenheiten in Sprechsituationen ganz gezielt eingesetzt werden.

Es ist nicht einfach, die verschiedenen klanglichen Aspekte der Stimmqualität, der Stimmgebung und Artikulation zuverlässig und anschaulich zu beschreiben.[11] Auf der einen Seite können die physiologischen Besonderheiten der Phonation und Artikulation, also die spezifischen Sprech- oder Gesangstechniken erläutert werden (vgl. unten). Um diese Erläuterungen und die damit verbundenen klanglichen Resultate nachvollziehen zu können, ist allerdings eine gute Kenntnis des Aufbaus und der Funktionsweise des Stimmapparats erforderlich. Auf der anderen Seite wird immer wieder versucht, die subjektiven Klangempfindungen verschiedener Stimmqualitäten mit Adjektiven systematisch zu erfassen. Wertvolle Vorschläge zu einer Charakterisierung der Stimmqualitäten, die sowohl physiologische Sachverhalte als auch klangliche Wahrnehmungsqualitäten umfassen, finden sich in einer Liste des US-amerikanischen *National Center for Voice and Speech* (NCVS), die 1994 von Ingo Titze zusammengestellt worden ist.[12] Außerdem

---

Analysefenster (0,1 s Fensterlänger), einem Dynamikbereich von -40dB und einer Frequenzdarstellung bis 2 kHz.

[11] Zur Bedeutung von ‚Sound' in populärer Musik und zu verschiedenen Ansätzen der Sound-Beschreibung vgl. auch Pfleiderer 2003.

[12] Die Liste ist im Internet zugänglich, vgl. http://www.ncvs.org/ncvs/tutorials/voiceprod/tutorial/quality.html (URL vom 22.11.2008). Die Vorschläge zur verbalen Beschreibung von Henrich et al. (2008) und Rapoport (1996a, 1996b) beziehen sich ausschließlich auf den klassischen Gesang. Die sechs bei Kayes (2004) genannten Stimmqualitäten setzen bei der introspektiven Beschreibung verschiedener Gesangstechniken an und können daher für Laien nur schwer auf wahrgenommene

gibt es eine Reihe klinischer Messverfahren, die zur Diagnostik von Stimmgebungsanomalien eingesetzt werden, so die Visualisierung der Stimmlippenschwingungen durch Stroboskopie und verschiedene Glottographie-Verfahren (Wendler/Seidner/Eysholdt 2004, S. 113ff.). Schließlich können aus Messungen der spektralen Energieverteilung bei Ausschnitten von Klangaufnahmen Rückschlüsse auf die Stimmqualität und Klanglichkeit einer Stimme gezogen werden. Aus Spektrogrammen wird der zeitliche Verlauf von Lage und Stärke der Formanten sowie der Anteil von geräuschhafter Rauigkeit und Hauchen ersichtlich. Während Spektrogramme die verschiedenen harmonischen Teiltöne eines gesprochenen oder gesungenen Vokals als dünne, parallel verlaufende Linien darstellen, werden bei einer behauchten bzw. rauen Stimmqualität aufgrund der Unregelmäßigkeiten der Schwingungen bzw. aufgrund der erhöhten Luftanteile relativ breite Frequenzbänder dargestellt. Zudem können aufgrund von digitalen Stimmaufnahmen verschiedene Maße für Schwingungsirregularitäten errechnet werden, und zwar hinsichtlich der Unregelmäßigkeiten in der Lautstärke (Shimmer-Maße) und in der Tonhöhe (Jitter-Maße) sowie hinsichtlich der Rauschanteile innerhalb eines periodischen Signals, etwa einer gesungenen Tonhöhe (z.B. harmonics-to-noise Ratio, HNR).[13]

Ich möchte nun die physiologischen und klanglichen Eigenheiten verschiedener Phonationsweisen und der entsprechenden Stimmqualitäten beschreiben und diese an Beispielen aus der populären Musik sowie anhand der Stimmqualitäten in „Come Back, Baby“ von Ray Charles erläutern. Hartwig Eckert und John Laver (1994) unterscheiden zwischen einer neutralen Stimmgebung, dem Falsett (auch Kopfstimme und, bei Frauen, Pfeifstimme genannt), der behauchten Stimmgebung (engl.: breathy) und der Flüsterstimme, der knarrenden Stimmgebung (engl.: creaky) sowie einer rauen und heiseren Stimmgebung (engl.: rough, harsh, hoarse). Auch wenn es zahlreiche Kombinationen und fließende Übergänge zwischen diesen Phonationstypen gibt, so sind sie dennoch ein hilfreicher Ausgangspunkt für die Beschreibung von Stimmqualitäten.

Wie bereits geschildert, wird bei der neutralen oder „normalen“ Stimmgebung die Stimmritze in jedem Schwingungszyklus einmal komplett geschlossen. Die neutrale Stimmgebung wird oft als Bruststimme bezeichnet. Die Abgrenzung zwischen einer sog. Brust- und Kopfstimme beruht zwar zu großen Teilen auf Unterschieden in der Tonlage und auf dem subjektiven Vibrationsempfinden des Vokalisten. Sie lässt sich aber auch durch die Messung der relativen Dauer des Anteils der Phasen mit geschlossenen Stimmlippen an einem Schwingungszyklus vornehmen. So definieren etwa Schutte und Miller (1993) die Bruststimme durch relativ lange geschlossene Phasen (> 50% des Schwingungszyklus'), Kopfstimme dagegen durch relativ kurze geschlossene Phasen (< 40% des Schwingungszyklus').

Statt „Kopfstimme“ wird vielfach der Ausdruck „Falsett“ (von italienisch „falsetto“ = falsch) verwendet. Beim Falsett wird in den Stimmlippen eine starke adduktive und mediale Spannung erzeugt. Die Längsmuskeln werden jedoch nicht angespannt, sodass die Stimmlippen viel dünner sind als bei der neutralen Stimmgebung und sich zudem nie

---

Stimmqualitäten bezogen werden. Ein Hörversuch von Henrich et al. (2008) zeigt außerdem, dass die Übereinstimmung von Einschätzungen zur Stimmproduktion selbst unter professionellen Sängern und Gesangspädagogen nicht sehr hoch ist.

[13] Die Klanganalyse-Software Praat (vgl. Anmerkung 10) bietet verschiedene Shimmer-, Jitter- und HNR-Maße zur Bestimmung der Stimmqualität anhand digitalisierter Stimmaufnahmen.

völlig schließen (Sundberg 1987, S. 63). Die Form der Schwingung ist stärker sinusförmig und somit ärmer an Obertönen. Wegen der geringeren Masse können die Stimmlippen zudem schneller schwingen, sodass höhere Töne erzeugt werden können als bei der neutralen Stimmgebung. Die charakteristische Klangqualität des Falsett wird auch dann nicht mit einer neutralen Stimmgebung verwechselt, wenn die Tonlage identisch ist. In der Alltagskommunikation wird das Falsett nicht verwendet, da es unnatürlich – eben ‚falsch' – klingt. Falsettgesang von Männern wird jedoch in vielen populären Musikgenres eingesetzt.

In „Come Back, Baby" von Ray Charles dominiert eine neutrale, bei einzelnen Silben bisweilen leicht gepresst wirkende Stimmgebung. Bei einer gepressten Stimmgebung wird das Schwingen der Stimmlippen durch einen hohen subglottalen Druck eher behindert, die Stimme wird dabei aufgrund der relativ lauten Partialtöne oft als grell empfunden. An vielen Stellen lässt Ray Charles am Schluss von Phrasen seine Stimme ins Falsett-Register „hochschnappen", so in der ersten Strophe bei „lover *you*", „know-*ow*", „ba-*by*". Darüber hinaus setzt Charles an mehreren Stellen Falsett-Rufe („huh") ein. Im Spektrogramm (Abb. 3)[14] sind die klanglichen Unterschiede zwischen der neutralen Stimmgebung und dem Falsett klar erkennbar: Die im Falsett gesungenen Töne verfügen über weit weniger Partialtöne, ablesbar an der weit geringeren Zahl der durchgezogenen horizontalen Linien, als die mit neutraler Stimme gesungenen Wörter.

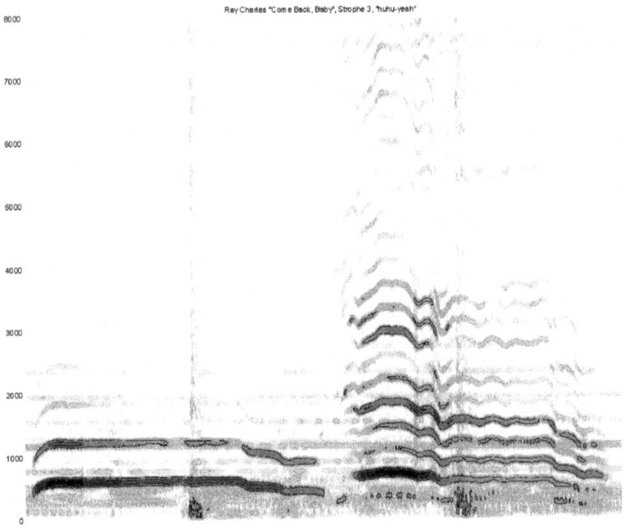

Abb. 3: Spektraldarstellung eines Ausschnitts aus Ray Charles „Come Back, Baby": Falsett „hu" (links) und neutrale Stimmgebung „yeah" (rechts).

---

[14] FFT, Gauß-Analysefenster mit 0,03 s Fensterlänge, Dynamik von -40 dB und bis 8 kHz; diese Optionen gelten auch für die Abbildungen 4 und 5.

Für den Gebrauch der Falsett-Stimme gibt es in verschiedenen Stilen der populären Musik zahlreiche Beispiele. Ein kurzes Hochschnappen ins Falsett und Falsett-Rufe sind im Blues und Soul-Gesang immer wieder anzutreffen. Seltener, doch in frühen Blues-Aufnahmen durchaus nicht ungewöhnlich, ist ein schneller, regelmäßiger Wechsel zwischen normaler Stimme und Falsett. Besonders ausgeprägt findet sich dieser jodelartige Gesangsstil in Aufnahmen der Country Music, so etwa bei Jimmie Rodgers.[15] Im Soul gibt es eine Reihe von Sängern, die vorwiegend oder ausschließlich mit einer „weichen", entspannten Falsett-Stimme singen, so etwa Curtis Mayfield oder Smokey Robinson. Auf ganz andere Weise wird das Falsett im Hardrock und später im Heavy Metal eingesetzt. So singt etwa Robert Plant von der einflussreichen Hardrock-Band *Led Zeppelin* mit einer lauten, sehr gepresst und angestrengt wirkenden Kopfstimme, wodurch laut Alan Moore der Eindruck von hoher emotionaler Intensität erzeugt wird (Moore 2001, S. 48). Auch manche Sängerinnen verwenden die Kopfstimme, die dann jedoch aufgrund der ohnehin höheren weiblichen Stimmlage keinen so drastischen klanglichen Unterschied zur neutralen Stimme darstellt wie bei Sängern.

Charakteristisch für die knarrende Stimme (auch: „Strohbass") ist ein geringer subglottaler Luftdruck und eine tiefe Tonlage. Bei der knarrenden Stimme sind die Stimmlippen durch die Stellknorpel fest zusammengezogen, sodass sie nur im vorderen Bereich frei schwingen können. Zugleich besteht eine geringe Längsspannung, sodass sich die Stimmlippen in einem relativ breiten Bereich berühren. Da weniger Luft verbraucht wird, neigen Sprecher vor allem am Schluss von Sätzen und Äußerungen dazu, in eine knarrende Stimmgebung zu fallen. Eine tiefe, knarrende Stimmgebung wird in der Regel als sehr gelassen und entspannt empfunden (Eckert/Laver 1994, S. 66ff.).[16]

Bei der behauchten Stimme werden die Stimmlippen nicht geschlossen, sondern bei mäßiger Längsspannung sowie geringer Adduktion und medialer Spannung einander nur angenähert, sodass im Bereich der Glottis Luftwirbel entstehen. Bei schnellem Luftstrom kommt es dennoch zu einer periodischen Verengung und Erweiterung der Stimmritze, sodass eine Tonhöhe hörbar wird. Dem Ton ist jedoch ein großer Anteil nicht-tönender, „wilder" Luft beigemischt, der als Hauchen, Rauschen oder Zischen (engl.: hissing) wahrgenommen wird. Eine behauchte Stimme wird aufgrund der geringen Spannung, Kraftanstrengung und Lautstärke als Ausdruck großer Intimität wahrgenommen (Eckert/Laver 1994, S. 72ff.).[17]

Bei einer rauen Stimmgebung werden die Stimmlippen durch adduktive und mediale Spannung stark gegeneinander gepresst, sodass unregelmäßige Schwingungen entstehen. Die schnellen Schwankungen der Periodenlänge werden jedoch nicht als Tonhöhenschwankungen, sondern als Rauigkeit wahrgenommen. Der Anteil der Frequenz-

---

[15] Timothy Wise unterscheidet in seiner Studie „Yodel Species: A Typology of Falsetto Effects in Popular Music Vocal Styles" (2007) verschiedene Jodeltypen und nennt zahlreiche Beispiele aus der populärer Musik.

[16] Mit der knarrenden Stimme verwandt ist eine weitere Stimmgebungsweise im tiefsten Register, das sog. „vocal fry" oder „vocal pulse register", bei der es zu regelmäßigen Unterbrechungen des Schwingungszyklus' der Stimmlippen kommt. Diese Stimmgebung wird in verschiedenen ostasiatischen Musikkulturen verwendet, z.B. beim Gesang tibetanischer Mönche, ist jedoch in westlicher populärer Musik ungebräuchlich.

[17] Eine weitere Sprechweise ist die Flüsterstimme, bei der die Stimmlippen durchweg geschlossen sind, jedoch ein sog. Flüsterdreieck geöffnet bleibt, durch welches die Luft entweicht (vgl. Eckert/Laver 1994, S. 77ff.).

schwankungen oder Unregelmäßigkeiten der Periodenlänge innerhalb eines stationären Klanges wird als Jitter bezeichnet und lässt sich akustisch messen; der Jitter einer normalen Stimme liegt unter 0,4% (vgl. Wendler/Seidner/Eysholdt 2004, S. 126). Es gibt verschiedene Ursachen für die Unregelmäßigkeiten der Stimmlippenschwingungen. So kann es vorkommen, dass die Stimmlippen zusammen mit den sog. falschen Stimmlippen, die sich oberhalb der eigentlichen Stimmbänder befinden, oszillieren. Aufgrund des starken Verschlusses und der größeren Masse ist dann ein hoher Atemdruck notwendig, der Unregelmäßigkeiten in der Periodenlänge und Amplitude eines jeden Schwingungszyklus‘ erzeugt bzw. weiter verstärkt. In seltenen Fällen schwingen ausschließlich die falschen Stimmlippen (engl.: ventricular folds), was zu einer extrem rauen Stimmqualität führt; vermutlich ist die überaus raue Stimmqualität des Gesangs von Louis Armstrong hierauf zurückzuführen. Doch auch krankhafte Veränderungen der Stimmlippen – geschwollene oder belegte Stimmlippen bei einer Erkältung oder Unregelmäßigkeiten der Stimmlippen durch organische Veränderungen oder Polypen – können zu einer rauen, heiseren oder gepresst wirkenden Stimmgebung führen (vgl. Sundberg 1987, S. 182ff.). Oft ist einer heiseren Stimmgebung zusätzlich ein Anteil Zischen beigemischt. Das Sprechen oder Singen mit rauer Stimme stellt für die Stimmbänder immer eine starke Belastung dar. Auch Schreien, Brüllen und Kreischen geht aufgrund der hohen Lautstärke und des hohen subglottalen Drucks zumeist mit einer rauen Stimmqualität einher.

In der klinischen Diagnose wird heute bei der Beurteilung der Stimmqualität zwischen dem Grad der Rauigkeit (roughness), dem Grad der Behauchtheit (breathiness) und dem Grad der Heiserkeit (hoarseness) unterschieden, die jeweils auf einer vierstufigen Skala bewertet und zum sog. RBH-System zusammengefasst werden (vgl. Wendler/Seidner/Eysholdt 2004, S. 124f.). Insbesondere Rauigkeit und Behauchtheit haben sich als zwei voneinander unabhängige Beurteilungskriterien erwiesen, die hoch mit den Maßen für die Irregularität der Stimmlippenoszillation (Rauigkeit) bzw. für Anteile von Rauschen (Behauchtheit) korrelieren. Dagegen kann Heiserkeit als Kombination einer rauen Stimmgebung mit einem Anteil glottalen Rauschens angesehen werden.

Ray Charles verwendet bei vielen seiner Rufe und Schreie („yeah“, „huh“, „ooh“) eine ausgesprochen raue Stimmgebung. In allen vier Strophen schreit er ein „yeah, yeah“ vor der Zeile „let’s talk it over“. Weitere Schreie finden sich bei „that *I* was wrong“ (Strophe 2) und „*moun*tain jack“ (Stophe 3). In der Spektraldarstellung der ersten Strophe lässt sich das geschriene „yeah“ durch die weit höheren Geräuschanteile klar von dem unmittelbar darauf gesungenen „yeah“ unterscheiden (Abb. 4). Während bei letzterem die Partialtöne eindeutig als parallel geschwungene Linien zu erkennen sind, verschwinden sie bei der rauen und geräuschhaften Klanggebung des geschrienen „yeah“ fast völlig in einem breiten, bis ca. 4 kHz reichenden Frequenzband. Auffällig ist die große gesangstechnische Kontrolle des Schreies. Es handelt sich hierbei offensichtlich um einen willkürlichen vokalen Ausbruch, denn nach dem Schrei singt Charles sofort mit neutraler Stimmgebung weiter.

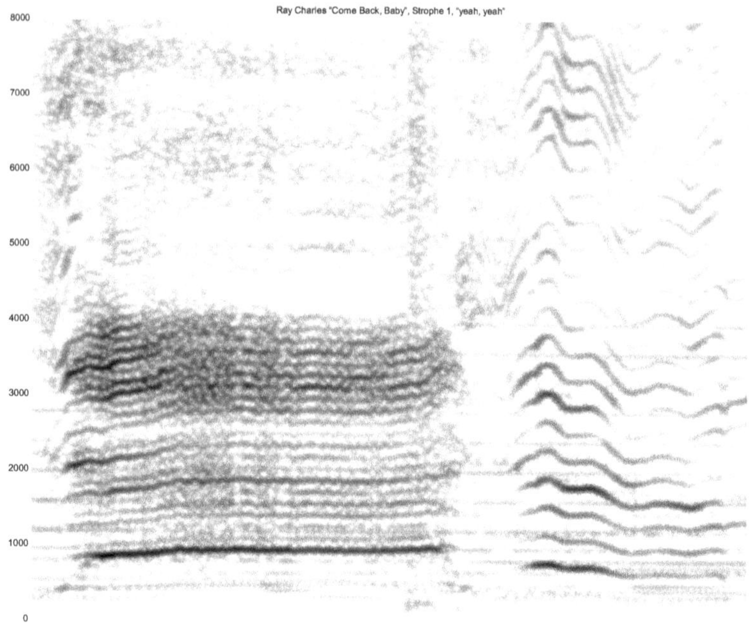

Abb. 4: Spektraldarstellung eines Ausschnitts aus Ray Charles „Come Back, Baby“: Geschrienes „yeah“ (links) und gesungenes „yeah“ (rechts).

In der dritten Strophe kombiniert Charles einen hohen Falsett-Ruf mit einer rauen schreiartigen Stimmgebung (Abb. 5). Die Spektraldarstellung zeigt sowohl die relativ wenigen Teiltöne (horizontale Linien, links) als auch die großen Geräuschanteile dieses „geschrienen Falsetts“. Es folgt ein gesungenes, jedoch leicht geräuschhaftes „yeah“ (rechts).[18]

[18] Neben der hier gewählten Möglichkeit, klangliche Merkmale der Stimmgebung und Stimmqualität mit Hilfe von Spektrogrammen darzustellen, können diese natürlich auch in einer konventionelleren Notationsweise durch zusätzliche Symbole festgehalten werden. So notiert Mike Daley (1998) in seiner Transkription einer Aufnahme der amerikanischen Sängerin Patti Smith – ihrer Version des Van Morrison-Songs „Gloria“ (1975) – sowohl den spektralen Klangcharakter des Gesangs als auch verschiedene Phonationsweisen parallel zum Notentext mit zusätzlichen Zeichen. Unterhalb der Melodielinie, bei der Daley zusätzlich Abweichungen der Tonhöhe mit Pfeilen notiert, ist in einem weiteren Notensystem die Klangfarbe der Vokale notiert (von einem dunklen „u“ mit wenigen Obertönen und tiefem zweiten Formanten in der untersten Linie des Notensystems bis zum hellen „i“ mit vielen Obertönen und hohem zweiten Formanten in der höchsten Linie); zusätzlich weist Daley durch Symbole auf fünf Stimmqualitäten oder „vocal modifiers“ hin: „raspy“ (kratzig, krächzend), „hard/nasal“, „breathy“ (behaucht), „sob“ (schluchzend) und „creaky“ (knarrend).

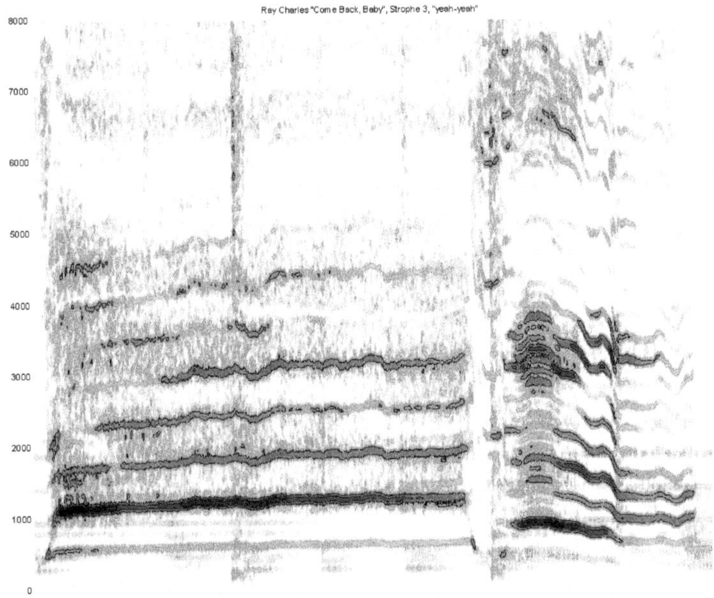

Abb. 5: Spektraldarstellung eines Ausschnitts aus Ray Charles „Come Back, Baby“: Falsett-Schrei „yeah“ (links) und gesungenes, leicht raues „yeah“ (rechts).

Für verschiedenen Arten der Rauigkeit und Heiserkeit sowie für Anteile von Luftwirbeln und Zischen lassen sich zahlreiche Beispiele aus der populären Musik heranziehen – von der charakteristischen Stimmgebung bei Louis Armstrong und bei frühen Blues-Sängern über die Schreie von Howlin' Wolf oder James Brown bis hin zum Schreien und Kreischen von Rock-, Punk- und Heavy Metal-Sängern usw. In den neuen Stilvarianten des Post-Punk/Hardcore und in verschiedenen Metal-Subgenres kommen weitere Vokaltechniken und Stimmgebungsarten wie z.B. Röcheln, Grunzen, Stöhnen zum Einsatz, die auf einen Grenzbereiche zwischen sprachlichen Äußerungen und einer Klanggebung jenseits der menschlichen Alltagskommunikation verweisen.

Zusätzlich zu den genannten Stimmqualitäten, die aus Eigenheiten der Phonation herrühren, unterscheiden sich manche Sprecher und ganze Sprachgemeinschaften durch unterschiedliche habituelle Dispositionen bezüglich der Form und Beweglichkeit des Vokaltraktes, z.B. der Zungenstellung, dem Grad der Nasalität und der Weite des Rachenraums. Diese anhaltenden Artikulationsdispositionen werden auch „Settings“ genannt (Eckert /Laver 1994, S. 91-130). Mit diesen Settings hängen gewisse habituelle Besonderheiten einer Sprechstimme zusammen, die mitunter auf die regionale, ethnische und soziale Herkunft eines Sprechers verweisen und als Akzent bezeichnet werden. Dialekte und Soziolekte zeichnen sich neben den klanglichen Charakteristika eines Akzents zusätzlich durch Besonderheiten der Grammatik und des Vokabulars aus. Allerdings unterscheiden sich das britische und das US-amerikanische Englisch nicht nur in manchen Aspekten der Grammatik und des Vokabulars, sondern bereits in der Aussprache, im Akzent. Artikulations-Settings können von Vokalisten auch gezielt eingesetzt werden, um eine regionale, ethnische oder auch soziale Zugehörigkeit zu verdeutlichen – oder aber vorzutäuschen. So bemühten sich viele britische Beat- und Rocksänger, die Mitte der 1960er Jahre den US-amerikanischen Popmarkt eroberten, darum, möglichst „amerikanisch“ zu klingen. Später sangen britische Punk-Sänger zum Teil in einen unterschichtstypischen Akzent, obwohl sie selbst der englischen Mittelschicht entstammten. John Potter schreibt hierzu: „(…) many working-class speakers of the British Isles and the USA use a relatively high larynx position for normal speech, whereas middle-class speakers of Received Pronuncination lower the larynx to produce the rounded vowels associated with RP. Because these settings can be learned they can also be used subversively, and the identification of accent purely as a marker of the singer's class may be extremely misleading (as in the case of Mick Jagger or numerous punk singers)“ (Potter 1998, S. 169).[19]

Gesangstechniken in populärer Musik

Fast jeder Mensch vermag ohne gesangstechnische Ausbildung zu singen. Doch erfordern die besonderen Belastungen, der die Stimme beim Singen über längere Zeiträume und bei größerer Lautstärke ausgesetzt ist, einen besonderen, schonenden Umgang mit dem Stimmapparat. Die Lautstärkeanforderungen bei einem Auftritt vor großem Publikum und in einem großen Raum, aber auch bestimmte handwerkliche Fähigkeiten der klanglichen Gestaltung – etwa das exakte Treffen von Tonhöhen, auch nach größeren Intervallsprüngen, oder eine rhythmisch-metrisch präzise Artikulation von Silben – erfordern eine gewisse Übung und können ein angeleitetes Training und den gezielten Einsatz von Gesangstechniken erforderlich machen. Während der klassische Opern- und Liedgesang über eine längere Geschichte der Gesangsausbildung verfügt, fehlt diese Ausbildungstradition in weiten Bereichen des populären Gesangs. Viele Sänger sind auf ihre eigenen Erfahrungen, auf Tipps von anderen Sängern, vor allem aber auf das Nachahmen von anderen Vokalisten angewiesen. Zwar gibt es insbesondere im Bereich des populären Musiktheaters und im Jazzgesang seit einiger Zeit eine akademisch verankerte Gesangsausbildung, und auch im professionellen zeitgenössischen Popgesang ist Stimm-Coaching inzwischen weit verbreitet. Das Bemühen um eine gute, d.h. sowohl den vokalen Erfordernissen angemessene als auch das Stimmorgan schonende Gesangstechnik hat sich jedoch noch nicht in allen Bereichen der populären Musik durchgesetzt.

Bis in die ersten Jahrzehnte des 20. Jahrhunderts hinein stellte die hohe Lautstärke, die bei Auftritten erforderlich ist und dort auch über einen längeren Zeitraum beibehalten werden muss, ohne das Stimmorgan zu schädigen, ein zentrales Problem für Sänger dar. Eine hohe Lautstärke wird in der Regel durch einen höheren subglottalen Druck erreicht, der zu einer stärkeren Schwingung und damit zu einer höheren Belastung der Stimmlippen führt. Die Herausbildung des klassischen Gesangs vor allem im 19. Jahrhundert hängt nicht nur mit einem bestimmten Ideal der Klangschönheit zusammen, sondern zugleich mit den wachsenden Erfordernissen an die Lautheit der Sänger (vgl. Potter 1998, S. 47ff.). In dieser Zeit wurden die Opernhäuser größer und die Orchester lauter, worauf die Sängerinnen und Sänger mit einer neuen Gesangstechnik reagierten,

[19] Potter (1998, S. 158ff.) geht ausführlich auf die soziale Bindung bestimmter Akzente und Gesangstechniken ein (vgl. dazu auch Middleton 2000 und Moore 2001, S. 49); er beruft sich dabei auf den Sozioliniguisten Peter Trudgill (1983).

die seither den klassischen Lied- und Operngesang („bel canto") prägt: Durch das Absenken des Kehlkopfs, teilweise auch durch ein Nachvorne-Stülpen der Lippen kommt es zu einer Verlängerung des Vokaltraktes, die wiederum zu einem Absenken der höheren Formanten führt. Indem auf diese Weise der dritte, vierte und fünfte Formant zum so genannten „Sängerformanten" verschmelzen, kommt es zur charakteristischen Verstärkung eines hohen Frequenzbereichs, der je nach Länge des Vokaltraktes und Registerlage zwischen 2,3 und 4 kHz liegt (Sundberg 1987, S. 118f.). Durch die Verstärkung dieses Frequenzbereichs erhöht sich die wahrgenommene Lautheit der entsprechenden Stimme, sodass sie sich auch in großen Räumen und gegenüber einem großen Orchester durchzusetzen vermag. Die durch den Sängerformanten geschaffene Stimmqualität wird bisweilen auch als „twang" bezeichnet (vgl. Kayes 2004, S. 110ff.).

Hinzu kommt im klassischen Gesang ein ausgeprägtes Vibrato, d.h. schnelle und regelmäßige Schwankung der gesungenen Tonhöhen (vgl. Sundberg 1987, S. 163ff., Fischer 1993, S. 141-166 u. 293-301), wodurch jedoch die Textverständlichkeit leidet. Sundberg betont, dass es eine Reihe unterschiedlicher Typen des Vibratos, u.a. auch ein „natürliches" Vibrato gibt, die sich von der Technik des Bel canto-Vibrato zum Teil unterscheiden.[20] Auch in vielen Bereichen des populären Gesangs, z.B. im Musical- oder Jazzgesang, wird mit Vibrato gesungen, oft jedoch ausschließlich auf gehaltenen Tönen, wodurch die Verständlichkeit des gesungenen Textes weit weniger leidet. Während beim Vibrato die Größe der Tonhöhenschwankungen abhängig von der Lautstärke gezielt beeinflusst werden kann, kann das Tempo der Tonhöhenschwankungen in der Regel nicht bewusst variiert werden und gehört somit zum persönlichen Stil eines Sängers.

Beim Sprechen und im populären Gesang versuchen Vokalisten in der Regel auf eine natürliche Weise, die Lautheit ihrer Stimme zu erhöhen: Lautes Sprechen wird zum Rufen, lautes Singen nähert sich dem Schreien oder Brüllen an (Klausmeier 1999). Aufgrund des höheren Atemdrucks und der höheren Belastung der Stimmlippen geht diese Lautstärkesteigerung nicht selten mit einer gepressten und rauen oder heiseren Stimmgebung einher. Allerdings schwingen bei einer Charakterisierung des populären Gesangs durch Ausdrücke wie „Schreien" oder „Brüllen" auch ideologische Komponenten mit. Denn gemäß des Anspruch auf Authentizität, wie er in der populären Rock- und Folk-Musik verbreitet ist, wird Authentizität gerade durch die Natürlichkeit und den emotionalen Einsatz der Interpreten eingelöst bzw. konstruiert (vgl. Moore 2001, S. 44, s. auch Moore 2002). Der Hinweis auf eine Nähe zu „rohen", also unkultivierten Gefühlen dient allerdings umgekehrt auch als Begründung für die ästhetische Abwertung populären Gesangs.

Im Bereich des populären Musiktheaters wurden einige „kultiviertere" Gesangstechniken entwickelt, so vor allem die Belting-Technik, durch welche die Lautstärke

[20] Nach Fischer (1993), der sich hauptsächlich auf Untersuchungen von Karl Hartlieb (1960/67) stützt, setzt sich das Vibrato im klassischen Gesang aus drei Komponenten oder „Wellen" zusammen, einer „Atemwelle" (Zwerchfell), einer „Kippwelle" der Schildknorpel und einer „Glottiswelle": „Das Vibrato, die ‚Vibratowelle' ist das Ergebnis der ausgewogenen Bewegungen des Zwerchfells, der äußeren Aufhänge- und Abfederungsmuskulatur des Kehlkopfes und schließlich jener Muskelgruppen, die ringförmig die Glottis umschließen und die Stellknorpel ‚rollend' bewegen (...). Diese Teilbewegungen müssen zum einen zeitlich exakt koordiniert sein, also synchron erfolgen, zum anderen spannungsmäßig und damit energetisch feinst (synerg) aufeinander abgestimmt sein" (Fischer 1993, S. 158). Vermutlich setzt sich ein „natürliches" Vibrato nicht aus allen dieser Komponenten zusammen, oder aber deren Koordination erfolgt nicht völlig synchron.

erhöht und die Tonlage der weiblichen Bruststimme um eine Quinte bis Sexte nach oben gedehnt werden kann (Banfield, 2000, S. 65f.). Schutte und Miller (1993) und Kayes (2004) charakterisieren das Belting übereinstimmend durch eine erhöhte Stellung des Kehlkopfs und einen starken Atemdruck. Allerdings kann die Belting-Technik durch den hohen Atemdruck zu einer Belastung der Stimmbänder führen.[21]

Schutte und Miller (1993) haben die verschiedenen physiologischen und akustischen Eigenheiten von vier Gesangstechniken bei mehreren Sängerinnen untersucht und miteinander verglichen: das Singen mit Bruststimme im klassischen Gesang und in der sog. „Legit"-Technik (von engl. legitimate), die „Belting"-Technik im Musical-Gesang sowie die „Pop-Stimme". Dabei stützen sie sich auf die Beobachtung bzw. Messung der Kriterien Stimmgebung („Falsetto" oder „Brust")[22], Position des Kehlkopfs (gesenkt, erhöht, normal), subglottaler Druck (hoch, mittel) und verwendete Tonlage. Die Pop-Stimme charakterisieren sie demzufolge als eine „Falsetto"-Stimmgebung mit mittlerer bis hoher Position des Kehlkopfs, mittlerem subglottalen Druck und einer vorwiegend tiefen bis mittleren Tonlage (vgl. Schutte/Miller 1993, S. 148). Diese sehr pauschale Beschreibung müsste freilich hinsichtlich der unterschiedlichen Singweisen in verschiedenen populären Musikgenres weiter differenziert werden. Detaillierte, empirische Studien zu den verschiedenen Singweisen und Gesangstechniken in populärer Musik fehlen leider bislang fast völlig.[23] Kayes (2004, S. 153ff.) liefert detaillierte Beschreibungen der physiologischen bzw. gesangstechnischen Grundlagen von mehreren Stimmqualitäten, wie sie heute im Musical-Gesang eingesetzt werden; ihre Klassifikation geht auf die US-amerikanische Sängerin und Gesangspädagogin Jo Estill zurück. Die sechs Stimmqualitäten „speech", „falsetto", „cry", „twang", „opera" und „belt" entstehen, so Kayes, aus einem je charakteristischen Zusammenwirken von subglottalem Druck, Aktivität der Stimmlippen, Position des Kehlkopfes, Abkippen der Stimmknorpel, Einsatz der Resonatoren im Vokaltrakt sowie der bevorzugten Stimmlage.[24] Barb Jungr hat protokolliert, wie Gillyanne Kayes zahlreiche Blues- und Gospel-Aufnahmen hinsichtlich der Stimmqualität der Vokalisten beurteilt: Kayes „identified *twang* as present in almost all examples of blues and gospel singers we heard. *Cry*, *belt* and *speech* were used by most singers in various combinations. *Falsetto* was rare and *opera* quality not in evidence (...)" (Jungr 2002, S. 107). Eine verlässliche Zuordnung der wahrgenommenen Klanglichkeit einer Gesangsstimme zu diesen sechs Stimmqualitäten – oder Kombinationen von ihnen – ist jedoch vermutlich nur Vokalisten möglich, die sie aus eigener Erfahrung zu unterscheiden wissen. Außerdem sind manche Bezeichnungen missverständlich – so ist mit „speech" nicht Sprechen, sondern durchaus eine Singweise

---

[21] Zum Belting vgl. auch die Studien von Miles/Hollien (1990) und Sundberg/Gramming/Lovetri (1993).

[22] Wie bereits erwähnt bezeichnen Schutte und Miller mit „Falsetto" und „Brust" den prozentualen Anteil der Verschlussphase der Stimmlippen innerhalb eines Schwingungszyklus': Bei der „Bruststimme" ist diese Verschlussphase > 50% des Zyklus'; bei Falsetto dagegen < 40%; wie der Bereich dazwischen klassifiziert werden soll, wird von den Autoren nicht problematisiert. In einer ähnlichen Bedeutung verwendet auch Kayes (2004) den Ausdruck „Falsetto".

[23] Eine Ausnahme bilden die zahlreichen Studien zu Eigenheiten des US-amerikanischen Country-Gesangs, die in den 1990er Jahren von einer Forschergruppe um Johan Sundberg durchgeführt wurden, vgl. Sundberg et al. (1999), Stone et al. (1999), und Cleveland et al. (1997).

[24] Einige physiologischen Aspekte dieser sechs Stimmqualitäten wurden von Estill et al. (1985) empirisch ermittelt.

gemeint, und „falsetto“ bezieht sich nicht auf die Kopfstimme, sondern auf eine behauchte Stimmgebung, bei der sich die Stimmlippen nie völlig schließen. Festzuhalten bleibt, dass für populären Gesang gerade eine Kombination verschiedener Stimmqualitäten und Gesangstechniken bzw. ein schneller Wechsel zwischen ihnen charakteristisch ist. Kayes schreibt hierzu: „It is impossible to generalise about, or even define, ‚pop‘. However, higher larynx, speech, breathy speech and twang certainly seem to be common features, with belting being used for heavy rock numbers. Shifts of voice quality are rapid in pop music, with changes often being made mid-note and phrase. Here the voice is being used very much as an instrument, and the text is often less important” (Kayes 2004, S. 174).

Technische Manipulation der Stimme

Die Einführung der Mikrophon- und Verstärkertechnik ermöglichte ab Mitte der 1920er Jahre eine Singweise, die sich nicht mehr um die physiologisch bedingten Lautstärkebegrenzungen der Stimme zu kümmern braucht. Die Mikrophontechnik, die auch leise, fast nur gehauchte Laute sowie Atemgeräusche hörbar macht, bildet allgemein die Grundlage dafür, dass auch nicht ausgebildete Stimmen – der Mann oder die Frau „von Nebenan“ – in populären Musikgenres durchaus erfolgreich werden können und vermittels Radio oder Tonträger in die Wohnungen der Hörer gelangen. Am Übergang vom Bühnen-Gesang der Vaudeville-Theater zum Mikrophon-Singen mit zurückgenommener Lautstärke in den späten 1920er Jahren steht das so genannte Crooning (bei Männern) und Torch Singing (bei Frauen) (Moore 1998, Lockheart 2003, Pitts/Hoffman 2002). Durch ein zurückhaltendes Singen mit Mikrophon – ein Singen mit „halber Stimme“ – wird Nähe und Intimität zwischen Vokalisten und Zuhörern suggeriert. Vor diesem Hintergrund bildete sich in den späten 1920er Jahren ein neuer Sängertyp heraus, der sich von den Bühnensängern und Komödianten des Musiktheaters grundlegend unterscheidet (vgl. Banfield, 2000, S. 72f., MacCracken 2001).

In der populären Musik seit den 1930er Jahren wird Gesang nur noch in seltenen Fällen ohne technische Verstärkung und Bearbeitung aufgeführt und gehört. Auch in Live-Situationen ist ein Anheben der Lautstärke durch Mikrophone und Verstärkeranlagen seither allgemein üblich geworden. Dabei können einerseits leises Singen, Flüstern und Atemgeräusche, andererseits lautes Singen und Schreien durch Variieren des Abstands zwischen Mikrophon und Sänger, durch Nachregeln am Mischpult sowie durch Dynamik-Kompressoren auf eine einigermaßen gleich bleibende Lautstärkeebene gebracht werden; Kompressoren heben automatisch die Lautstärke von leisen Vokalpassagen an, während sie besonders laute Passagen abschwächen. Darüber hinaus wird die klangtechnische Manipulation und Nachbearbeitung zunehmend zu einem wichtigen Charakteristikum von Stimmen populärer Musik. So wird der Klang der Stimme durch die Verwendung von Filtern (sog. Equalizer) und Hall (Reverb) den räumlichen Gegebenheiten angepasst oder gemäß ästhetischer Kriterien gezielt modifiziert. Ein vollerer Stimmklang wird dadurch künstlich erzeugt, dass entweder Vokalisten die selbe Gesangslinie mehrfach singen („doubling“), oder aber indem durch einen digitalen „Transposer“ oder „Harmonizer“ die gesungene Melodielinie in der Tonhöhe leicht erhöht oder erniedrigt und diese nur wenig transponierte Stimme sodann der ursprünglichen Stimme beigemischt wird; natürlich können auf diese Weise auch zusätzliche Stimmen im Abstand einer Oktave oder Quinte erzeugt werden. Die Anwendung des

„doubling" und „harmonizing" ist in populärer Musik sowohl bei Aufnahmen als auch, im Falle des digitalen „harmonizing", bei Live-Auftritten so selbstverständlich geworden, dass sie vielen Hörern nicht mehr auffällt – oder erst dann auffällt, wenn der Effekt fehlt und eine „nackte" Stimme hörbar wird. Weitere regelmäßig eingesetzte Effekte sind Echo und Delay, die dem Originalsignal ein verzögertes Signal beimischen. Im Falle des seltener eingesetzten Phasers und Flangers ist die Verzögerungszeit so kurz gewählt, dass aufgrund von Phasenverschiebungen kontinuierliche Klangspektrumsveränderungen entstehen. Seit Ende der 1990er Jahre wird das digitale Effektgerät *Auto-Tune* bei vielen Popmusikaufnahmen eingesetzt, um Intonationsabweichungen des Gesangs in Realzeit auf eine vorgegebene Tonhöhen zurecht zu rechnen und zu korrigieren. Der Einsatz dieser automatischen Intonationskorrektur wird freilich dann als Effekt hörbar, wenn die Sänger zwischen Tonhöhen gleiten und dieses Gleiten vom digitalen Gerät in schnelle Wechsel von diskreten Tonhöhen transformiert wird.

Manche Stimmen der populären Musik sind inzwischen untrennbar mit einer bestimmten technischen (Nach-)Bearbeitung verbunden. Populäre Stimmen oder auch die typische Gesangsweise in einem bestimmten Stilbereich definiert sich nicht mehr nur über musikalische Gestaltungsmittel, sondern auch über den stiltypischen Einsatz von Effekten wie Filter, Hall oder Kompressoren.

Gesangstile, vokale Stereotype und die „persönliche Stimme" in populärer Musik

Allgemein lassen sich die zahlreichen Genres und Stilrichtungen populärer Musik durch eine ganze Reihe von Faktoren beschreiben und voneinander abgrenzen: anhand der typischen Inhalte von Songtexten, anhand unterschiedlicher Präsentationsformen sowie anhand typischer Rezeptionsformen und Bewertungskriterien oder „ästhetischer Schemata". Zumeist unterscheiden sie sich zudem durch den genretypischen Einsatz unterschiedlicher musikalischer Gestaltungsmittel, durch die sich typische „ästhetische Formen" herausbilden.[25] Neben der instrumentalen Gestaltung spielt hier natürlich die vokale Gestaltung eine zentrale Rolle, vor allem die charakteristischen Eigenheiten der Melodiegestaltung und deren Bezug auf den tonal-harmonischen und rhythmisch-metrischen Rahmen, aber auch der Einsatz von bestimmten Gesangstechniken und Stimmqualitäten. Der typische Gesang im R&B und Soul ließe sich beispielsweise durch Offbeat-Akzentuierungen, Freiheiten der mikrorhythmischen Gestaltung, bisweilen auch durch eine melismatische Ausgestaltung der Melodielinien sowie eine Anreicherung durch Rufe und Schreie charakterisieren, die oftmals mit einer rauen Stimmqualität oder aber im Falsett erfolgen. „Come Back, Baby", gesungen von Ray Charles, wäre ein frühes Beispiel für den genretypischen Einsatz dieser Gestaltungsmittel.

Betrachtet man die unterschiedlichen stimmlichen Ausdrucksweisen und Gestaltungsmittel in populärer Musik, so lassen sich grob zwei Tendenzen unterscheiden: einerseits die Orientierung an einer instrumentalen Virtuosität, die in der Regel eine vokale Ausbildung und den gezielten Einsatz von Gesangstechniken voraussetzt, andererseits die Nähe zu „natürlichen", alltäglichen vokalen Ausdrucksformen. Die eher

[25] Ich folge hier der Unterscheidung zwischen ästhetischen Schemata und ästhetischen Formen, die der Kultursoziologe Rainer Diaz-Bone (2001, S. 162-173) vorgeschlagen hat. Ästhetische Schemata und ästhetische Formen stehen laut Diaz-Bone in einem kontingenten Verhältnis zueinander, d.h. sie wandeln sich bisweilen auch unabhängig voneinander.

instrumental und klanglich orientierte Singweise steht zum Teil in der Tradition des klassischen Gesangs („bel canto“ mit tiefer Kehlkopfstellung und starkem Vibrato), so vor allem im Musiktheater- und Vaudeville-Gesang. Auch im Crooning wird auf eine virtuose Vokalgestaltung Wert gelegt, etwa beim schnellen Wechsel zwischen eher gesprochenen bzw. „gehauchten“ und gesungenen Passagen. An einer instrumentalen Virtuosität orientiert sich vor allem der Scat-Gesang im Jazz (vgl. Bauer 2001/02, Stewart 1987). Doch auch bei der melismatischen Gestaltung in der afroamerikanischen Gospel Music und im Rhythm & Blues und Soul steht oftmals ein virtuoser Stimmgebrauch im Zentrum – nicht selten verbunden mit einer rauen Stimmqualität. Dagegen liegt ein einfacher, „natürlicher“ Stimmgebrauch nahe an der alltäglichen Sprechstimme und deren sozial oder ethnisch geprägten Artikulations-Settings, so etwa in der amerikanischen Country Music und im Blues, und zeichnet sich zudem durch nonverbale Ausdrucksweisen und stimmliche Ausbrüche sowie durch Anteile von Rauheit und Luftwirbel im Stimmklang aus. Emotional intensive Ausdrucksweisen wie Schreien, Schluchzen oder Weinen sind sowohl in vielen afroamerikanischen Musikrichtungen als auch im Rock und im Punk verbreitet. Sprechnaher Gesang und Sprechgesang findet sich, zumindest in einzelnen Passagen, in zahlreichen populären Musikgenres, und in ausgeprägter Form im Rap.

Richard Middleton (2000, S. 29) fügt diesen beiden idealtypisch Gesangstraditionen populärer Musik, die er als „vernacular derivatives of bel canto“ und als „natural technique“ bezeichnet, einen dritten Traditionsstrang hinzu, den er durch Techniken der Stilisierung und des performativen Spiels mit Rollen und Images charakterisiert und mit der Tradition von Theater, Cabaret und karnevalesken Rollenspielen in Verbindung bringt. Anders als bei der natürlichen Gesangstechnik geht es, so Middleton, bei diesen vokalen Stilisierungsstrategien gerade nicht um die Konstruktion von Authentizität im Sinne eines unmittelbaren emotionalen Involvements oder einer sozialen, regionalen oder ethnischen Gruppenzugehörigkeit. Der Popmusikforscher Philip Tagg hat darauf hingewiesen, dass diese klanglich stilisierten Ausdrucksmuster oder “vocal persona” für das Musikerlebnis vieler Popmusikhörer von zentraler Bedeutung sind. Vor dem Hintergrund seiner Erfahrung mit Popanalyse-Kursen schreibt Tagg:

> “When exhorted to speak the words of a particular vocal line emulating its approximate pitches, dynamics, timbre, duration, accentuation and rhythm, students are not slow to come up with words which pinpoint what might be called a *vocal persona*. The kinds of vocal persona heard in analysis classes tend to receive such vernacular epithets as: ‘little girl’, ‘nice boy’, ‘complete bitch’, ‘utter bastard’, ‘desperate rebel’, ‘macho git’, ‘sexy whisper’, ‘confident woman’, ‘evil child’, ‘worried man’, ‘death voice’, ‘voice of Satan’, ‘hound of hell’, ‘nervous hiccuping teenager’, ‘angry scream’, ‘Harpie shriek’, ‘frustrated squawk’, ‘bedside discoman’, ‘football hooligan’, ‘one of the lads’, ‘nudge-nudge, wink-wink’, ‘friend and confidente’, ‘tired and abused’, ‘resigned’, ‘depressed’, ‘demoralised’, ‘cynical’, ‘angry’, ‘hysterical’, ‘out of breath’, ‘stressed out’, etc. The list seems virtually endless and intersubjective agreement in analysis classes about the connotative traits of the vocalist(s) and vocal line in question is usually very high” (Tagg 2001, S. 13).

Bei den genannten vokalen Stereotypen handelt es sich offensichtlich um kulturelle Muster, also um „patterns, explicit or implicit, of and for behavior acquired and transmitted by symbols“ (Kluckhohn 1962, S. 73), über deren stimmliche Symbolisierungen eine hohe Übereinstimmung zwischen den Angehörigen einer kulturellen Gemeinschaft besteht. Vermutlich orientieren sich nicht nur Vokalisten an den bestehenden stimmlichen Ausdrucksmustern, sondern auch viele Hörer orientieren sich in der eigenen Imagination und Identitätsbildung an vokal vermittelten kulturellen Stereotypen.[26]

Bei der Erforschung der Stimmen populärer Musik stellt sich somit nicht nur die Aufgabe, vokale Gestaltungsmittel in verschiedenen Musikgenres und Stilrichtungen populärer Musik angemessen und schlüssig zu analysieren und zu beschreiben, sondern auch, die durch stimmliche Ausdrucksmittel vermittelten kulturellen Verhaltensmuster und Stereotype zu charakterisieren. Und es stellt sich noch eine weitere Aufgabe: Wie lässt sich der Personalstil einer Sängerin oder eines Sängers, deren oder dessen „persönliche Stimme" charakterisieren?

Christian Bielefeldt hat in seinen Überlegungen zur „Pop-Stimme" vorgeschlagen, „(...) im Falle populärer Stimmen zwischen interpretatorisch eingesetzten Gestaltungsmitteln auf der einen Seite und andererseits solchen Elementen zu trennen, die einer Stimme in anderer, persönlicherer Weise anhaften und zusätzlich zur stimmlichen Affektdarstellung und Songinterpretation zu Gehör kommen (können)" (Bielefeldt 2008, S. 211). Bielefeldt wendet sich dagegen, diesen zweiten Aspekt ausschließlich auf die Stimmphysiologie zu beziehen, da „(...) das Körperliche in den meisten populären Stilen selbst ein ebenso standardisiertes Ausdrucksmittel darstellt wie stimmliche Limits und Kontrollverluste" (ebd.). Vielmehr schlägt er vor, die persönliche oder, wie er schreibt, „personale" Stimme „als ein komplex verschaltetes Netzwerk" ihrer physiologischen Eigenheiten und biographischen Spuren zu verstehen – „als Echo erlebten Lebens, in das sich Organisches und Soziales, Materie, Körper und Geschlecht, Alter und Herkunft einschreiben" (Bielefeldt 2008, S. 212). Neben physiologischen Gegebenheiten wären in diesem Konzept einer persönlichen Stimme also u.a. auch regional und ethnisch sozialisierte Stimm-Settings sowie persönliche Dispositionen der Stimmqualität zu berücksichtigen. Zusätzlich können aber, so Bielefeldt weiter, auch audiotechnische Prozesse der Klangbearbeitung und bestimmte kulturelle Muster integraler Bestandteil einer persönlichen Stimme werden. Ähnlich wie Tagg zielt Bielefeldt mit diesen Mustern oder Images auf typische vokale Ausdrucksweisen, die von den Mitgliedern einer kulturellen Gemeinschaft unmittelbar als Ausdruck bestimmter Stereotype – dem Verführer, dem Prediger, dem Erzähler u.a. (vgl. Bielefeldt 2008, S. 213) – verstanden werden.

Versteht man die „persönliche Stimme" eines Vokalisten als Netzwerk von physiologischen Gegebenheiten und audiotechnischen Transformationen, von biographischen Spuren und kulturellen Mustern, so steht man vor der schwierigen Aufgabe, all diese Aspekte anhand konkreter Gesangsaufnahmen analytisch zu beschreiben. Dabei müssten auch mögliche Veränderungen der persönlichen Stimme im Laufe eines Künstlerlebens, etwa aufgrund des natürlichen Alterungsprozesses, aber auch aufgrund biographischer Einschnitte berücksichtigt werden – und die stilistischen Brüche und Neuorientierungen innerhalb des Schaffens eines Sängers, die mit biographischen Erfahrungen oder Verschiebungen der kulturellen Muster einhergehen können, aber nicht immer einhergehen müssen.[27]

---

[26] Zum Zusammenhang zwischen populärer Musik, Imagination und Prozessen der Identitätskonstruktion vgl. die Überlegungen von Frith (1996, S. 273ff.).

[27] Bielefeldt (2008) diskutiert diese Aspekte am Beispiel des umfangreichen musikalischen Schaffens von Prince.

Die Analyse und Interpretation des Klanggeschehens und der vokalen Klangproduktion stellt jedoch nur einen möglichen Zugang zur Erforschung des Phänomens Stimme in populärer Musik dar. Ein alternativer bzw. ergänzender Zugang lässt sich von der Frage leiten, wie die Stimmen von den Hörern wahrgenommen werden.

Emotion, Persönlichkeit und stimmlicher Ausdruck – empirische Forschungsansätze

Zum stimmlichen Ausdruck und zum Eindruck bzw. den Zuschreibungen von affektiven Zuständen wie Gefühlen, Stimmungen, Persönlichkeitseigenschaften usw. aufgrund von para- und extra-linguistischen Eigenheiten vokaler Äußerungen wurden zahlreiche empirische Untersuchungen durchgeführt – zumindest was die Äußerungen der Sprechstimme betrifft (vgl. Scherer 1982a, Tischer 1993, S. 83ff., Juslin/Laukka 2003, Juslin/Scherer 2008).[28] Patrik Juslin und Klaus Scherer geben in ihrem Beitrag zum *New Handbook of Methods in Nonverbal Behavior Research* konkrete Empfehlungen zur Konzeption, Durchführung und Auswertung von empirischen Studien zum stimmlichen Ausdruck von Affekten (Juslin/Scherer 2008, S. 92-121), die sich auch für eine empirische Erforschung des Zusammenhangs von vokalem Ausdruck und Affektwahrnehmungen in populärer Musik fruchtbar machen lassen. Juslin und Scherer nennen die folgenden konzeptionellen Schritte und Entscheidungen: Auswahl der zu untersuchenden affektiven Zustände; Auswahl der Vokalbeispiele sowie Segmentierung der Beispiele; Beurteilung der phonatorisch-artikulatorischen Vorgänge und Messung der akustischen Stimmeigenschaften; Beurteilung der Klangempfindungen; Einschätzung der affektiven Zustände; Anwendung statistischer Methoden der Modellbildung, durch welche affektive Zustände, Stimmmerkmale, Klangempfindungen und Einschätzungen der affektiven Zustände aufeinander bezogen werden.

Im Hintergrund dieser methodischen Empfehlungen steht ein mehrstufiges heuristisches Modell, das Scherer bereits in den 1980er Jahren vorgeschlagen hat (vgl. Scherer 1982a, 1982b, Juslin/Scherer 2008, S. 83ff.). Scherer unterscheidet zwischen

(1) den affektiven Zuständen des Sprechers,
(2) den akustischen und phonatorisch-artikulatorischen Eigenheiten der vokalen Äußerungen („voice cues“),
(3) der Empfindung dieser klanglichen Äußerungen seitens des Hörers („proximal cues“) sowie
(4) der Einschätzung oder Zuschreibung der affektiven Zustände der Sprecher, die hieraus hergeleitet werden.

Gelingt es, alle vier Instanzen des Modells exakt zu beschreiben, so lassen sich mittels statistischer Verfahren die Zusammenhänge zwischen ihnen modellieren und unter Umständen Vorhersagen über die Zuschreibung von affektiven Zuständen aufgrund von klanglichen Eigenheiten der Stimmäußerungen machen. Ich möchte auf einige Aspekte des Modells etwas genauer eingehen.

[28] Während Bernd Tischer 1993 ca. 50 Studien nennt, so ist die Zahl der Studien, die sich allein auf den Ausdruck der Grundemotionen beziehen, zehn Jahre später bereits auf über 100 angewachsen (vgl. Juslin/Laukka 2003).

Affektive Zustände (1) können nach Juslin und Scherer anhand mehrerer Kriterien in die Bereiche Emotionen und Stress, Stimmungen und zwischenmenschliche Haltungen sowie affektive Einstellungen und Persönlichkeitsdispositionen unterteilt werden (Juslin/Scherer 2008, S. 69ff., s. auch Scherer/Zentner 2001, S. 362f.). Emotionen und Stress sind kurze, aber intensive Reaktionen mit hoher Bedeutung für das Individuum. Die verschiedenen Reaktionskomponenten (Kognition, körperliche Erregung, Bewegungsausdruck usw.) sind miteinander synchronisiert und beeinflussen allgemein das Ausdrucksverhalten und insbesondere die stimmlichen Äußerungen. Ein Großteil der Forschungen zum stimmlichen Ausdruck hat sich bislang auf die Faktoren Stress (durch Überarbeitung oder Zeitdruck) und die Emotionen Wut, Furcht, Traurigkeit, Freude und Zärtlichkeit konzentriert (Juslin/Laukka 2003). Im Gegensatz zu diesen Basisemotionen und zum Stress lassen sich Stimmungen sowie bestimmte affektive zwischenmenschliche Haltungen (z.B. höflich, aggressiv, abweisend, kühl), die sich auch im stimmlichen Interaktionsverhalten ausdrücken, nur selten auf eine ganz bestimmte Ursache zurückführen. Stimmungen können zahlreiche Ursachen haben, und die Gründe für affektive zwischenmenschliche Haltungen ergeben sich aus der spezifischen Interaktionssituation, oder aber rühren von Dispositionen her, die länger anhalten bzw. generell die Interaktion mit bestimmten Personen prägen. Diese affektiven Phänomene sind bislang kaum untersucht worden, obwohl die Auswirkungen auf die Stimmäußerungen vermutlich beträchtlich sind. Schließlich kommt es zu längerfristigen affektiv besetzten Bewertungen von Objekten oder Menschen sowie zu anhaltenden affektiven Dispositionen, die zu Persönlichkeitsmerkmalen werden können (vgl. Eckert/Laver 1994, S. 149ff.).

Bei der Beurteilung und Klassifizierung der stimmlichen Äußerungen (2) sind verschiedene Erhebungsmethoden erprobt worden. Zur Klassifizierung phonatorisch-artikulatorischer Vorgänge wurden in den Studien zur Sprechstimme in der Regel Phonetiker herangezogen. Im Kontext der Musik wären zudem professionelle Sänger, Gesangspädagogen oder sonstige Stimmexperten zu Rate zu ziehen. Daneben wurden verschiedene Verfahren zur Messung der akustischen Eigenschaften der „voice cues“ vorgeschlagen, die sich in die Bereiche Tonhöhe (z.B. Grundtonhöhe und deren Streuung, Tonhöhenkontur, Jitter), Lautheit oder Intensität (z.B. Intensität, Stärke des Einsatzes, Shimmer), Stimmqualität (hochfrequente Energie, Formanten, Präzision der Artikulation, relative Zeit des Stimmlippenverschlusses pro Periode) und zeitliche Aspekte (Geschwindigkeit, Pausen, Regelmäßigkeit) unterteilen lassen.[29] Bei der Einschätzung von Gesang und Sprechgesang sollten darüber hinaus immer auch genuin musikalische Eigenschaften wie Tonalität und Harmonik, Metrum, Verwendung von Synkopen und Offbeats, Grad der mikrorhythmischen Exaktheit usw. herangezogen und möglicherweise in quantitative Kennwerte übersetzt werden. Dies erfordert eine genaue Analyse der musikalisch-klanglichen Gestaltung.

Sicherlich ist die von Scherer vorgenommene Unterscheidung in Wahrnehmungsqualitäten (3) und Affektzuschreibungen (4) sinnvoll. Allerdings stellt sich die Frage, ob diese beiden Einschätzungen sich aneinander anschließen, wie es Scherers Modell suggeriert, oder nicht vielmehr parallel vollzogen werden. Eine gesonderte Erhebung von Wahrnehmungsqualitäten ist meiner Ansicht nach nicht unbedingt erforderlich, solange es um Zusammenhänge zwischen Affektzuschreibungen und klanglichen Eigen-

[29] Vgl. Juslin/Laukka (2003, S. 790ff.) und Juslin/Scherer (2008, S. 102-104). Siehe außerdem die sehr differenzierten Kennwerte bei Tischer (1993, S. 200ff.).

heiten der Stimme („voice cues“) geht. Bei der Entwicklung einer verlässlichen deskriptiven Terminologie der stimmlichen Gestaltung kann die Erhebung von Wahrnehmungsqualitäten jedoch zweifelsohne äußerst hilfreich sein.

Es ist hier nicht der Ort, auf die zahlreichen bisherigen Forschungsergebnisse zum Zusammenhang zwischen Affektzuschreibungen und Eigenheiten der Sprechstimme en detail einzugehen (s. dazu Tischer 1993, Juslin/Laukka 2003, Juslin/Scherer 2008). Während in manchen Fragen große Übereinstimmung herrscht, sind andere Untersuchungsergebnisse doch eher uneinheitlich. Juslin und Scherer räumen ein, dass die „voice cues“ vermutlich nur unsichere Indikatoren für affektive Zustände sind, da derselbe affektive Zustand sich in Stimmäußerungen mit unterschiedlichen akustischen und physiologischen Eigenschaften äußern kann – die aber dennoch zum Teil interkorrelieren und daher teilweise redundant sind. Bei der Einschätzung des affektiven Zustands eines Sprechers werden von den Zuhörern somit wahrscheinlich eine Reihe von Stimmeigenschaften herangezogen, die jeweils für sich genommen weder notwendig noch hinreichend sind: „(...) multiple cues that are partly redundant yield a robust communicative system that is ‚forgiving‘ towards deviation from optimal code usage" (Juslin/Scherer 2008, S. 85).

Außerdem muss mit Dritt- und Störvariablen gerechnet werden: den individuellen Besonderheiten eines Sprechers, den Einflüssen der Semantik bei sprachlichen Äußerungen, den Veränderungen des stimmlichen Ausdrucks und Eindrucks im Verlauf einer Äußerung (vgl. Tischer 1993) sowie schließlich den Wechselwirkungen und Verbindungen zwischen mehreren affektiven Zuständen. Diese intervenierenden Variablen haben im musikalischen Bereich besondere Relevanz. Es sind vor allem zwei Faktoren, durch die sich der Einsatz der Stimme im künstlerisch-performativen Rahmen der Musik vom alltäglichen Sprechen unterscheidet: einerseits der Bezug auf die Inhalte der Songtexte und die Vielfalt der Personen und Stimmen, die dabei zur Geltung kommen[30], andererseits kommt es zu Rückwirkungen des durch die kompositorische Vorlage und die Begleitband vorgegebenen musikalischen Rahmens auf die stimmliche Gestaltung und auf den emotional-affektiven Gesamteindruck. Denn der Gesang lässt sich oft nicht vom musikalischen Kontext und dem gesamten Ausdrucksgehalt eines Musikstücks isolieren.

Der Umstand, dass es sich bei vielen künstlerischen Äußerungen von Schauspielern oder Sängern vermutlich nicht um den Ausdruck von „realen“ Emotionen und Stimmungen handelt, die tatsächlich in diesem Augenblick spontan gefühlt werden, sondern vielmehr um deren stilisierte Äußerung in einem Aufführungskontext, stellt dagegen nicht unbedingt ein methodisches Problem für die empirische Stimmausdrucksforschung dar. Denn Menschen können ja generell ihre stimmlichen Äußerungen bis zu einem gewissen Grade auch kontrolliert und strategisch einsetzen. Nach Scherer und Wallbot kommt es bei emotionalen Äußerungen sowohl zu „Push“-Effekten als auch zu „Pull“-Effekten (Scherer/Wallbot 1990): Einerseits bewirken affektive Zustände natürlich unmittelbare körperliche Veränderungen, z.B. Veränderung der Atmung und der Muskelspannung, die wiederum die stimmlichen Äußerungen hörbar beeinflussen und

[30] Frith (1996, S. 183ff.) hat darauf hingewiesen, dass bei der Aufführung vieler Popsongs nicht nur die tatsächlich erklingende Stimme der Sängerin oder des Sängers, sondern eine Reihe weiterer Stimmen eine Rollen spielen können, so die Stimmen des Erzählers und des Protagonisten des Songtextes sowie die Stimmen des Textautors oder des Komponisten.

verändern („Push“-Effekte). Andererseits kann stimmlicher Ausdruck aber auch strategisch und für manipulative Zwecke eingesetzt werden („Pull“-Effekte), z.B. zur Selbst-Präsentation – also um einen bestimmten Eindruck beim Gegenüber zu erzeugen – oder als bewusste Konventionalisierung und Stereotypisierung des Ausdrucks, wodurch die Verständlichkeit und Eindeutigkeit kommunikativer Botschaften optimiert werden kann. Nicht anders werden in der Musik bestimmte vokale Ausdrucksmittel im Gesang zur gezielten Übermittlung affektiver Botschaften eingesetzt.

Allerdings finden musikalische Aufführungen in der Regel unter anderen Vorzeichen statt als die zumeist zweckgerichtete Alltagskommunikation, nämlich in einem performativen, ästhetischen Rahmen. Vokale Qualitäten und die klanglichen Gestaltungseigenheiten der Stimme können hier zu eigenen sinnlichen Qualitäten werden, die zweckfrei erlebt und genossen werden können. Stimmen werden in einem künstlerischen Kontext mitunter so eingesetzt, dass die Aufmerksamkeit der Hörer allein auf deren Klanglichkeit gerichtet wird, oder dass die Imagination einer fiktiven Welt oder irrealer Personen anregt wird.[31] Scherer und Zentner (2001) weisen ebenfalls auf die Möglichkeit hin, dass affektive Zustände beim Musikhören womöglich nichts mit alltäglichen Emotionen und Affekten zu tun haben könnten, sondern dass es sich etwa bei der Ergriffenheit oder Rührung angesichts eines Musikerlebnisses um „ästhetische Emotionen“ handelt, die sich allerdings bislang gegen eine empirische psychologische Erforschung sperren.

Affektzuschreibungen zu Stimmen populärer Musik – eine Pilotstudie

Welche affektiven Zustände – Emotionen, Stimmungen und Persönlichkeitsdispositionen – werden Sängerinnen und Sängern anhand ihrer vokalen Äußerungen und der von ihnen verwendeten stimmlichen Gestaltungsmittel zugeschrieben? Sind die Affektzuschreibungen bei verschiedenen Hörerinnen und Hörern einheitlich? Um diese Fragen am Beispiel des oben analysierten Songs „Come Back, Baby“ von Ray Charles zu klären, wurden vierzehn Probanden gebeten, den Höreindruck der ersten Strophe des Songs auf eine Reihe von Adjektiven zu beziehen.[32] Die konkrete Aufgabe der Probanden lautete: „Hören Sie sich bitte die Stimme im Musikbeispiel aufmerksam an. Wie wirkt der Sänger auf Sie? Er ist ....“ gefolgt von 44 Adjektiven in alphabetischer Reihenfolge, wobei der Grad der Zustimmung zu jedem Item auf einer fünfstufigen Skala mit den Polen „überhaupt nicht“ (1) und „sehr stark“ (5) angegeben werden sollte. Die Versuchsteilnehmer hatten zusätzlich die Möglichkeiten, bei Adjektiven, die ihnen als unpassend erschienen, auf eine Antwort zu verzichten.

Als Ausgangsbasis für diese Item-Liste wurden zunächst zum einen Basisemotionen, zum anderen Persönlichkeitsmerkmale nach dem Persönlichkeits-Kreis von Eysenk und dem in der Persönlichkeitspsychologie gebräuchlichen Fünf-Faktoren-Modell (vgl. Zimbardo/Gerrig 2004, S. 606f.) herangezogen. Aus diesem umfangreichen Pool von zunächst knapp 100 Adjektiven wurden sodann jene Items ausgesondert, die sich, wie eine Voruntersuchung zeigte, nur schwer auf Musik, Gesang und Stimme beziehen

---

[31] Zu den ästhetischen Erfahrungsdimensionen populärer Musik hat sich ausführlich Ralf von Appen (2007) geäußert, der wiederum an die philosophische Ästhetik von Martin Seel (2003) anknüpft.

[32] Die Befragung wurde vom Autor im April 2008 am musikwissenschaftlichen Institut der Universität Basel im Rahmen der Lehrveranstaltung „Einführung in die Musikpsychologie“ durchgeführt.

lassen, sowie einige Adjektive, die einen Pol von gegensätzlichen Adjektivpaaren darstellen (z.B. „glücklich“ – „unglücklich“, „freundlich“ – „unfreundlich“). Ergänzt wurden einige weitere Adjektive, die von den Teilnehmern des Pretests vermisst wurden, z.B. „ironisch“, „verführerisch“, „erotisch“, sodass schließlich die 44 in der Untersuchung verwendeten Adjektive übrig blieben.

Affektzuschreibung	n	Mittelwert	Standardabweichung
aus sich herausgehend	14	4,36	,842
ehrlich	13	4,08	1,115
natürlich	14	3,86	,663
nachdenklich	14	3,86	1,027
energiegeladen	14	3,79	1,311
niedergeschlagen	14	3,79	,699
offen	13	3,77	1,013
durchsetzungsfähig	14	3,71	,994
mitfühlend	14	3,71	,726
verzweifelt	14	3,64	1,216
freundlich	14	3,50	,760
erregt	14	3,50	,941
lässig	14	3,50	1,160
impulsiv	13	3,46	1,050
aktiv	14	3,43	1,158
friedlich	13	3,38	1,121
traurig	14	3,29	1,139
verführerisch	13	3,15	1,068
ruhig	13	3,15	1,214
beherrscht	14	3,14	,770
kontrolliert	14	3,00	,877
intellektuell	10	3,00	,816
emotional instabil	12	2,92	,793
launisch	14	2,71	1,139
ausgeglichen	14	2,71	,914
euphorisch	13	2,69	,947
erotisch	13	2,69	1,251
optimistisch	13	2,62	1,193
freudig	13	2,46	,877
wütend	14	2,36	1,082
ironisch	12	2,33	1,371
glücklich	13	2,31	,751
ängstlich	14	2,29	1,139
oberflächlich	14	2,14	,770
unintelligent	10	2,00	,816
reserviert	14	1,93	,616
aggressiv	14	1,86	,663
schüchtern	14	1,86	,663
gleichgültig	14	1,79	,975
gelangweilt	14	1,79	,802
sorglos	14	1,64	,497
verachtend	13	1,54	,519
kalt	14	1,50	,650

Tabelle 1: Affektzuschreibungen bei „Come Back, Baby“ von Ray Charles (erste Strophe) auf einer fünfstufigen Skala (1 = „überhaupt nicht“, 5 = „sehr stark“).

Tabelle 1 zeigt die Ergebnisse der Befragung, wobei die Adjektive nach dem Grad der Zustimmung angeordnet sind. Demnach wird der Sänger der ersten Strophe von „Come Back, Baby“ – der Name Ray Charles wurde gegenüber den Probanden nicht erwähnt – in hohem Maße als „aus sich herausgehend“, „ehrlich“, „natürlich“, „energiegeladen“, aber auch als „nachdenklich“ und „niedergeschlagen“ eingeschätzt, dagegen nicht oder nur in geringem Maße als „kalt“, „verachtend“, „sorglos“, „gelangweilt“, „gleichgültig“, „schüchtern“ oder „aggressiv“. Es muss daher davon ausgegangen werden, dass in Musik generell und insbesondere durch stimmliche Äußerungen eine Vielfalt von Gefühlsebenen vermittelt werden, die sich dann, oberflächlich betrachtet, auch widersprechen können – wie hier die Zuschreibungen „nachdenklich“ und „energiegeladen“.

Der Grad der Übereinstimmung der Einschätzungen zwischen den Versuchsteilnehmern ist mit Standardabweichungen zwischen 0,497 („sorglos“) und 1,371 („ironisch“) erstaunlich hoch; zudem sind die Ratings mit wenigen Ausnahmen normalverteilt. Im Falle des Ray Charles-Songs kann also davon ausgegangen werden, dass Hörer in der Tat relativ einheitlich vom stimmlichen Ausdruck auf affektive Zustände des Sängers schließen. Allerdings lässt sich nur schwer rekonstruieren, welcher Stellenwert dem Songtext bei den Einschätzungen der Probanden zukommt und welche Merkmale und Passagen des Gesangs letztendlich zu den einzelnen Beurteilungen geführt haben. Genau dies ist aber die entscheidende Frage: Auf welche klanglichen und musikalischen Eigenheiten der stimmlichen Gestaltung lassen sich die verschiedenen Affektzuschreibungen zurückführen?

Mit einer explorativen Pilotstudie habe ich versucht, diese Fragen einzugrenzen und konkrete Hypothesen zu Zusammenhängen zwischen vokalen Ausdrucksmitteln und bestimmten Affektzuschreibungen zu gewinnen. Außerdem sollte mit der Studie die Einheitlichkeit von affektiven Zuschreibungen überprüft und ein mögliches statistisches Vorgehen zur Ermittlung der Zusammenhänge zwischen stimmlichen Merkmalen und der Zuschreibung von Gefühlen und Persönlichkeitsmerkmalen erprobt werden.

Die Untersuchung besteht aus zwei Teilen. Im ersten Teil wurde den 32 Teilnehmerinnen und Teilnehmern einer Einführungsveranstaltung im Bachelorstudiengang Populäre Musik und Medien an der Universität Paderborn[33] zunächst Zielsetzung und Ablauf der ca. 40 Minuten dauernden Befragung folgendermaßen erläutert: „Mit dem folgenden Hörversuch soll untersucht werden, wie verschiedene Stimmen und Vokalstile in populärer Musik auf Hörerinnen und Hörer wirken. Ich spiele Ihnen mehrere Musikausschnitte aus verschiedenen Songs vor. Ihre Aufgabe ist es, bei jedem Ausschnitt anzugeben, wie die Sängerin / der Sänger auf Sie wirkt. Hierzu finden Sie auf den folgenden Seiten jeweils eine lange Liste von Adjektiven. Bitte kreuzen Sie bei jedem Adjektiv je nach Ihrer Empfindung den Grad der Ausprägung zwischen ‚sehr stark' und ‚überhaupt nicht' an. Nutzen Sie bitte auch die Zwischenstufen.“

[33] Da die Studierenden dieses Studiengangs sich sowohl im Rahmen ihres Studiums als auch in ihrer Freizeit als Musiker intensiv mit populärer Musik beschäftigen, erschienen sie mir hervorragend geeignet, populäre Stimmen in einer differenzierten Weise zu beurteilen. Die Befragung fand im Wintersemester 2007/08 statt.

Affektzuschreibung	n	Minimum	Maximum	Mittelwert
glücklich	10	,70	,98	,8456
aus sich herausgehend	10	,55	1,11	,8628
freundlich	10	,61	1,06	,8736
aktiv	10	,64	1,21	,8869
optimistisch	10	,72	1,07	,9027
friedlich	10	,66	1,03	,9201
energiegeladen	10	,49	1,22	,9380
freudig	10	,81	1,06	,9390
schüchtern	10	,56	1,24	,9406
unintelligent	10	,73	1,14	,9439
ausgeglichen	10	,81	1,09	,9469
ruhig	10	,57	1,23	,9514
aggressiv	10	,54	1,37	,9523
mitfühlend	10	,66	1,27	,9617
durchsetzungsfähig	10	,80	1,19	,9632
ehrlich	10	,74	1,19	,9691
ängstlich	10	,56	1,25	,9741
intellektuell	10	,82	1,10	,9855
erotisch	10	,52	1,48	,9868
reserviert	10	,81	1,36	1,0075
wütend	10	,72	1,43	1,0104
niedergeschlagen	10	,69	1,30	1,0108
nachdenklich	10	,80	1,28	1,0120
impulsiv	10	,76	1,14	1,0121
oberflächlich	10	,89	1,14	1,0157
verzweifelt	10	,80	1,22	1,0191
traurig	10	,76	1,18	1,0239
ironisch	10	,56	1,37	1,0397
offen	10	,86	1,18	1,0398
gelangweilt	10	,57	1,22	1,0409
sorglos	10	,87	1,26	1,0437
emotional instabil	10	,77	1,23	1,0442
verachtend	10	,72	1,20	1,0499
lässig	10	,92	1,37	1,0554
erregt	10	,80	1,23	1,0564
kalt	10	,84	1,36	1,0674
gleichgültig	10	,67	1,26	1,0684
verführerisch	10	,38	1,57	1,0716
beherrscht	10	,98	1,32	1,0751
kontrolliert	10	,90	1,24	1,0884
launisch	10	,90	1,26	1,0889
euphorisch	10	,90	1,42	1,1261
natürlich	10	,88	1,33	1,1447

Tabelle 2: Minimum und Maximum sowie Mittelwert der Standardabweichungen der 44 Affektvariablen bei den zehn Musikbeispielen.

Sodann wurden zehn Musikausschnitte[34] mit einer Dauer zwischen 20 und 30 Sekunden je zwei Mal vorgespielt, wobei die Versuchsteilnehmer jeweils ein Blatt mit der folgenden Anweisung ausfüllten: „Hören Sie sich bitte die Stimme im Musikbeispiel aufmerksam an. Wie wirkt die Sängerin / der Sänger auf Sie? Sie / er ist ....:“, gefolgt von der bereits erläuterten Liste der 44 Adjektive, zu denen der Grad der Zustimmung („sehr stark“) bzw. Ablehnung („überhaupt nicht“) auf einer fünfstufigen Skala angegeben werden sollte. Anschließend wurde danach gefragt, ob die Probanden die Sängerin bzw. den Sänger sowie den Song bereits gekannt hatten. Am Schluss der Befragung wurden zusätzlich einige personenbezogene Daten zu Musikpräferenzen, Hörgewohnheiten sowie Alter und Geschlecht erfragt.

Die Zuschreibungen der 44 Adjektive durch die 32 Versuchspersonen sind auch hier erstaunlich einheitlich. Abhängig vom Musikbeispiel und vom jeweiligen Adjektiv lag die Standardabweichung zwischen $0{,}491 < s < 1{,}556$, wobei sie zumeist bei etwa einem Skalenpunkt der fünfstufigen Ratingskala lag. Demnach werden Sängerinnen und Sängern anhand von kurzen Musikausschnitten bestimmte Emotionen und Persönlichkeitsmerkmale erstaunlich einheitlich zugeschrieben. In Tabelle 2 sind für alle 44 Items jeweils der größte und der kleinste Wert der zehn Standardabweichungen (bezogen auf die zehn Musikbeispiele, $n = 10$) sowie der Mittelwert der Standardabweichungen aufgeführt.

Während manche Items bei allen zehn Musikbeispielen relativ einheitlich eingeschätzt wurden (z.B. „glücklich“, „freudig“, „ausgeglichen“, „durchsetzungsfähig“, „ehrlich“), schwankt die Streuung der Einschätzungen bei anderen Items abhängig vom Musikbeispiel doch recht stark, so etwa bei „verführerisch“ oder „ironisch“. In Folgestudien mit weiteren Musikbeispielen und anderen Versuchsteilnehmer müsste näher untersucht werden, in wie weit die Einheitlichkeit der Urteile eher auf die Auswahl der Musikbeispiele bzw. der Probanden zurückzuführen ist, oder ob ihr tatsächlich intersubjektiv gültige Affektwahrnehmungen zu Grunde liegen.

An dieser Stelle kann nicht weiter auf die verschiedenen Affektzuschreibungen bei den einzelnen Musikbeispielen eingegangen werden, denn eine nachvollziehbare Diskussion dieser Ergebnisse setzt die Kenntnis der Musikbeispiele durch die Leser voraus. Stattdessen möchte ich mich der Frage widmen, auf welche Eigenheiten der musikalischen Gestaltung sich die Affektzuschreibungen ursächlich zurückführen lassen und wie die Zusammenhänge zwischen „voice cues“ und Affektzuschreibungen empirisch ermittelt werden können. Hierzu wurde der zweite Teil der explorativen Pilotstudie durchgeführt, in dem nun die zehn im ersten Versuchsteil verwendeten Musikbeispiele bezüglich einer Reihe von Eigenheiten der stimmlichen Gestaltung eingeschätzt werden sollten. Hieran waren sechs Teilnehmer eines Seminars „Stimme und Gesang in populärer Musik“ beteiligt, die aufgrund der inhaltlichen Vorarbeiten in der Lehrveranstaltung als „Experten“ hinsichtlich stimmlicher Gestaltungs- und Ausdrucksmittel

[34] Verwendet wurden Ausschnitte aus den folgenden Songs: 1. Muse „In your world“, 2. TokTokTok „The Breakdown“, 3. Amanda Woodward “Binaire Et Lisible“, 4. Primus “Seas of Cheese“, 5. Shivaree „Goodnight Moon”, 6. Sizzla „Boom and Go through“, 7. Christina Aguilera feat. Herbie Hancock “A Song for You”, 8. Dead Kennedys “Bleed for Me“, 9. DFA1979 “Sexy Results”, 10. Cypress Hill “Make a Move”.

gelten können.[35] Als Merkmale des stimmlichen Ausdrucks dienten die verschiedenen Stimmgebungsarten nach Eckert/Laver (1994) sowie einige der sog. Cantometrics-Skalen nach Alan Lomax (1962, 1966) – und zwar all jene, die sich auf Eigenheiten des solistischen Gesangs beziehen.[36] Weitere Variablen wurden ergänzt, so vor allem Items zur studiotechnischen Bearbeitung der Stimme. Auf diese Weise wurden 36 siebenstufige Beurteilungsskalen zusammengestellt. Bei der Mehrzahl der Items sollte von den Probanden eingeschätzt werden, in welchem Grade ein Merkmal im Musikausschnitt vorhanden ist (von „nicht vorhanden“ = 1, bis „vorherrschend“ = 7). Die Items beziehen sich auf die folgende Bereiche: auf Stimmgebung und Stimmqualität (z.B. „neutrale Stimmgebung“, „Falsett“, „Rauigkeit“, „Schreien“), auf Lautstärke bzw. Stimmaufwand und deren Variation, auf verschiedene Verzierungsarten („Vibrato“, „Melismen“, „Glissando“), auf die rhythmische Gestaltung („einfach“ vs. „komplex“), die mikrorhythmische Gestaltung („durchgehend rubato“ vs. „metrisch präzise“), das Tempo („schnell“ vs. „langsam“) und die vorherrschenden Notenwerte („kurz“ vs. „lang“), auf die Tonhöhen-, Melodie- und Phrasengestaltung[37] sowie schließlich auf die studiotechnische Bearbeitungen der Stimme („Stellung der Stimme im Gesamtmix“, „studiotechnische Verfremdung“, „Hall“ sowie „Echo / Delay“). Anders als bei Lomax‘ Cantometrics-Projekt (Lomax 1976) konnten im Rahmen der Pilotstudie der jeweilige Ausprägungsgrad der einzelnen Variablen nicht durch Musikbeispiele veranschaulicht werden – dies bleibt einer zukünftigen Untersuchung vorbehalten.

Ähnlich wie im ersten Teil der Studie wurde zunächst die Schwierigkeit bzw. Streuung der einzelnen Items einer Prüfung unterzogen. Bei manchen Items und Musikbeispielen gab es eine völlige Übereinstimmung aller Probanden hinsichtlich der siebenstufigen Einschätzungen. Andere Merkmalen wurden jedoch sehr unterschiedlich eingeschätzt (maximal mit $s = 2{,}87$), was als Indiz dafür gewertet werden könnte, dass bei der Wahrnehmung der einzelnen Stimmmerkmale große Unterschiede bestehen, oder aber dass manche Items unklar definiert sind bzw. Raum für unterschiedliche Interpretation offen lassen (z.B. das Item „präzise Artikulation“ mit $0{,}41 < s < 2{,}87$). Auf die Optimierung der Beurteilungsskalen, die gegenüber den Versuchsteilnehmern möglichst durch eindeutige Referenzbeispiele illustriert werden sollten, muss bei zukünftigen Untersuchungen ein Hauptaugenmerk gerichtet werden.

---

[35] Das Seminar fand ebenfalls im Wintersemester 2007/08 im Masterstudiengang Populäre Musik und Medien an der Universität Paderborn statt. Den Teilnehmerinnen und Teilnehmern sei ausdrücklich für ihre konstruktive Mitarbeit gedankt; ein besonderer Dank gilt Johannes Bert, der die Rohdaten beider Befragungen in die SPSS-Tabellen eingegeben hat.

[36] Der amerikanische Ethnomusikologe Alan Lomax hat Anfang der 1960er Jahre ein umfassendes und Aufsehen erregendes Forschungsprojekt initiiert, in dem es um eine empirische Klärung der Zusammenhänge zwischen Vokalstilen und sozialen Strukturen geht (Lomax 1962, zu den Ergebnissen des Projekts vgl. Lomax 1968). In unserem Zusammenhang von Interesse sind die von Lomax vorgeschlagenen 37 Cantometrics-Skalen, mit denen zahlreiche musikalische Merkmale der Vokalmusik in den verschiedenen Musikkulturen der Welt eingeschätzt wurden. Neben den zahlreichen Items zur Stimmqualität und zu verschiedenen vokalen Gestaltungsmitteln, die in die Pilotstudie übernommen wurden, beziehen sich einige der Cantometric-Skalen allerdings auch auf Merkmale des mehrstimmigen Gesangs und der Ensemblebegleitung, die im vorliegenden Zusammenhang von eher untergeordneter Bedeutung sind.

[37] Die Angaben zu einer nominal skalierten Variable „Melodiekontur“ wurden nicht in die folgende Auswertung einbezogen.

Um exemplarisch zu testen, ob es möglich ist, die Affektzuschreibungen und Persönlichkeitseigenschaften aus dem ersten Untersuchungsteil sinnvoll mit den Einschätzungen stimmlicher Eigenheiten aus dem zweiten Untersuchungsteil zu verknüpfen, wurden für alle 44 Adjektive multiple lineare Regressionsrechnungen durchgeführt. Die Regressionsrechnung ist ein statistisches Verfahren auf der Basis der Korrelationsrechnung, deren Ziel es ist, das beste Vorhersagemodell für eine abhängige Variable (Kriteriumsvariable) aufgrund einer oder mehrerer gewichteter unabhängiger Variablen (Prädikatorvariablen) zu finden. Als Kriteriumsvariablen wurden die Mittelwerte der Affekteinschätzungen zu den zehn Musikbeispielen gewählt, als Prädikatorvariablen fungierten die Mittelwerte der Einschätzungen zu stimmlichen Merkmalen aus dem zweiten Versuchsteil.[38] Gesucht war also jeweils jene Kombination aus gewichteten „voice cues", durch die am besten eine bestimmte Affektzuschreibung vorhergesagt werden kann. Die Güte eines Modells zeigt sich daran, welcher Anteil der Varianz (standardisiertes r^2) der Kriteriumsvariablen durch das ermittelte Vorhersagemodell tatsächlich erklärt werden kann. Ich möchte im Folgenden einige Ergebnisse nennen.

Die Affektzuschreibung „aggressiv" lässt sich mit einer Varianzaufklärung von knapp 68% ($r^2 = 0{,}68$) auf die Stimmqualität „Schreien" (standardisierter Gewichtungskoeffizient $b = 0{,}85$, hoch signifikant) zurückführen. Mit anderen Worten: Aufgrund der vorliegenden Datenbasis können durch die Einschätzungen des siebenstufigen Grades von „Schreien", multipliziert mit dem Faktor 0,85, mehr als zwei Drittel der Varianz bei der fünfstufigen Einschätzung der Variablen „aggressiv" bei einem vergleichbaren Musikbeispiel vorhergesagt werden; auf weitere Prädikatorvariablen muss hierbei nicht zurückgegriffen werden. Dieser relativ eindeutige Zusammenhang zwischen dem Ausmaß des Schreiens und dem Grad der empfundenen Aggressivität leuchtet unmittelbar ein. Ähnlich eindeutig ist der Zusammenhang zwischen „verführerisch" und der Variable „behauchte Stimmgebung" ($b = 0{,}91$, $r^2 = 0{,}815$, hoch signifikant) sowie zwischen „erotisch" und „behauchte Stimmgebung" ($b = 0{,}85$, $r^2 = 0{,}68$, hoch signifikant) – was auf eine weitgehende Überschneidung zwischen den beiden Affektvariablen schließen lässt; die Korrelation zwischen den Einschätzungen zu den Items „verführerisch" und „erotisch" ist in der Tat mit $r = 0{,}96$ extrem hoch. Weitere einfache und einleuchtende Zusammenhänge ergeben sich zwischen den Variablen „durchsetzungsfähig" und „laut" ($b = 0{,}88$, $r^2 = 0{,}75$, hoch signifikant) sowie „gelangweilt" und „hoher Stimmaufwand" ($b = -0{,}91$, $r^2 = 0{,}80$, hoch signifikant). Das negative Vorzeichen bei der Gewichtung der Variablen bedeutet hier, dass der Eindruck „gelangweilt" um so höher ist, je geringer der Stimmaufwand eingeschätzt wird. Ein vergleichbarer Zusammenhang besteht zwischen „schüchtern" und „hohem Stimmaufwand" ($b = -0{,}89$, $r^2 = 0{,}76$, hoch signifikant).

Weitere Affektzuschreibungen lassen sich durch die Kombination mehrerer Prädikatorvariablen erklären. So ergibt sich bei der Affektzuschreibung „beherrscht" ein Modell ($r^2 = 0{,}92$) mit negativen Gewichtungen für „Schreien" ($b = -0{,}84$) und „rubato"

[38] Als statistisches Verfahren wurde in SPSS 14 ein Regressionsverfahren mit schrittweisem Einschluss der Prädikatorvariablen gewählt; als Kriterien für die Aufnahme bzw. den Ausschluss einer Variablen in das bzw. aus dem Modell wurden übliche F-Wahrscheinlichkeiten verwendet, für die Aufnahme $F < 0.05$ und für den Ausschluss $F > 0.1$. In zukünftigen Studien sollten auch alternative Verfahren erprobt werden, so die logistische Regression und sog. Regressionsbäume, wodurch auch ordinal bzw. nominal skalierte Variablen einbezogen werden können.

(b = - 0,312). Demnach gilt ein Vokalist dann als beherrscht, wenn seine Äußerungen weder Anzeichen von Schreien noch von einer freieren rhythmischen Feingestaltung aufweisen. Noch differenzierter ist das mit $r^2 = 0{,}998$ extrem exakte Regressionsmodell für Erregung („erregt"). Es setzt sich aus den Vorhersage-Variablen „Phrasenenden auf hohem Ton" (b = 1,31), „schnelles Tempo" (b = 1,27), „neutrale Stimmgebung" (b = 0,89) und „entspannte Stimmgebung" (b = 0,484) zusammen sowie einer geringen negativen Gewichtung beim Item „Echo / Delay" (b = - 0,143). Die absolute Größe der jeweiligen Gewichtungen (b) lässt Rückschlüsse über den relativen Stellenwert der einzelnen Prädikatorvariablen zu. Entscheidend für den Eindruck des Erregt-Seins ist somit eine in der Tonhöhe ansteigende Melodielinie und ein schnelles Tempo, doch auch eine neutrale und entspannte Stimmgebung spielen eine gewisse Rolle. Das ebenfalls gut nachvollziehbare Modell für die Variable „lässig" ($r^2 = 0{,}98$) setzt sich aus den Vorhersagevariablen „kurze Notenwerte" (b = 0,76), den jeweils negativ gewichteten Variablen „Phrasenenden mit hohem Ton" (b = - 0,57) und „gepresst" (b = - 0,58) sowie einem geringen Anteil technischer Verfremdung (b = 0,173) zusammen. Der Eindruck von „Lässigkeit" geht demnach mit eher kurzen Notenwerten, einer nicht gepressten Stimmgebung und Phrasenenden bei tiefen Tönen einher.

Wie aufgrund der mit nur zehn Musikbeispielen doch relativ kleinen Datenbasis nicht anders zu erwarten war, konnte nicht bei allen der 44 Affektvariablen überzeugende und interpretierbare Regressionsmodelle mit befriedigender Varianzaufklärung ermittelt werden. So gab es mehrere Affektzuschreibungen, die sich anhand der vorhandenen Daten nicht sinnvoll auf Merkmale der stimmlichen Gestaltung beziehen ließen – vor allem jene Items, über die unter den Versuchsteilnehmern nur relativ geringe Übereinstimmung herrschte, z.B. „intellektuell", „ehrlich", „mitfühlend", „nachdenklich", „niedergeschlagen", „oberflächlich" und „offen".

Natürlich lassen sich die statistischen Vorhersagemodelle zunächst nur auf vokale Gestaltungsmittel beziehen, die in den zehn bislang verwendeten Musikausschnitten eingeschätzt worden sind. Vermutlich deckt die doch relativ kleine Auswahl von zehn Musikbeispielen nur einen Teil der großen Bandbreite möglicher stimmlicher Ausdrucksmittel in populärer Musik ab. Dennoch kam es bei einer Reihe von Affekten zu interessanten Ergebnissen, die Licht auf möglicherweise verallgemeinerbare Zusammenhänge zwischen Merkmalen der stimmlichen Gestaltung und Zuschreibungen von Emotionen und Persönlichkeitsmerkmalen werfen. Mit der hier vorgestellten Vorgehensweise ist somit prinzipiell zu sinnvollen Aussagen über Zusammenhänge zwischen der Empfindung und Zuschreibung affektiver Zustände und stimmlichen Ausdrucksmitteln zu gelangen, die, je nach Auswahl der Musikausschnitte, für das Repertoire eines bestimmten Vokalisten, eines Musikgenres oder Musikstils allgemein gültig sind.

Bei einer Weiterführung des vorgestellten empirischen Forschungsansatzes müssten allerdings zunächst Wege gefunden werden, wie die klanglichen und musikalischen Eigenheiten von Stimmen objektiv, valide und reliabel eingeschätzt oder gemessen werden können. Vielleicht wäre es sinnvoll, sich zunächst auf die Erhebung jener Vokalvariablen zu beschränken, die besonders eindeutig und verlässlich gemessen oder beurteilt werden. Außerdem sollte berücksichtigt werden, ob und wie der affektive Eindruck sich im Laufe eines Musikausschnittes verändert, wodurch unter Umständen die Vielschichtigkeit bzw. Widersprüchlichkeit mancher Einschätzungen erklärt werden könnte. Eine weitere mögliche Vorgehensweise bestünde darin, passende Musikbeispiele mit professionellen Sängerinnen und Sängern selbst zu produzieren, bei denen einerseits den Vokalisten bestimmte zu realisierende Ausdrucksqualitäten vorgegeben werden und andererseits aufgrund des Aufnahmeverfahrens die klanglichen Eigenschaften eindeutiger bestimmbar und miteinander vergleichbar wären. Zudem wäre es sinnvoll, die mit 44 Adjektiven noch relativ hohe Zahl von Affektvariablen, die zum Teil in hohem Maße miteinander korrelieren, durch eine Faktorenanalyse oder Variablenclusteranalyse auf eine geringere Anzahl zu verkleinern. Schließlich müsste näher untersucht werden, welche intervenierenden Variablen einen Einfluss auf die Einschätzungen der Versuchsteilnehmer haben. Denn es ist nicht auszuschließen, dass die Urteile der Probanden in einem gewissen Maße durch die Semantik der Songtexte, durch ihre individuellen musikalischen Vorlieben und Hörgewohnheiten sowie durch die Assoziation von Musikbeispielen mit bestimmten Sängerinnen und Sängern und deren persönlichen Images geprägt werden. Denkbar wäre zum Beispiel, verschiedene Aufnahmen desselben Songs und somit desselben Liedtextes, gesungen jedoch von verschiedenen Vokalisten, heranzuziehen, oder aber bei der Auswahl der Probanden stärker deren Präferenzen und Hörgewohnheiten zu berücksichtigen.

Ausblick

Die Erforschung des Phänomens Stimme und Gesang in populärer Musik steht vor einer Reihe von Herausforderungen. Es bieten sich mehrere Herangehensweisen und damit verbunden verschiedene Forschungsmethoden an, sich diesem Phänomen wissenschaftlich zu nähern. Die unterschiedlichen Herangehensweisen ergänzen und befruchten sich, meiner Ansicht nach, gegenseitig und können so zu einem umfassenden Verständnis der Stimmen populärer Musik, der stimmlichen Gestaltungsmittel und ihrer Rezeption beitragen.

Zunächst gilt es, alle verfügbaren Quellen und Tondokumente zu sammeln und zu erschließen, um auf diese Weise ein umfassendes Archiv von Tonaufnahmen und Informationen zu den Stimmen populärer Musik aufzubauen. Auf dieser Grundlage sollten sich Analysen der Gesangsstile in verschiedenen Epochen und Bereichen der populären Musik anschließen. Dabei gilt es zunächst, adäquate und schlüssige Beschreibungs-, Analyse- und Darstellungsmöglichkeiten für stimmliche Gestaltungsmittel zu entwickeln, die sowohl melodische, rhythmische, harmonische Gestaltungsweisen als auch klangliche Besonderheiten, verschiedene Stimmqualitäten und Gesangstechniken sowie schließlich die technische Bearbeitungen von Stimmaufnahmen berücksichtigen. Darüber hinaus wäre es hilfreich und erhellend, durch verschiedene Messverfahren die physiologischen Grundlagen von Stimmen und Gesangstechniken (Eigenheiten der Stimmlippen und der Phonation, subglottaler Druck, spektrale Energieverteilung u.a.) in verschiedenen Stilrichtungen und bei verschiedenen Vokalisten zu ermitteln.

Auf dieser musikanalytischen Grundlage ließe sich sodann eine vergleichende musikhistorische Charakterisierung von verschiedenen Vokalstilen vornehmen – hinsichtlich der gemeinsamen Gestaltungsmittel von Sängerinnen und Sängern innerhalb bestimmter musikalischer Genres und Stile, aber auch hinsichtlich des Personalstils und der „persönlichen Stimme“ eines Vokalisten. Letztere hängt vermutlich mit physiologischen und biographischen Eigenheiten der jeweiligen Person, gegebenenfalls auch mit aufnahmetechnischen Besonderheiten sowie schließlich mit kulturellen Mustern zusammen. Stimmäußerungen verweisen immer auch auf kulturelle Muster, Stereotype, Images und

Identitäten, auf die der Vokalist zurückgreift und die er umgekehrt durch sein künstlerisches Schaffen prägt und formt. Stimmliche Ausdrucksmuster stehen zugleich in engem Zusammenhang mit nicht-klanglichen kulturellen Mustern, sodass die stimmlichen Eigenheiten in einem weiteren Analyseschritt auf visuelle und performative Aspekte populärer Musik bezogen werden müssen, so etwa auf die (Selbst-)Präsentation von Sängerinnen und Sängern bei Konzerten und in den Medien.

Schließlich erscheint mir eine Verknüpfung dieser stimm-, musik- und kulturanalytischen Ansätze mit empirischen Untersuchungen zur Rezeption der Stimmen populärer Musik als unumgänglich, will man nicht Gefahr laufen, von den subjektiven Empfindungen, Einschätzungen und Interpretationen eines einzelnen Forschers möglicherweise unzutreffende Rückschlüsse auf die Rezeptionsweisen in anderen Hörerkreisen zu ziehen. Eine mögliche empirische Vorgehensweise, die sich dem Zusammenhang von Eigenheiten der vokalen Gestaltung und den Zuschreibungen von Emotionen, Stimmungen und Persönlichkeitsmerkmalen widmet, habe ich skizziert. Daneben bieten sich zahlreiche weitere Vorgehensweisen an. So können die Verknüpfung von stimmlichen Gestaltungsmitteln mit bestimmten Popmusikgenres sowie deren jeweilige Relevanz für die Hörer durch Ähnlichkeitseinschätzungen von Gesangsbeispielen und mit Methoden der multidimensionalen Skalierung ermittelt werden. Ein möglicher Ansatz, sich der Frage des Personalstils und der „persönlichen Stimme“ empirisch anzunähern, bestünde darin, unterschiedliche Aufnahmen desselben Sängers oder derselben Sängerin hinsichtlich der „voice cues“ und Affektzuschreibungen beurteilen zu lassen und Gemeinsamkeiten dieser Einschätzungen zu bestimmen.

Vermutlich lassen sich viele Ergebnisse aus einer so verstandenen Stimmforschung auch über den Bereich der populären Stimmen hinaus weiterführen. Ein Vergleich mit klassischem Opern- und Liedgesang wäre ebenso von Interesse wie die Untersuchung von Stimmen und vokalen Gestaltungsmitteln in asiatischen, afrikanischen und lateinamerikanischen Musiktraditionen (vgl. Födermayr 1971). Da die Klanggestaltung bei vielen Musikinstrumenten an den Klang und die Ausdrucksmittel der menschlichen Stimme anknüpft, ließe sich zudem auch das Spiel auf Musikinstrumenten – etwa der elektrischen Gitarre im Rock (vgl. Schneider 2001) oder den Blasinstrumenten im Jazz – mit Ansätzen der Stimmforschung in einem neuen Licht betrachten.

Literatur

Appen, Ralf von (2007): *Der Wert der Musik. Zur Ästhetik des Populären*, Bielefeld: transcript.

Ashley, Richard (2002), Do[n't] Change a Hair for Me. The Art of Jazz Rubato, in: *Music Perception*, 19, 311-332.

Banfield, Stephen (2000), Stage and Screen Entertainers in the Twentieth Century, in: *The Cambridge Companion to Singing*, hg. v. John Potter, Cambridge: Cambridge University Press, 63-81.

Bauer, William R. (1993), Billie Holiday and Betty Carter: Emotion and Style in the Jazz Vocal Line, in: *Annual Review of Jazz Studies*, 6, 99-152.

Bauer, William R. (2001/02): Scat Singing: A Timbral and Phonemic Analysis, in: *Current musicology*, 71/73 (Special Issue Jazz Studies: A Commemorative Festschrift in Honor of Mark Tucker), 303-323.

Bielefeldt, Christian (2006), Hiphop im Candyshop. Überlegungen zur populären Stimme, in: *Cut and paste. Schnittmuster populärer Musik der Gegenwart* (= Beiträge zur Popularmusikforschung, Bd. 34), hg. von Dietrich Helms und Thomas Phleps, Bielefeld: transcript, 135-152.

Bielefeldt, Christian (2008), Voices of Prince. Zur Popstimme, in: *Popmusicology. Perspektiven der Popmusikwissenschaft*, hg. von Christian Bielefeldt, Udo Dahmen und Ralf Grossmann, Bielefeld: transcript, 201-219.

Brackett, David (1995), *Interpreting Popular Music*, Cambridge: Cambridge University Press.

Bradby, Barbara / Brian Torode (1984), Pity Peggy Sue, in: *Popular Music, 4, Performers and Audiences*, hg. von Richard Middleton und David Horn, Cambridge: Cambridge University Press, 183-205.

Cleveland, Thomas F., R.E. Stone, Johan Sundberg, Jenny Iwarsson (1997), Estimated Subglottal Pressure in Six Professional Country Singers, in: *Journal of Voice*, 11, 403-409.

Crowther, Bruce, Mike Pinfold (1986), *The Jazz Singers. From Ragtime to the New Wave*, Poole: Balndford.

Cubitt, Sean (1984), 'Maybellene'. Meaning and the Listening Subject, in: *Popular Music, 4, Performers and Audiences*, hg. von Richard Middleton und David Horn, Cambridge: Cambridge University Press, 207-224.

Daley, Mike (1998), Patti Smith's 'Gloria'. Intertextual Play in Rock Vocal Performance, in: *Popular Music*, 16, 235-253.

Diaz-Bone, Rainer (2002), *Kulturwelt, Diskurs und Lebensstil. Eine diskurstheoretische Erweiterung der bourdieuschen Distinktionstheorie*, Opladen: Leske + Budrich.

Eckert, Hartwig, John Laver (1994), *Menschen und ihre Stimmen. Aspekte der vokalen Kommunikation*, Weinheim: Beltz.

Epping-Jäger, Cornelia, Erika Linz (Hg.) (2003), *Medien/Stimmen*, Köln: DuMont.

Estill, Jo, Thomas Baer, Kiyoshi Honda, Katherine Harris (1985), Voice Quality in Singing: An Acoustic, Electromyographic, and Glottographic Study, in: *Proceedings of the Stockholm Music Acoustics Conferende 1983*, Vol. 1, Stockholm: The Royal Swedish Academy of Music, Publ. No. 46, 157-174.

Felderer, Brigitte (Hg.) (2004), *Phonorama. Eine Kulturgeschichte der Stimme als Medium*, Berlin: Matthes & Seitz.

Fischer, Julia (2006), Tierstimmen, in: *Stimme*, hg. v. Doris Kolesch und Sybille Krämer, Frankfurt: Suhrkamp, 172-190.

Fischer, Peter-Michael (1993), *Die Stimme des Sängers. Analyse ihrer Funktion und Leistung – Geschichte und Methodik der Stimmbildung*, Stuttgart/Weimar: Metzler.

Födermayr, Franz (1971), *Zur gesanglichen Stimmgebung in der außereuropäischen Musik. Ein Beitrag zur Methodik der vergleichenden Musikwissenschaft*, 2 Bde., Wien: Institut für Völkerkunde.

Friedwald, Will (1990), *Jazz Singing: America's Great Voices from Bessie Smith to Bebop and Beyond*, New York: Scribner.

Frith, Simon (1996), *Performing Rites. Evaluating Popular Music*, Oxford: Oxford University Press.

Göttert, Karl-Heinz (1998), *Geschichte der Stimme*, München: Fink.

Hartlieb, Karl (1960), Stimmbildung als Wissenschaft, in: *Beiträge zur Musikwissenschaft*, 1960, H. 2, 41-57.

Henrich, Nathalie et al. (2008), Towards a Common Terminology to Describe Voice Quality in Western Lyrical Singing: Contribution of a Multidisciplinary Research Group, in: *Journal of Interdisciplinary Music Studies*, 2, 71-93.

Hudson, Richard (1994), *Stolen Time. The History of Tempo Rubato*, Oxford: Clarendon Press.

Jungr, Barb (2002), Vocal Expression in the Blues and Gospel, in: *The Cambridge Companion to Blues and Gospel Music*, hg. v. Allan Moore, Cambridge: Cambridge University Press, 102-115.

Juslin, Patrik N., P. Laukka (2003), Communication of Emotions in Vocal Expression and Music Performance: Different Channels, Same Code? In: *Psychological Bulletin*, 129, 770-814.

Juslin, Patrik N., Klaus R. Scherer (2008), Vocal Expression of Affect, in: *The New Handbook of Methods in Nonverbal Behavior Research*, hg. von Jinni A. Harrigan, Robert Rosenthal und Klaus R. Scherer, Oxford: Oxford University Press, 65-135.

Kayes, Gillyanne (2004), *Singing and the Actor*, 2. Aufl., London: A. & C. Black.

Kittler, Friedrich, Sigrid Weigel, Thomas Macho (Hg.) (2002), *Zwischen Rauschen und Offenbarung. Zur Kultur- und Mediengeschichte der Stimme*, Berlin: Akademie Verlag.

Klausmeier, Friedrich (1999), *Belconato und Pop. Zwei Arten des Singens*, Augsburg: Wißner

Kluckhohn, Clyde (1962): *Culture and Behavior. The Collected Essays of Clyde Kluckhohn*, hg. v. Richard Kluckhohn, New York: Free Press.

König, Ekkehard, Johannes G. Brandt (2006), Die Stimme – Charakterisierung aus linguistischer Perspektive, in: *Stimme*, hg. v. Doris Kolesch und Sybille Krämer, Frankfurt: Suhrkamp, 111-129.

Kolesch, Doris, Sybille Krämer (Hg.) (2006), *Stimme*, Frankfurt: Suhrkamp.

Krims, Adam (2000), *Rap Music and Politics of Identity*, Cambridge: Cambridge University Press.

Laver, John (1980), *The Phonetic Description of Voice Quality*, Cambridge: Cambridge University Press.

Lockheart, Paula (2003), A History of Early Microphone singing, 1925-1939: American Mainstream Popular Singing at the Advent of Electronic Microphone Amplification, in: *Popular Music and Society*, 26, 367-385.

Lomax, Alan (1962), Song Structure and Social Structure, in: *Ethnology*, 1, 425-251.

Lomax, Alan (1968), *Folk Song Style and Culture*, Washington: American Association for the Advancement of Science.

Lomax, Alan (1976), *Cantometrics. An Approach to the Anthropology of Music,* Berkeley.

MacCracken, Allison (2001), Real Men Don't Sing Ballads: The Radio Crooner in Hollywood, 1929-1933, in: *Soundtrack Available*, hg. von Pamela Robertson Wojcik and Arthur Knight, Durham: Duke University Press, 105-133.

Middleton, Richard (2000), Rock Singing, in: *The Cambridge Companion to Singing*, hg. v. John Potter, Cambridge: Cambridge University Press, 28-41.

Middleton, Richard (2006), *Voicing the Popular. On the Subjects of Popular Music*, New York: Routledge.

Miles, Beth, Harry Hollien (1990), Whither Belting? In: *Journal of Voice*, 4, 64-70.

Moore, John (1989), 'The Hieroglyphics of Love'. The Torch Singers and Interpretation, in: *Popular Music*, 8, 39.

Moore, Allan F. (2001), *Rock. The Primary Text. Developing a Musicology of Rock*, 2. Aufl., Aldershot: Ashgate.

Moore, Allan F. (2002), Authenticity and Authentication, in: *Popular Music*, 21, 209-223.

National Center for Voice and Speech (1994), *Tutorials. Voice Quality*, http://www.ncvs.org/ncvs/tutorials/voiceprod/tutorial/quality.html (URL vom 22.11.2008).

Pétursson, Magnús, Joachim Neppert (2002), *Elementarbuch der Phonetik*, 3. Aufl., Hamburg: Buske.

Pfleiderer, Martin (2003), Sound. Anmerkungen zu einem populären Begriff, in: *PopSounds. Klangtexturen in der Pop- und Rockmusik*, hg. von Thomas Phleps und Ralf van Appen, Bielefeld: transcript, 19-29.

Pfleiderer, Martin (2006a): *Rhythmus. Psychologische, theoretische und stilanalytische Aspekte populärer Musik*, Bielefeld: transcript.

Pfleiderer, Martin (2006b), Brown Sugar. Tendenzen im Rhythm'n'Blues, in: *Cut and paste. Schnittmuster populärer Musik der Gegenwart* (= Beiträge zur Popularmusikforschung, Bd. 34), hg. von Dietrich Helms und Thomas Phleps, Bielefeld: transcript, 47-60.

Pitts, Michael, Frank Hoffman (2002), *The Rise Of The Crooners. Gene Austin, Russ Columbo, Bing Crosby, Nick Lucas, Johnny Marvin, and Rudy Vallee*, Scarecrow Press.

Pleasants, Henry (1974), *The Great American Popular Singers*, New York: Gollancz.

Potter, John (1998), *Vocal Authority. Singing Style and Ideology*, New York: Cambridge University Press.

Potter, John (Hg.) (2000), *The Cambridge Companion to Singing*, Cambridge: Cambridge University Press.

Putschögl, Gerhard (1998), Zu den rhythmisch freien Gestaltungsformen der afroamerikanischen Musik am Beispiel John Coltranes, in: *...und der Jazz ist nicht von Dauer. Aspekte afro-amerikanischer Musik. Festschrift für Alfons Michael Dauer*, hg. v. Bernd Hoffmann und Helmut Rösing (= Forum Jazz Rock Pop 1), Karben: CODA, 333-369.

Rapoport, Elizier (1996a), Emotional Expression Code in Opera and Lied Singing, in: *Journal of New Music Research*, 25/2, 109-149.

Rapoport, Elizier (1996b), Temporal Aspects of Timbre and Emotional Expression in Opera and Lied Singing. An FFT Study, in: *Systematische Musikwissenschaft*, 4, 311-329.

Scherer, Klaus, R. (1982a), Methods of Research on Vocal Communication. Paradigms and Parameters, in: *Handbook of Methods in Nonverbal Behavior Research*, hg. v. Klaus R. Scherer und Paul Ekman, Cambridge: Cambridge University Press 1982, 136-198.

Scherer, Klaus R. (1982b), Stimme und Persönlichkeit – Ausdruck und Eindruck, in: *Vokale Kommunikation. Nonverbale Aspekte des Sprachverhaltens*, hg. v. Klaus R. Scherer, Weinheim/Basel: Beltz, 188-210.

Scherer, Klaus R., Harald G. Wallbott (1990), Ausdruck von Emotionen, in: *Psychologie der Emotionen*, hg. v. Klaus R. Scherer (= Enzyklopädie der Psychologie, Themenbereich C, Serie 4, Bd. 3), Göttingen: Hogrefe, 345-422.

Scherer, Klaus R., Marcel R. Zentner (2001), Emotional Effects of Music. Production Rules, in: *Music and Emotion. Theory and Research*, hg. von Patrik N. Juslin u. John A. Sloboda, Oxford: Oxford University Press, 361-392.

Schneider, Albrecht (2001), Klanganalyse als Methodik der Popularmusikforschung, in: *Musikwissenschaft und populäre Musik: Versuch einer Bestandsaufnahme*, hg. v. Helmut Rösing, Albrecht Schneider u. Martin Pfleiderer (= Hamburger Jahrbuch für Musikwissenschaft, Bd. 19). Frankfurt: Peter Lang, 107-129.

Schutte, Harm K., Donald G. Miller (1993), Belting and Pop, Nonclassical Approaches to the Female Middle Voice: Some Preliminary Considerations, in: *Journal of Voice*, 7, 142-150.

Seel, Martin (2003), *Ästhetik des Erscheinens*, Frankfurt: Suhrkamp.

Seidner, Wolfram, Jürgen Wendler (2004), *Die Sängerstimme. Phoniatrische Grundlagen für die Gesangsausbildung*, 4. Aufl, Berlin: Henschel.

Shapiro, Peter (2000), *Soul. 100 Essential CDs. The Rough Guide*, London: Rough Guides.

Stewart, Milton Lee (1987), Stylistic Environment and the Scat Singing Styles of Ella Fitzgerald and Sarah Vaughan, in: *Jazzforschung/Jazz Research*, 19, 61-76

Stone, R.E., Thomas F. Cleveland, Johan Sundberg (1999), Formant Frequencies in Country Singers' Speech and Singing, in: *Journal of Voice*, 13, 161-167.

Sundberg, Johan (1987), *The Science of the Singing Voice*, Dekalb: Northern Illinois University Press.

Sundberg, Johan (1994), Perceptual Aspects of Singing, in: *Journal of Voice*, 8, 106-122.

Sundberg, Johan, Patricia Gramming, Jeanette Lovetri (1993), Comparisons of Pharynx, Source, Formant, and Pressure Characteristics in Operatic and Musical Theatre Singing, in: *Journal of Voice*, 7, 301-310.

Sundberg, Johan, Thomas F. Cleveland, R.E. Stone, Jenny Iwarsson (1999), Voice Source Characteristics in Six Premier Country Singers, in: *Journal of Voice*, 13, 168-183.

Tagg, Philip (2001), *Music analysis for 'non-musos'. Popular perception as a basis for understanding musical structure and signification*, Paper for Conference on Popular Music Analysis, University of Cardiff, 17 November 2001 (http://www.tagg.org/articles/xpdfs/cardiff01.pdf, URL vom 22.11.2008).

Tischer, Bernd (1993), *Die vokale Kommunikation von Gefühlen*, Weinhein: Beltz.

Titon, Jeff Todd (1977), *Early Downhome Blues. A Musical and Cultural Analysis*, Urbana: University of Illinois Press.

Trojan, Felix (1975), *Biophonetik*, hg. von Herbert Schendl, Mannheim: Bibliographisches Institut.

Trudgill, Peter (1983), *On Dialect. Social and Geographical Perspectives*, Oxford.

Wendler, Jürgen, Wolfram Seidner, Ulrich Eysholdt (2005), *Lehrbuch der Phoniatrie und Pädaudiologie*, , 4. Aufl., Stuttgart: Thieme.

Wise, Timothy (2007), Yodel Species: A Typology of Falsetto Effects in Popular Music Vocal Styles, in: *Radical Musicology*, 2.

Zimbardo, Philip G., Richard J. Gerrig (2004), *Psychologie*, 16. Aufl., München: Pearson Studium.

Helmut Rösing

Was leistet Musik für die Gesellschaft? Anmerkungen aus rezeptions- und sozialpsychologischer Sicht

Mehr als Anmerkungen sind zu der zentralen Frage, was Musik für die Gesellschaft zu leisten vermag, kaum zu erwarten. Entsprechend der Komplexität der Thematik sind meine Anmerkungen notgedrungener Weise sehr abstrakt gehalten. Denn Konkretion und Vollständigkeit lassen sich hier nur durch die Analyse von Einzelfallstudien erzielen. Mir geht es im Folgenden aber darum, übergeordnete Faktoren des Zusammenspiels von Musik mit Gesellschaft und Individuum zu benennen, anders gesagt, die Sichtweise auf das Ganze zu richten und es dann jedem Leser selbst zu überlassen, die Markierungspunkte der ganzheitlichen Sichtweise als Analyseraster am Einzelfall zu erproben. Diese Vorgehensweise entspricht der Empfehlung von Bruno Nettl (1983), Musik als Bestandteil eines organischen Ganzen zu sehen und dann zu fragen, welche Funktion(en) sie im soziokulturellen Gesamtsystem hat.

Zu allen Zeiten und in allen Ethnien ist Musik von besonderer Bedeutung für die Menschheit gewesen. Ein gesellschaftliches Zusammenleben ohne Musik gibt es nicht und scheint es auch nie gegeben zu haben. Das kann nur zu einer Schlussfolgerung führen: Musik hat zu jeder Zeit für die kulturelle Evolution gesellschaftsschaffende und -begleitende Funktion gehabt. In seinem Buch „Der musizierende Mensch. Eine Anthropologie der Musik“ (1984, S. 180) fasst Wolfgang Suppan diesen Sachverhalt folgendermaßen zusammen: „Die biologische Disposition zum Musikgebrauch in entscheidenden Phasen des menschlichen Zusammenlebens ist in allen Gesellschaften dieser Erde [...] in derselben Weise gegeben, doch hat die Fülle unterschiedlicher kultureller Evolutionen diese Disposition jeweils anders genutzt“. Man kann also Musik der verschiedensten Erscheinungsformen und Stile durchaus als klingendes Alphabet der Gesellschaft bezeichnen. Allerdings erweist es sich immer wieder als schwierig, die vielschichtig-komplexen, auf symbolischen Codes beruhenden musikalischen Bedeutungsebenen dieses klingenden Alphabets zu entschlüsseln, vor allem dann, wenn man die entscheidenden Codes nicht von frühester Kindheit an hat verinnerlichen können.

Jede Musik, die erklingt, hat Funktionen und erfüllt Funktionen. Unterschieden werden sollte aber – wie das Hans Heinrich Eggebrecht 1973 bündig dargelegt hat – zwischen drei Ebenen der Funktionalität. Die musikimmanente Funktionsebene regelt das Funktionieren der musikalischen Strukturen. Intendierte Funktionen beruhen auf der Zweckbestimmung eines musikalischen Produkts durch Komponisten, Musiker, Produzenten und Vermittler. Tatsächlich realisierte Funktionen aber sind ein Ergebnis des Rezeptionsprozesses. Allein um diese dritte Funktionsebene mit ihren personen- und gesellschaftsbezogenen Faktoren soll es im Folgenden gehen. Denn hier vor allem entscheidet sich, welche Bedeutung Musik für jeden Einzelnen von uns hat bzw. haben kann.

Eine Arbeitshypothese von Georg Knepler (1982, S. 36 f.) lautet, dass die Menschheitsgeschichte durch zwei wesentliche Entwicklungszüge gekennzeichnet ist, (1) die Ausprägung interner kognitiver Schemata, um dem Ansturm neuer, unbekannter Objekte, Situationen und Handlungen gewachsen zu sein, und (2) die Schaffung interner emotionaler Schemata, um den sich daraus ergebenden psychischen Anforderungen standhalten zu können. Folgt man dieser These, dann liegt es nahe, den beiden akustischen Kommunikationsmedien Sprache und Musik die folgende Funktionsaufteilung zuzuordnen: Sprache steht relativ frei von Emotionalität als Verständigungssystem für den kognitiven Bereich zur Verfügung; mit Musik dagegen können die durch Sprache nicht mehr zu leistenden emotionalen Zustände hörbar, nachvollziehbar und – dank verschiedener Formen symbolisch-ritueller Überhöhung – häufig überhaupt erst erträglich gemacht werden. Damit wird Musik zu einem allein schon aus sozialpsychologischen Gründen (über)lebensnotwendigen Gebrauchsgegenstand menschlichen Daseins. Eine ausschließliche Luxusfunktion, wie sie der Kunstmusik so gerne unterstellt wird, hat sie allenfalls in einigen wenigen, ästhetisch hochstilisierten Erscheinungsformen.

Die sozialpsychologische Kraft von Musik wird offensichtlich, wenn man Musik – wie unlängst in meinem Artikel zu der Frage „Wie politisch kann Musik sein?" (2004) vorgeschlagen – als mehrdimensionales Bezugssystem versteht. Jeder musikalische Zirkulationsprozess umfasst eine Vielzahl von aufeinander bezogenen und voneinander abhängigen Stationen: (1) die musikalische Produktionshandlung (Komposition, Improvisation), (2) verschiedene Vermittlungsschritte bis hin zur klingenden Realisation des musikalischen Produkts (die Umsetzung von notierten musikalischen Substraten, Soundfiles oder Texturen in spezifischen Aufführungs- und Wiedergabekontexten) und (3) die Rezeption durch Hörer mit unterschiedlichen musikalischen Erfahrungsinventaren, Musikerwartungen und durch Musik hervorgerufenen Verhaltensweisen. Diese Komplexität von Musik als einem mehrdimensionalen Bezugssystem hat in der traditionellen Musikwissenschaft und Musikkritik eine all zu oft unangemessene Einengung hin auf das musikalische Produkt erfahren. Alles, was nicht in ihm, also der „reinen" Musik selbst enthalten ist, das läge, so Carl Dahlhaus und Hans Heinrich Eggebrecht in ihren Essays zur Frage „Was ist Musik?" (1985, S. 68 u. 139 ff.), außerhalb der Musik. Musikalischer Sinn sei allein in der Form bzw. im „Formsinn" der Musikstrukturen verdinglicht. Bei dieser Sichtweise wird außer Acht gelassen, dass Musik immer erst in einem situativ geprägten gesellschaftlichen Umfeld Wirkung entfalten kann. Sie muss von Individuen wahrgenommen werden – im Rahmen einer Live-Darbietung, während der Wiedergabe über Lautsprecher, oder, im Extremfall, beim stumm-imaginativen Lesen der Noten. Ihre Bedeutung beim Rezipienten erzielt sie auf der Grundlage von Beschriftungen, Konnotationen, biographischen Erfahrungen, die den gesamten Zirkulationsprozess des Bezugssystems Musik umfassen.

Musikalische Bedeutungen, musikbezogene Wirkungen und musikgenerierte Funktionen sind so gut wie immer gebunden an konkrete Situationen kultureller Praxis, und sie sind eingebunden in einen lebensweltlichen Bezug. Hier fungieren die jeweiligen sozialen, politischen, ökonomischen, technischen, und räumlichen Ressourcen innerhalb eines gesellschaftlichen Systems als objektive Bedingungen für musikkulturelles Handeln und Verhalten. Das haben Kurt Blaukopf (1989) und Alfred Smudits (2002) mit dem Begriff der „Mediamorphose“ quer durch die europäische Musikgeschichte beschrieben und am Beispiel der aktuellen elektronischen und digitalen Mediamorphose im Einzelnen belegt. Technische Entwicklungen verändern die Ressourcen eines gesellschaftlichen Systems und üben nachhaltigen Einfuß auf Musik als Bestandteil von Kommunikationskultur aus: Sie beeinflussen die Musikproduktion, die Musikverbreitung, den Umgang mit Musik, das Musikverständnis und die Funktionen von Musik in der Gesellschaft.

Die objektiven Ressourcen einer Gesellschaft stehen in einer dynamischen Wechselbeziehung mit den subjektiven Strukturen der Individuen, die die Gesellschaft konstituieren. Zu den subjektiven Strukturen zählen u. a. Veranlagung und psychische Sensibilität, kognitive Disposition, emotional-affektive Gegebenheiten und die daraus resultierenden persönlichen Vorlieben. Aus den Übereinstimmungen und Differenzen zwischen objektiven Ressourcen und subjektiven Strukturen ergeben sich nicht nur die Strategien für kreativ-kulturelles Handeln, die z. B. im Bereich der Musik zur Ausprägung der verschiedenen kunst- und popmusikalischen Stile geführt haben. Sie konstituieren und bedingen auch die Bedeutungen, Bewertungen und Funktionen von Musik in der Gesellschaft und für die Gesellschaft – teilweise auf recht direkte Art, teilweise aber auch vielfach vermittelt und gebrochen.

Mit Blick auf die soziokulturellen Ressourcen im Europa der Neuzeit hat Max Weber in seinen Abhandlungen zur Musik (1921) vier Primärfunktionen von Musik in der Gesellschaft benannt, die individuell-psychische und gesellschaftlich-kommunikative Aspekte beinhalten: (1) affektbestimmte emotionale Funktionen wie psychische Resonanz, Projektion oder Abreaktion von Gefühlen, (2) traditionsorientierte Funktionen – rituell, geschichtsbezogen, überliefernd – (3) wertabhängige Funktionen wie z. B. gute vs. schlechte, anspruchsvolle vs. triviale, unterhaltende vs. ernste Musik und (4) zweckrationale Funktionen mit politischer, wirtschaftlicher, erzieherischer oder ideologischer Komponente. Und aus vergleichend-musikanthropologischer Sicht glaubte Alan P. Merriam (1964) zehn gute Gründe dafür anführen zu können, warum Menschen Musik brauchen. Sie decken sich in mancherlei Hinsicht mit den von Max Weber propagierten Primärfunktionen: (1) emotionaler Ausdruck, (2) psychische Reaktion, (3) ästhetischer Genuss, (4) Unterhaltung, (5) Kommunikation, (6) symbolische Repräsentation, (7) soziale Normierung, (8) rituelle und institutionelle Überhöhung, (9) kulturelle Stabilität und Kontinuität sowie (10) Integration in gesellschaftliche Gruppenprozesse.

Wie unlängst Paul Riggenbach in seiner umfassenden empirischen Studie über „Funktionen von Musik in der modernen Industriegesellschaft“ (2000) zeigen konnte, sind nahezu alle diese Funktionen auch bei Popmusikhörern der Gegenwart anzutreffen: Musik beeinflusst deren Gefühle (F 1) und bewirkt psychische Reaktionen (F 2), sie ist Medium der Unterhaltung (F 4) und Trägerin von Botschaften (F 5). Musik verändert Realitätsbezüge durch symbolische Repräsentation (F 6), sie konstituiert Lebensstile und kann hierbei sozial normierend sein (F 7), sie befördert Identitätsprozesse und führt zu sozialen Gruppenbildungen (F 9 u. 10), wie sie Victor Turner (2005) unter dem Begriff der Communitas wiederholt beschrieben hat.

Die im Anschluss an die Popmusik-Funktionen von Paul Riggenbach konstatierten Tendenzen zielen allerdings in eine Richtung, die den musikbezogenen Funktionsfächer um so mehr einengen und begrenzen, wie die digitale Mediamorphose voranschreitet. Musik ist allgegenwärtig geworden. Das führt häufig zu erhöhtem passivem oder gar ungewolltem Musikkonsum bis hin zur Abstumpfung gegenüber Musik. Sie verkümmert zudem durch die technische Übertragungskette immer mehr zu einer einseitigen Kommunikationsform. Dementsprechend wurden die durch Musik ausgelösten Gefühle und Handlungen von Riggenbachs Interviewpartnern zur Jahrtausendwende als zunehmend fremdbestimmter erfahren. Der Bezug zur Realität wurde gegenüber den 1960er – 1990er Jahren als abgeschwächt empfunden und der reine Unterhaltungs- oder gar Berieselungsaspekt von Musik trat immer stärker in den Vordergrund.

Live dargebotene Musik kann diesen Tendenzen der Einengung und Abschwächung musikalischer Funktionen aber nach wie vor entgegenwirken. Das zumindest belegen die Auswertungen von mehreren hundert Fragebögen der Besucher verschiedener Rockkonzerte, die Roland Hafen 1998 vorgelegt hat. Psychophysiologische und sozialpsychologische Wirkungen motivieren zum Besuch eines Rockkonzertes. Zum psychophysiologischen Funktionsfeld gehören vor allem die Qualität und Intensität des Körpergefühls gemäß der Maxime „Das Glück ist körperlich". Dieses Gefühl umfasst das Verlangen nach Rhythmus, Sound und Lautstärke, das Spiel mit dem Körper, den Wunsch nach Nähe in einer Gruppe Gleichgesinnter („ein Bad in der Menge nehmen") und die körperliche Verausgabung bis zur Erschöpfung. Das sozialpsychologische Funktionsfeld wird abgesteckt durch die Demonstration von Haltungen (u. a. durch Bekleidung und Accessoires), das Zur-Schau-Stellen von Einstellungen und Lebensstilen (Rockmusik als Lebensgefühl) und das Verlangen nach Atmosphäre und Authentizität (Echtheit der Botschaft).

Naturgemäß etwas anders sind allerdings die Erwartungen und Wünsche gelagert, die die Besucher z. B. von André Rieu- und Helmut Lotti-Konzerten haben. Nina Polaschegg fasst die Ergebnisse ihrer qualitativen Studie über „Populäre Klassik – Klassik populär" (2005, S. 187) folgendermaßen zusammen: Die Konzertgänger „möchten einen harmonischen und besonderen Konzertabend verbringen, sie möchten unterhalten werden, lachen und schwelgen können und den Alltag für diese Zeit weit hinter sich lassen. Der Abend soll ausschließlich angenehme, schöne und lustige Erlebnisse versprechen. Unbekanntes hat in diesem Kontext keinen Platz und wird [...] negativ und ablehnend beschrieben". Im Hinblick auf die Funktionen dieser Musik im Konzertsaal dominieren emotionale Kompensation, Einsamkeitsüberbrückung, Konfliktbewältigung und Stimulierung zu Handlungen, die sich im Beifallsritual ebenso wie im körperlichen Mitvollzug (Schunkeln, Takt klatschen, Tanzen) manifestieren. Der Communitas-Wirkung von Musik, dem Sich-Geborgen-Fühlen im Harmoniemilieu alltagsästhetischer Schemata (Gerhard Schulze, 1992) kommt hierbei der Stellenwert einer Primärfunktion zu.

Derartige Untersuchungen bei unterschiedlichen Konzertpublika – wobei nachdrücklich auf die nach wie vor Richtung weisende Publikation des Autorenteams Dollase, Rüsenberg, Stollenwerk aus dem Jahr 1986 hingewiesen sei – belegen die verschiedenen

Funktionen von Musik in Abhängigkeit vom musikalischen Produkt, von konkreten Darbietungssituationen und von personengebundenen Variablen der Musikhörer. Trotz unterschiedlicher Gewichtung der einzelnen Funktionskomponenten im gesellschaftlich-kommunikativen und individuell-psychischen Funktionsfächer lassen sich letztlich immer wieder zentrale Funktionskomplexe dingfest machen, wie sie bereits von Max Weber und Alan P. Merriam benannt worden sind.

Zur Veranschaulichung seien einige weitere Beispiele aus unserem westlichen ‚Kulturkreis' benannt. Zum gesellschaftlich-kommunikativen, durch soziale Normen und Regeln geprägten Funktionsfächer gehören folgende musikbezogene Settings und Inhalte: religiöse oder magisch-rituelle Transzendenz (Kult- und Kirchenmusik), Repräsentation und Glorifizierung (Militärmusik), Gestaltung von Festen und Kundgebungen (Musik als akustisches Ornament für die besonderen Ereignisse des Lebens), Bewegungsaktivierung und -koordination (Tanz- und Marschmusik), Gruppenstabilisierung und -kontrolle (Nationalhymne als akustische Flagge kultureller Identität), Erziehung (Musik der Jugend zur Selbstfindung), Gesellschaftskritik (Musik gegen Krieg, Rassismus, Sexismus) und Gesellschaftsvisionen (Musik gegen Krieg, für die Freiheit), zwischenmenschliche Kontaktaufnahme und Communitas-Erfahrung in einer Gruppe von Gleichgesinnten.

Der individuell-psychische, auf die jeweilige persönliche Bedürfnislage abzielende Funktionsfächer umfasst vornehmlich die folgenden Teilkomponenten: emotionale Kompensation (Projektion oder Abreaktion von Gefühlen), Einsamkeitsüberbrückung (parasozialer Kontakt), Konfliktbewältigung (Stressabbau durch Musik), Entspannung (Musik zum Einschlafen), Aktivierung (Musik als psychisches Stimulanzmittel), Unterhaltung (Spaß- bzw. Lustgewinn durch Musik), ästhetische Befriedigung (Transzendenz-Erfahrung) und Freiraum für Träume.

Alle die hier angedeuteten Funktionen von Musik sind erst einmal wertneutral, und jede von ihnen hat eine mehr oder weniger direkt nachvollziehbare gesellschaftliche Relevanz. Musikpädagogik versucht deren positiven Seiten zu stärken. Aber bereits hier können interessengeleitete Konflikte entstehen, etwa wenn der Referenzrahmen der Unterrichtenden nicht deckungsgleich mit dem der Unterrichteten ist, wenn also z. B. einzig die kognitiv-strukturelle Hörweise als musikadäquat angesehen wird, nicht aber emotionale, assoziative, imaginative oder vegetativ-motorische Umgangsformen mit Musik. So können psychische Abreaktion schnell als Eskapismus, träumerische Visionen als Realitätsverlust und Lustgewinn durch Musik als Hedonismus interpretiert werden, bis hin zu dem Eindruck, Musik habe die Funktion einer Droge. Positiver Nutzen und Missbrauch liegen also immer eng beieinander.

Die Bewertung musikalischer Funktionen stellt somit ein Problem dar. Eine ideologiefreie Verwendung des Systems Musik mit seinen wie auch immer gearteten Zirkulationsbedingungen gibt es so gut wie nicht. Die gesellschaftlichen Beschriftungen von Musik lassen sich nicht ungeschehen machen, wie die l'art pour l'art-Ästhetik ab dem 19. Jahrhundert unterstellt hat. Die europäische Musikgeschichte liest sich so gesehen als Ansammlung von Beispielen für den ideologischen Musikgebrauch. Walter von der

Vogelweides „Palestina-Lied“, die deutsche Nationalhymne oder Hanns Eislers Kampflieder der 1920er Jahre markieren nur drei beliebig genannte Eckpunkte von ideologischer Vereinnahmung mit den Mitteln der Musik. Wie man Musik für gesellschaftspolitische Zwecke funktionalisieren und missbrauchen kann, das lässt sich ihrem Einsatz in Diktaturen wie dem nationalsozialistischen Deutschland oder dem stalinistischen Russland unschwer entnehmen (Geiger 2004). Wie man sie zur kritischen und freiheitlichen Meinungsbildung nutzen kann, das wurde Theodor W. Adorno zu fordern und zu beschreiben nicht müde.

Wenn auch die Macht von Musik und ihre Wunderwirkungen sicher nicht so groß sind, wie der Zusammenbruch der Mauern von Jericho vorgibt, so ist ihre mitreißende, ja durchaus massensuggestive Kraft bei der Instrumentalisierung politischer, kultisch-religiöser oder kriegerischer Ambitionen keineswegs von der Hand zu weisen. Das funktioniert im Guten wie im Schlechten, weil die funktionalen Auslösemechanismen immer die gleichen bleiben. Selbst Art und Stil der Musik können kein Garant gegen Missbrauch sein. Denn die Bedeutung eines Musikstücks im Zirkulationsprozess von historischer Zeit, gesellschaftlicher Referenzbildung und symbolhaltiger Konditionierung unterliegt der Veränderung. Das lässt sich am Beispiel der sinfonischen Dichtung „Les Préludes“ von Franz Liszt unschwer belegen. Eine neue, vom Komponisten so niemals beabsichtigte Beschriftung im zweiten Weltkrieg hat zu einer Umfunktionierung und Aufladung des musikalischen Produkts mit einem politischen Inhalt geführt, der, je nach der Selbstpositionierung der Hörer im Nationalsozialismus, Freude oder Entsetzen hervorrief.

Die bisherigen Ausführungen haben deutlich werden lassen, dass Musik und Gesellschaft nicht unabhängig voneinander betrachtet werden dürfen. Es handelt sich hier um zwei Systeme, die sich gegenseitig bedingen und im musikalischen Rezeptionsvorgang miteinander verschmelzen. Das geschieht auf eine eher implizite Art, die darum auch dem Hörenden üblicher Weise eher verborgen bleibt und sich erst rationaler Analyse erschließt. Im Folgenden möchte ich den Versuch wagen, zumindest die wichtigsten der komplexen Phänomene musikalischer Bedeutungszuweisungen systematisierend einzugrenzen.

Die vielen psychischen, sozialpsychologischen und gesellschaftlichen Wirkungen, die durch das Bezugssystem Musik ausgelöst werden können, sind abhängig von einigen wenigen funktionalen Universalien, die sich als individuell-psychische und gesellschaftlich-kommunikative Komponenten beschreiben lassen (Rösing 1992, S. 315 f.). Will man diesen Wirkungen und ihren gesellschaftlichen Implikationen im Detail nachspüren, dann müsste in Zukunft weit mehr als bislang das komplette Bezugssystem Musik im soziokulturellen Zusammenhang gesichtet und gedeutet werden. Als Untersuchungsleitfaden böten sich kulturwissenschaftlich fundierte, übergeordnete Kriterien gesellschaftlichen Handelns und Verhaltens an. Kriterien, die auch für Musik gleich welcher Herkunft und welchen Stils von großer Bedeutung sind und es erlauben, die Komplexität der Wirkung stiftenden bzw. Wirkung beeinflussenden Faktoren für Forschungszwecke so zu operationalisieren, dass keine unangemessene Einengung und Ergebnisverfälschung die Folge ist.

Sieben dieser Kriterien seien hier kurz skizziert:

1. Andersartigkeit und Differenz – hier wird Musik als ein Kultur schaffendes und prägendes künstlerisches Ausdrucksmedium verstanden. Dieser Ansatz berücksichtigt die Dynamik menschlichen Handelns im gesellschaftlichen Umfeld und fragt nach der Andersartigkeit gleichzeitig bestehender Musikszenen, vor allem nach ihrer häufig nicht unerheblichen Differenz im Vergleich zu den etablierten musikkulturellen Systemen. Auf diese Weise kann die Bandbreite aller das gegenwärtige Musikleben bestimmenden Richtungen – populäre ebenso wie elitäre – als ein Gefüge wechselseitiger Verzahnung mit jeweils spezifischen gesellschaftspolitischen, ökonomischen, technischen und lebensstilbezogenen Implikationen durchleuchtet werden.
2. Präsentation und Praxis – Kultur ist immer auch (so Erving Goffman 1976) Bestandteil von Inszenierungen. Diese bestimmen die soziale und künstlerische Praxis. Die Betonung spielerischer oder theatralischer Elemente lässt sich bei jeder Musikaufführung als ein Mittel von Selbstdarstellung beobachten. Dabei bilden musikalische Kommunikationsinhalte in der Form symbolischer Codes und die visuell nachvollziehbare Art der Präsentation eine Einheit. Typische Verhaltensweisen von Musikern und analog dazu typische Verhaltensrituale der Zuhörer geben oft mehr als die konkreten Strukturen des musikalischen Produkts Aufschluss über Wirkungen und Funktionen von Musik im gesellschaftlichen Umfeld.
3. Diskursivität und Textualität – Menschen belegen ihre Lebenswelt mit einer Fülle von Bedeutungen, und diese Bedeutungen bestimmen als kulturelles Gut die Handlungen anderer. Die durch Musik bedingten textuellen Bedeutungszuweisungen lassen sich aber nur in Verbindung mit einer Analyse des gesamtgesellschaftlichen Diskurses in ihrer Komplexität erfassen. Dies vor allem deswegen, weil es zu schwerwiegenden Differenzen der Bedeutungs- und Wertzuweisung bei verschiedenen gesellschaftlichen Gruppierungen kommen kann. Das drückt sich z. B. in der unterschiedlichen Einschätzung von atonaler Musik und Techno, politischer Musik und HipHop, volkstümelnder Popmusik und Klassik aus.
4. Urbanität und Rustikalität – Kultur in städtischen Ballungsräumen wird unter anderen Bedingungen geschaffen, präsentiert und rezipiert als in ländlichen Regionen. Urbanität war seit Jahrhunderten für die Ausprägung musikalischer Stile von großer Bedeutung. Die Entwicklung der Oper ist ohne städtischen Kontext kaum denkbar, die musikalische Klassik ist nachhaltig an die Ressourcen der Stadt Wien gebunden, New Orleans war erster Kristallisationspunkt des Jazz, Beat entstand im Umfeld von Liverpool usw. Andererseits sind ungarische Bauernmusik oder die Musik der Griots und selbst die Anfänge des ländlichen Blues ohne die jeweils typischen ruralen Lebensräume kaum denkbar. Wirkungen, Funktionen und gesellschaftliche Bedeutung sind an diese Gegebenheiten gebunden und nur im Zusammenspiel mit ihnen zu benennen. Auf die Gegenwart gewendet beinhaltet dieses Kriterium auch die Differenz zwischen globaler, nach Gesetzen des Marktes kreierter Popmusik und einer von persönlichem Mitteilungsbedürfnis getragener Musik der nahen Lebensräume.

5. Funktionalität – die Verlagerung des Musikkonsums von der Live-Darbietung zur medialen Übertragung hat zu einer zunehmenden Funktionalisierung von Musik in Alltagskontexten geführt. Immer weniger wird sie um ihrer selbst willen gehört und immer häufiger zur Erfüllung musikfremder Zwecke genutzt. Funktionelle Musik für die Verwendung in Einkaufszentren, Gaststätten, Wartesälen oder am Arbeitsplatz markiert diesbezüglich den Extremfall. Hier ist eine aktive und gar kognitiv gesteuerte Musikrezeption gar nicht mehr erwünscht. Anders und durchaus kontradiktorisch gesagt: Der selbstbestimmte Umgang mit fremdbestimmter Musik nimmt zu, und so mancher hat sich hier zum eigenen Musiktherapeuten ernannt.
6. Medialität – die technischen Medien sind fester Bestandteil der Kultur des 20. und 21. Jahrhunderts. Die Speicherung von Musik auf Tonträger und ihre Wiedergabe von Tonträgern haben nicht allein die Rezeptionserwartungen und das Rezeptionsverhalten deutlich verändert, sondern auch die Komposition und Distribution. Im Gefolge der Elektrifizierung wurden neue Musikinstrumente konstruiert und die Klänge der herkömmlichen Musikinstrumente durch Tonabnehmer oder Mikrophonübertragung abgewandelt und verstärkt. Synthetische Klangerzeugung und Arbeit im Tonstudio, schließlich die Digitalisierung der Musikproduktion zeigen, auf wie direkte Art Musik in den Kreislauf der technischen Ressourcen, die eine Gesellschaft zur Verfügung stellt, eingebunden ist, und wie stark dadurch die Wirkung von Musik in der Gesellschaft modelliert wird (vgl. Rösing 2000).

7. Korporalität – kulturelle Prozesse wirken auf das Bewegungsrepertoire und die Bewegungsabläufe des Körpers ein. Musik und Tanz sind dafür ein bündiger Beleg. In der Oper „Don Giovanni" hat das Wolfgang Amadeus Mozart durch die gleichzeitige Darbietung von Tänzen aus der niederen, mittleren und hohen Ständegesellschaft zum Ausdruck gebracht, wobei dem brüchigen Gefüge der Klassengesellschaft die Missklänge im Tonsatz entsprechen. In der verschiedenartigen Bezugnahme von Jazz, Rock und Pop zu Tanz und Körperlichkeit finden sich Anschauungsmodelle für die Gegenwart. Sie bedürfen einer höchst detaillierten Analyse, um in ihnen die Spuren gesellschaftlicher Prozesse aufzudecken.

Die Frage, was Musik für die Gesellschaft leistet, welche Funktionen sie im einzelnen erfüllt, lässt sich also nicht pauschal, sondern lediglich am Einzelfall konkret nachweisen. Und das auch nur, sofern die verschiedenen Stationen im Zirkulationsprozess des Systems Musik sehr genau berücksichtigt und analysiert werden. Die hier aufgezählten sieben Untersuchungskriterien – die sich unschwer noch ergänzen und auch weiter differenzieren ließen – dürften davon eine Vorstellung vermittelt haben. Eine unlängst von Alenka Barber-Kersovan vorgelegte Arbeit „Vom ‚Punk-Frühling' zum ‚slowenischen Frühling'. Der Beitrag des slowenischen Punk zur Demontage des sozialistischen Wertesystems" (2005) zeigt prototypisch, wie die auf kulturgesellschaftliche Phänomene bezogene Auswertung des Musiksystems „Slowenischer Punk" funktionieren kann, und sie belegt zugleich auch, dass die Ergebnisse nicht auf andere Gesellschaften und andere Zeiten übertragbar sind. Denn wie im "Wechselspiel zwischen Provokation und Reaktion [...] der Punk zu einem politischen Faktor hochstilisiert und mit einem zweifachen Symbolwert ausgestattet" wurde, das ist und bleibt gebunden an eine bestimmte zeitgeschichtliche Situation und eine Gesellschaft bzw. Nation mit ihren typischen politischen, ökonomischen, gesellschaftlichen, technischen und kulturellen Ressourcen.

Dem Einzelfall liegen jedoch – aus übergeordnet-abstrahierender Sichtweise betrachtet – durchaus gewisse Universalien zugrunde: Musik und Gesellschaft erweisen sich als aufeinander bezogene Systeme. Objektiv gegebene Ressourcen und individuelle Strukturen greifen bei der Produktion, Distribution und Rezeption von Musik ineinander. Sie prägen musikalische Zirkulationsprozesse und sind ablesbar an den verschiedenen Funktionen, die das System Musik in gesellschaftlich-kommunikativer wie individuell-psychischer Hinsicht auszulösen vermag. Ihre sozialpsychologische Kraft und ihre Communitas-Wirkung entfalten sich dann besonders nachhaltig, wenn Musik gemeinschaftlich in Live-Kontexten gehört wird.

Aus diesem wiederholt belegten Sachverhalt (s. Rösing 1998) ergeben sich musikbezogene Forderungen an die Gesellschaft, die heute nicht weniger aktuell sind als früher. Allein eine fundierte musikalische Unterweisung für jeden bewirkt jene Offenheit und jenes Verständnis gegenüber Musik, das einer ideologischen Vereinnahmung von Musik durch Politik und Ökonomie zumindest Grenzen auferlegen kann. Eigenes Musikmachen und ein vielfältiges musikalisches Live-Angebot ermöglichen mehr noch als technisch vermittelte Musik einen sinnvollen Bezug zu nahen Lebenswelten. Dies schafft ein Gegengewicht zu globalen, allein nach dem Gesetz der Ware konzipierten Musikproduktionen ohne authentische Botschaft.

Musikalische Mündigkeit durch Musikkenntnisse kommt nicht von selbst, wie neoliberale Argumentation vielleicht glauben machen will. Die gesellschaftlichen Ressourcen für einen selbst bestimmten und mündigen Umgang mit Musik müssen mit erheblichem ideellem und finanziellem Einsatz von der Gemeinschaft zur Verfügung gestellt werden. Das beginnt bei musikalischer Früherziehung, setzt sich fort über einen kontinuierlichen Musikunterricht an allgemein bildenden Schulen und führt zu Lebensstilen, in denen Musik nicht nur nebenbei konsumiert sondern mit Bewusstheit und aktiv rezipiert wird. Nur so lässt sich verhindern, dass Musik nicht ihrer gegenwärtigen Gesellschaft entgleitet und auch in Zeiten der medialen Allverfügbarkeit ihren Wert als kommunikatives, gesellschaftsbezogenes und gesellschaftsformendes Medium beibehält.

Literatur

Adorno, Theodor W. (1962), *Einleitung in die Musiksoziologie. Zwölf theoretische Vorlesungen*, Frankfurt am Main: Surkamp.

Alenka Barber-Kersovan (2005), *Vom ‚Punk-Frühling' zum ‚Slowenischen Frühling'. Der Beitrag des slowenischen Punk zur Demontage des sozialistischen Wertesystems*, Hamburg: Krämer.

Blaukopf, Kurt (1989), *Beethovens Erben in der Mediamorphose. Kultur- und Medienpolitik für die elektronische Ära*, Heiden (CH): Niggli.

Dahlhaus, Carl, Hans Heinrich Eggebrecht (1985), *Was ist Musik?*, Wilhelmshaven: Noetzel.

Dollase, Rainer, Micheal Rüsenberg, Hans J. Stollenwerk (1986), *Demoskopie im Konzertsaal*, Mainz: Schott.

Eggebrecht, Hans Heinrich (1973), Funktionale Musik, *Archiv für Musikwissenschaft* 30, 1 - 25.

Geiger, Friedrich (2004), *Musik in zwei Diktaturen. Verfolgung von Komponisten unter Hitler und Stalin*, Kassel: Bärenreiter.

Goffman, Erving (1976), *Wir alle spielen Theater. Die Selbstdarstellung im Alltag*, München: Piper (orig. 1959).

Hafen, Roland (1998), Rockmusik-Rezeption in Live-Konzerten, In: Dieter Baacke (Hg.), *Handbuch Jugend und Musik*, Opladen: Leske u. Budrich, 369 – 380.

Knepler, Georg (1982), *Geschichte als Weg zum Musikverständnis. Zur Theorie, Methode und Geschichte der Musikgeschichtsschreibung*, 2. Auflage, Leipzig: Reclam.

Merriam, Alan P. (1964), *Anthropology of music*, Evanston: North Western University Press.

Nettl, Bruno (1983), *The study of ethnomusicology*, Urbana: University of Illinois Press.

Polaschegg, Nina (2005), *Populäre Klassik – Klassik populär. Hörerstrukturen und Verbreitungsmedien im Wandel*, Köln: Böhlau.

Riggenbach, Paul (2000), *Funktionen von Musik in der modernen Industriegesellschaft. Eine Untersuchung zwischen Empirie und Theorie*, Marburg: Tectum.

Rösing, Helmut (1992), Musik als Lebenshilfe? Funktionen und Alltagskontexte, In: Wolfgang Lipp (Hg.), *Gesellschaft und Musik. Wege zur Musiksoziologie*, Berlin: Duncker & Humblot, 311-331 (= Sociologica Internationalis, Beiheft 1).

Rösing, Helmut (1998), Musikgebrauch im täglichen Leben – Musikalische Lebenswelten, In: Herbert Bruhn, Helmut Rösing (Hg.), *Musikwissenschaft. Ein Grundkurs,* Reinbek: Rowohlt, 107 – 152.

Rösing, Helmut (2000), Digitale Musik und Medien. Zwölf Thesen, In: Bernd Enders, Joachim Stange-Elbe (Hg.), *Musik im virtuellen Raum*, Osnabrück: Rasch, 13 – 23.

Rösing, Helmut (2004), Wie politisch kann Musik sein?, *Beiträge zur Popularmusik-Forschung* 32, 155 – 168.

Schulze, Georg (1992), *Die Erlebnisgesellschaft. Kultursoziologie der Gegenwart*, Frankfurt am Main: Campus.

Smudits, Alfred (2002), *Mediamorphosen des Musikschaffens. Kunst und Kommunikationstechnologien im Wandel*, Wien: Braunmüller.

Suppan, Wolfgang (1984), *Der musizierende Mensch. Eine Anthropologie der Musik*, Mainz: Schott.

Turner, Victor (2005), *Das Ritual: Struktur und Anti-Struktur*, Frankfurt am Main: Campus (orig. 1969).

Weber, Max (1921), *Die rationalen und soziologischen Grundlagen der Musik*, München: Drei Masken.

www.ingramcontent.com/pod-product-compliance
Ingram Content Group UK Ltd.
Pitfield, Milton Keynes, MK11 3LW, UK
UKHW021845100926
8556IPUK00019B/1058

9 783631 599914